MANAGEMENT TECHNIQUES APPLIED TO THE CONSTRUCTION INDUSTRY

FOURTH EDITION

R. Oxley MSc, MCIOB, MBIM
and
J. Poskitt MCIOB, MBIM

OXFORD
BSP PROFESSIONAL BOOKS
LONDON EDINBURGH BOSTON
MELBOURNE PARIS BERLIN VIENNA

BSP Professional Books
A division of Blackwell Scientific
Publications Ltd
Editorial Offices:
Osney Mead, Oxford OX2 0EL
25 John Street, London WC1N 2BL
23 Ainslie Place, Edinburgh EH3 6AJ
3 Cambridge Center, Cambridge,
MA 02142, USA
54 University Street, Carlton
Victoria 3053, Australia

Other Editorial Offices:
Librairie Arnette SA
2, rue Casimir-Delavigne
75006 Paris
France

Blackwell Wissenschafts-Verlag
Meinekestrasse 4
D-1000 Berlin 15
Germany

Blackwell MZV
Feldgasse 13
A-1238 Wien
Austria

First published by Crosby Lockwood
& Son Ltd 1968
Second edition 1971
Reprinted by Granada Publishing Ltd
in Crosby Lockwood Staples 1976, 1979
Third edition published by Granada Publishing
1980
Fourth edition published by Collins
Professional and Technical Books 1986
Reprinted by BSP Professional Books
1989, 1990, 1992

Printed and bound in Great Britain by
The Alden Press, Oxford

DISTRIBUTORS

Marston Book Services Ltd
PO Box 87
Oxford OX2 0DT
(*Orders*: Tel: 0865 791155
Fax: 0865 791927
Telex: 837515)

USA
Blackwell Scientific Publications, Inc.
3 Cambridge Center
Cambridge, MA 02142
(*Orders*: Tel: 800 759-6102
617 225-0401)

Canada
Oxford University Press
70 Wynford Drive
Don Mills
Ontario M3C 1J9
(*Orders*: Tel: 416 441-2941)

Australia
Blackwell Scientific Publications
(Australia) Pty Ltd
54 University Street
Carlton, Victoria 3053
(*Orders*: Tel: 03 347-0300)

British Library
Cataloguing in Publication Data

Oxley, Raymond
Management techniques applied to the
construction industry.—4th ed.
1. Construction industry.—Management
I. Title II. Poskitt, Joseph
658′.99 HD 9715.A2

ISBN 0–632–02487–9

Contents

Acknowledgements

We are indebted to our friends and colleagues, James Ekins, Jim Padden, Alan Dolman, Ian Cooper, Graham Stoneman, Julian Smith, Steve Westgate and Mark Fields for the help received when writing the various editions of this book.

We would also like to express our gratitude for the help and encouragement received from our wives.

Trademark acknowledgements: '1-2-3' is the registered trademark of the Lotus Development Corporation, 'dBase II' is the registered trademark of Ashton-Tate, 'Pertmaster' is the registered trademark of Abtex Computer Systems Ltd, 'CAMM' is the registered trademark of Vega Software Ltd.

Introduction

The book is concerned with the application of management techniques to problems occurring in the industry. It is assumed that readers are familiar with the basic techniques and no excuse is made for omitting some of the background information in some chapters.

The examples used have been simplified essentially to avoid over complication, and adapted to get the important points over to the reader. The techniques used are applied in a simple form avoiding the more sophisticated extensions applied by some firms. The methods of presentation have evolved from our experience with students studying the various topics.

The techniques in this book have been considered separately for convenience. It must be appreciated, however, that if a firm is to work efficiently, it needs an integrated information system which will meet the needs of management. For this to be possible, it is necessary to cross the normal departmental boundaries. Information which is gathered for one department such as production control, may be relevant to other departments such as surveying, cost control, etc. In all firms a certain amount of integration exists but there are few firms indeed which can claim to have a fully integrated system and this includes firms in industry generally.

There seems very little point in departments within a firm collecting information or processing information which is identical to that collected or processed by another department but this happens all too often in practice. In addition to this, much information is collected which is never used or is not presented in an acceptable manner for different levels of management. The system should avoid these deficiencies. It is a fact that many systems in a firm happen more by accident than by design. If an information system is to be designed for a whole business the 'traditional organisation structure' will be replaced by a 'functional structure'. This is necessary because information must flow vertically and horizontally and in a traditional structure information tends to flow vertically, but information flow across the lines of authority is not always present and this results in the problem of duplication of information mentioned earlier.

Management techniques are the tools of management, and the techniques considered in this book can and should be used by all firms, irrespective of size. They enable decisions to be taken with greater knowledge of the facts; ignorance of the techniques available to management leads to inefficient operation and may eventually lead to bankruptcy in the competitive years ahead. The use of scientific methods is essential to effective management.

The book is aimed mainly at students studying management as part of degree courses, higher technician courses and courses leading to the chartered Institute of Building examinations. It will also be of value to professionals in the construction industry who need to be aware of the scientific methods essential to effective management.

R. OXLEY and J. POSKITT

Preface to the Fourth Edition

In preparing the fourth edition of this book we have included a considerable amount of additional material. This includes extension of the work on cost optimisation, line of balance method and cash flows, and new material in programme evaluation and review technique, cost of working capital, tabular presentation of contract budgets and variance analysis. A completely new chapter has been added on computer applications.

R. Oxley and J. Poskitt

1 Construction Planning and Control

1.1 INTRODUCTION

Planning can be applied in varying degrees of detail, depending on the stage at which it is being carried out. For construction work it is usually divided into pre-tender planning, pre-contract planning, and short term planning on site.

Pre-tender planning is done to allow the estimator to arrive at an estimate of cost based on the proposed methods of working and an estimate of the time required to carry out the work. Programming at the pre-tender stage is usually in an outline form to consider only the phasing of the main operations, since much information is not at that time available.

Pre-contract planning is carried out when the contract has been won and the project is considered more fully. Planning at this stage includes the overall programme, labour schedule, plant schedule, materials schedules, etc. The overall programme should not break the operations down excessively or it will become unrealistic.

Short term planning on site is done in greater detail and the programmes at this stage are broken down much further.

Control must be carried out to make planning effective, as without control planning loses much of its value. It must be applied continuously to up-date the plans and to enable reconsideration of work ahead in the light of what has already taken place. The example which follows could be programmed using Network techniques. The resulting overall programme, converted to a bar chart, would then have presented much more information, e.g. float times on all activities could be shown. However, it is argued by some that for small projects, bar charts may be drawn direct and are quite satisfactory. Many contractors use bar charts as they are thought to be less involved. In addition it would not be possible in examination conditions to produce a programme showing movement of resources starting with a network as this would involve drawing a network, analysing it and then scheduling which would involve resource allocation. This could then result in an extension of the time first calculated. Whichever method of programming is used it should be appreciated that when progressing is done the aim is to adhere to the programme.

1.2 PRE-TENDER PLANNING

Pre-tender planning will not be considered in detail but it is felt that the relevance of the pre-tender method statement is too important to omit.

1.2.1 Pre-tender Method Statement

This is one of the most important pre-tender documents because it is used by the estimator as the basis for arriving at the cost of many of the major items of work. It follows that the method statement should include all work of any consequence where alternative methods or choice of plant are possible.

The method statement takes the form of a schedule (see Fig. 1.1) stating the method to be adopted or plant to be used for such activities as excavation, formwork, concreting, etc., and is compiled at the same time as the pre-tender programme and site layout drawing, having considered in detail each problem and having arrived at the best solution with regard to resources required and cost. Having arrived at the resources required, it is common practice to record these in the schedule for future reference. Together with the pre-tender programme it forms the basis for the overall programme. The schedule is also very useful for informing site staff of the method the estimator has used when arriving at a price.

Time spent in the preparation of a well thought out method statement usually results in a more competitive tender and when considered with a programme it can result in the solution of some potential problems before they arise.

A pre-tender Method Statement (Fig. 1.1) for the Amenity Centre and Office Block is shown in Figs 1.2, 1.3 and 1.4. This method statement is used as the basis for the overall programme, see Fig. 1.8.

Fig. 1.1 Pre-tender method statement

CONTRACT: AMENITY CENTRE & OFFICE BLOCK   SHEET NO: COMPILED BY:

NO. OF SHEETS   DATE

Item	*Quantity*	*Remarks*	*Labour & Plant*	*Time Req'd Days*
Hoarding	45 linear metres	Erected along N.E. boundary of site 2·5 metres high.	2 carpenters	2
Access	–	Vehicle access via works entrance		
	–	Cross over in S.W. footpath	2 labourers	1

Fig. 1.1 *continued*

Item	*Quantity*	*Remarks*	*Labour & Plant*	*Time Req'd Days*
	–	Workmen access via 1 metre wide gate in hoarding (position shown on site layout drawing).	–	–
Accommodation	–	All accommodation to be firm's standard sectional huts (position shown on site layout drawing) Toilets provided in form of mobile unit.	2 carpenters	8
Temp. Services	–	Provide electrical, water & W.C. services as shown on site layout drawing.	3 labourers	3
			Plumber & mate	3
			Electrician & mate	2
Excavation	105 cubic metres	*Reduced level dig* Front bucket of J.C.B. 4 – easy going.	J.C.B. 4 & operator & 3 labourers	1
	140 cubic metres	*Tie beam trenches* Rear back-acter bucket of J.C.B. 4. *Internal drain trenches* Rear back-acter bucket of J.C.B. 4.	J.C.B. 4 & operator & 3 labourers	4
		Disposal of surplus spoil Remove to tip 400 metres distance using lorries – within works.	2 lorries & drivers	5
General Hoisting & Transportation	–	The tower crane will be used for lifting all materials until frame is complete when it will be replaced by a 750 kg platform hoist which will be built adjacent to T.C. before it is removed. N.B. all bricks & blocks	Tower crane & driver	80
			Hoist	80

Fig. 1.1 *continued*

Item	*Quantity*	*Remarks*	*Labour & Plant*	*Time Req'd Days*
		lifted onto floor before crane removed. Platform hoist used for lifting materials used in finishings.		
Concreting	420 cubic metres	*Mixing* The concrete will be mixed in a 300/200 litre electrically powered reversing drum mixer, having a built-in weigh batcher, the aggregate being delivered to the weigh batcher by a hand scraper. All located at south corner of site. Cement is to be stored in a 20 tonne silo located directly over the mixer.	2 labourers 300/200 litre mixer set up	75
	420 cubic metres	*Transportation* Mixed concrete is to be transferred to all levels by means of the tower crane using a 400 litre skip.	Tower crane 500 kg capacity 21 metre max. reach + driver	–
	420 cubic metres	*Placing* The concrete is to be compacted by means of vibrator.	3 labourers Immersion vibrator	–
Reinforcement	37 tonnes	*Work Area* All reinforcing steel to be cut & bent on site in the area set aside at the east corner of the site, adjacent to the existing building. The equipment in this area consists of a scaffold rack for storage of steel when delivered,	3 steel fixers	}

Fig. 1.1 *continued*

Item	*Quantity*	*Remarks*	*Labour & Plant*	*Time Req'd Days*
		and a scaffold rack for storage of steel when processed.		$62\frac{1}{2}$
	–	*Plant* Cutting bench, bending bench and power bender all on hard standing and covered with tarpaulin on a scaffold frame.	Power bender	
	–	*Transport* Steel lifted to the required level using the tower crane.	Tower crane	
Form-work	–	*Working Area* All formwork is to be produced on site in the area set aside at the S.E. end of the building. A joiners' shop to be provided consisting of a scaffold tube frame covered with a tarpaulin and having a concrete hardstanding.		
	–	*Plant* Electrically powered circular saw.	Circular saw	
	–	*Transport* Formwork lifted to required level using tower crane.	Tower crane	
	–	*Foundations* Rough boarded form-work to bases and edge of floor slab at ground floor.	4 carpenters	$63\frac{1}{2}$

Fig. 1.1 *continued*

Item	*Quantity*	*Remarks*	*Labour & Plant*	*Time Req'd Days*
	–	*Columns* 12 mm plywood sheet with 100 mm × 50 mm sawn backing, secured by column cramps (see drawing no.) supported by telescopic props.	8 carpenters	
	–	*Beams* 12 mm plywood sheet with 100 × 50 mm sawn backing to soffite and sides, supported by Telescopic props.		
	–	*Slabs* 12 mm plywood sheet with 100 mm × 50 mm sawn bearers on expanding steel floor centres.		
Mortar-Mixing	–	Mortar for bricklaying will be mixed in a 150/100 litre mixer located on hard standing adjacent to the T.C. or hoist. A bagged cement store will be located alongside.	1 labourer	
Scaffolding	–	Scaffolding will be of independent type 1 m wide with staging at 2 m intervals and be constructed of tubular steel. A ramp will be provided at the N.W. end of the building to provide access to all platform levels.	2 scaffolders	

1.3 PRE-CONTRACT PLANNING

1.3.1 Basis of calculations

The information to be used as a basis for calculating productive hours should be obtained from operational records of past contracts and from work study synthetics if these are available. The bill of quantities can be used for the master programme providing it is accurate, but if it is employed many items will have to be collected together under the heading of the programme operations (e.g. the operation 'brickwork' will include such items as 'extra over for facings' and 'bedding cills').

The bill rates should be broken down to show labour and plant content so that the labour and plant contents of operations can be easily ascertained.

1.3.2 Breakdown into operations

At this stage the project should be broken down into the major operations only and the temptation to reduce it to detailed operations must be resisted.

1.3.3 Description of project

The project consists of a five storey building with a reinforced concrete frame cast *in situ*. It contains an amenity centre and office accommodation (see Figs. 1.2, 1.3 and 1.4), the amenity centre being confined to the lower three floors whilst the office accommodation is on the upper two floors.

1.3.4 Construction of the building

1.3.4.1 Foundations

Reinforced concrete bases and concrete strip foundations.

1.3.4.2 Superstructure

Frame: reinforced concrete columns, beams, floors and roof.

Ground floor: reinforced concrete on D.P.M. on blinding on hardcore filling.

Staircases and lift walls: reinforced concrete cast *in situ*.

External facing: aluminium curtain walling secured to reinforced concrete frame.

Concrete block backing to cill level on inside of curtain walling and 300 mm brick cladding where shown on drawing.

Covering to roofs and canopy: screed laid to falls with asphalt finish.

1.3.4.3 Internal finishes and services

Internal partition walls: 100 mm clinker blocks.

Rainwater goods and drainage: in service ducts.

Wall finishes: partial tiling to all toilets, washrooms, locker rooms and drying rooms. Timber panelling to entrance halls and office accommodation.

Plastering elsewhere with eggshell finish paint.

Ceiling finishes: suspended ceilings to office accommodation. Plastering elsewhere. All ceilings have eggshell paint finishes.

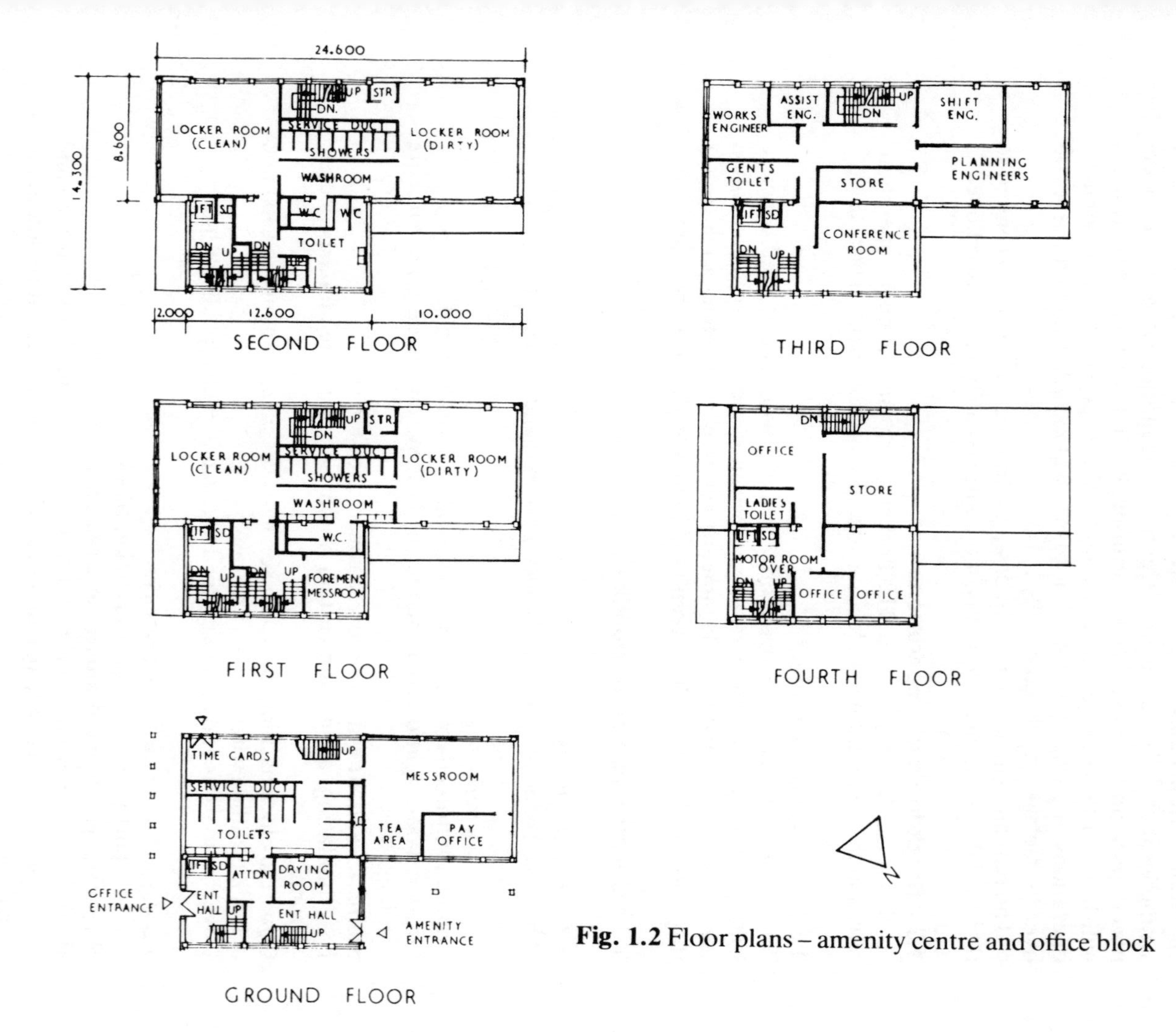

Fig. 1.2 Floor plans – amenity centre and office block

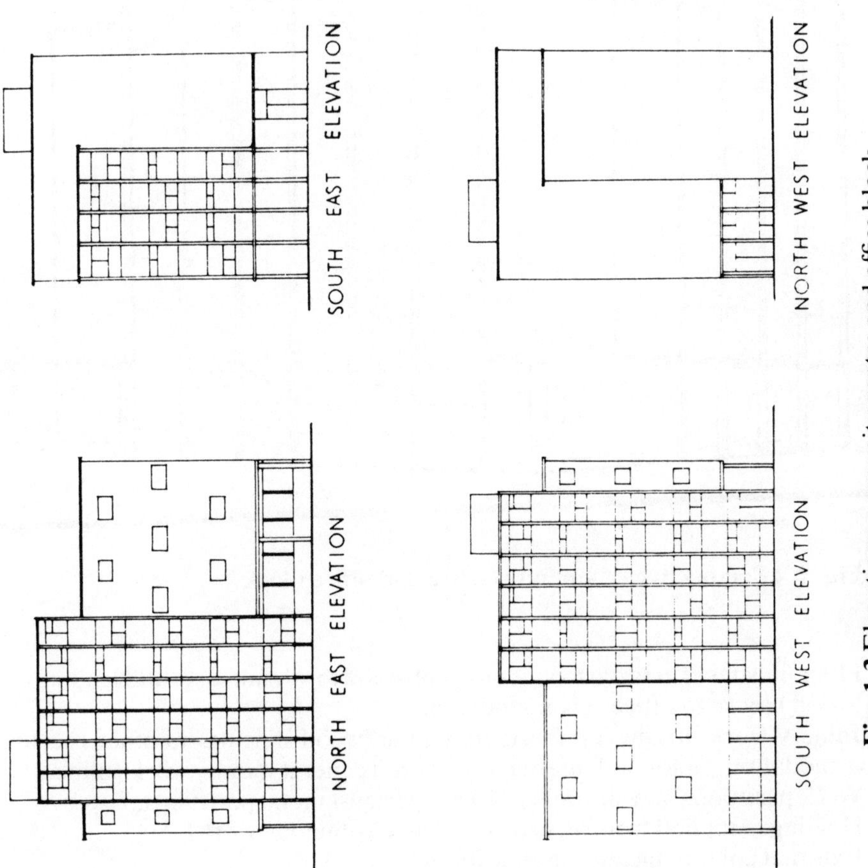

Fig. 1.3 Elevations – amenity centre and office block

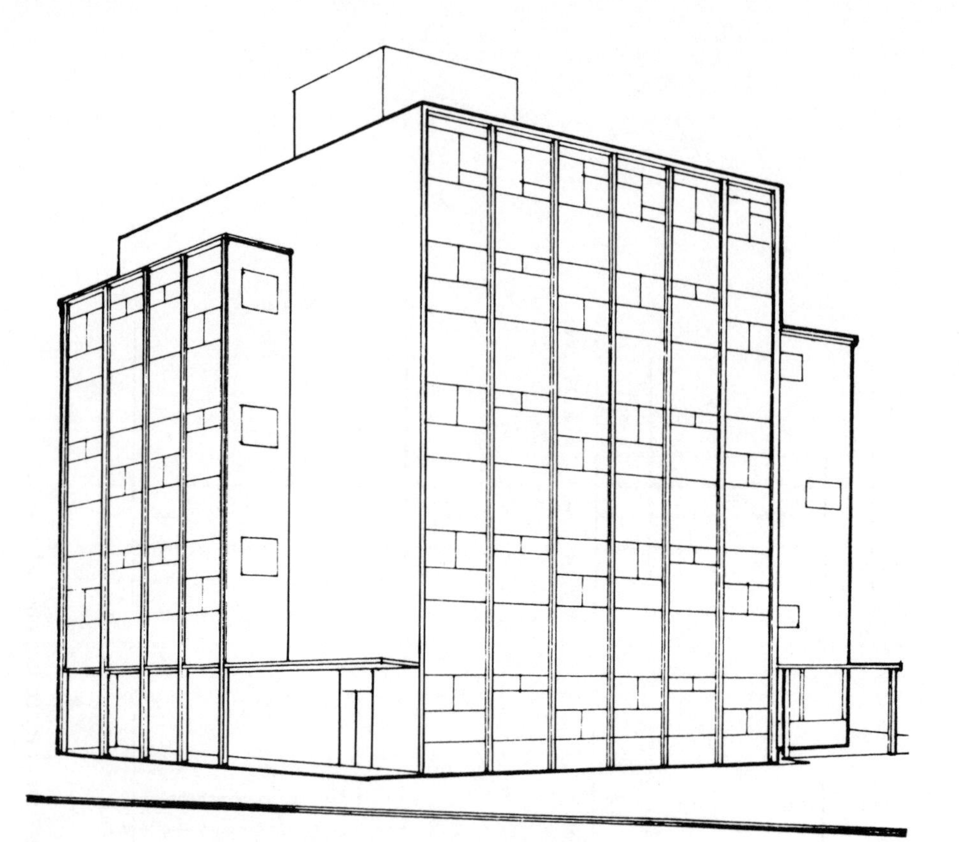

Fig. 1.4 Perspective of amenity centre and office block

Floor finishes: Terrazzo to amenity centre and to all staircases and landings.
Wood blocks to office accommodation.
Joinery work: hardwood faced doors, hardwood skirtings to office accommodation. Softwood internal fittings and general joinery work painted.
W.C. partitions and shower cubicles: specialist metal partitions.
Heating: supplied from an external source within the works.
External column facings: mosaic finish.
Lifts: to office accommodation only.

1.3.4.4 External works
Drainage as shown on drawings.
Boundary walls.
Services.
Tarmac paving on blinding on hardcore.

1.3.5 The site

The site is located 6.00 km from the contractors head office and is within the boundaries of a chemical works; access is via an existing works entrance.

The ground is level and consists of loose filling to a depth of 1.2 m. Below this is a subsoil of firm gravel.

1.3.6 Contract period

The contract period is 46 weeks, the starting date is February 19th.

1.3.7 Working week

A 40-hour, 5-day week has been used throughout.

1.3.8 Plant

The major plant has been selected at the pre-tender stage as follows:

500 kg static tower crane on concrete base
300/200 litres reversing drum concrete mixer, with loading hopper and power scoop
20 tonne cement silo
J.C.B. 4 for all excavation work by machine
500 kg platform hoist for lifting with the tower crane leaves site.

1.3.9 Productive hours

Plant-hours and man-hours (for the productive operatives only) have been established for all operations (Fig. 1.5).

1.3.10 Sub-contract work

Sub-contractors have been contacted and have agreed periods on site and notice required (Fig. 1.6).

1.3.11 Procedure for programming

1. Study the drawings before starting the programme.
2. Estimate the time required for major phases of the work, e.g.
   Set up site
   Sub-structure
   R.C. frame
   Structural brickwork and curtain walling
   Roof covering
   Internal trades and finishings
   External works
   Cleaning up and tidying site.

This will act as a guide when preparing the overall programme and will indicate in good time whether sufficient resources are being used to complete on time.

The planned completion date should be set before the contract period to allow for unforeseen delays.

3. List the operations to be used in the overall programme in approximate order of starting time, grouping the operations shown in the schedules as necessary. This list will contain the operations shown on the calculation sheet (Fig. 1.7) and the overall programme chart (Fig. 1.8).

4. Fill in the total plant-hours or productive man-hours for the programme operations.

5. Decide on the number of operatives or items of plant to be used on the first operation and calculate the time required. Enter this on the programme chart before proceeding to the next operation. Attention must be paid to providing continuity of work for plant and men where possible.

Fig. 1.5 Schedule of operations

Operation	*Plant-hours*	*Man-hours (productive trades)*
Erect hoarding		28
Cast tower crane base		13
Set up mixer, including all hard standings		40
Site accommodation		110
Excavate for drains and services		63
Water service and W.C.		20
Electric service		13
Reduce level dig	7 (J.C.B.)	21
Excavate trenches and bases, including internal drains	28 (J.C.B.)	84
Reinforcement in bases, foundations, lift base, and floor slab		98
Rough formwork to bases and edge of floor slab		196
Concrete in bases and trenches	22 (300/200 litre mixer)	
Internal drainage		96
Reinforced concrete frame:		
reinforcement		1402
formwork		4071
concrete	213 (300/200 litre mixer)	
Hardcore filling and blinding		152
D.P.M. and concrete ground floor slab	17 (300/200 litre mixer)	
Brickwork: to first floor level		185
first to second floor		264
second to third floor		258
third to fourth floor		264
fourth floor to roof		258

Operation	*Plant-hours*	*Man-hours (productive trades)*
External works:		
reduce level dig	14 (J.C.B)	
excavate for drains and manholes	10 (J.C.B.)	
excavate for services	7 (J.C.B.)	
excavate for foundations to boundary walls	7 (J.C.B.)	
concrete to foundations; drains and ducts	28 (300/200 litre mixer)	
brickwork to manholes and ducts, and lay drains		21
brickwork to boundary walls		85
backfilling, hardcore and blinding		205
clear huts from site		68
clear up		208
Blockwork backing to curtain walling:		
to first floor level		78
first to second floor		45
second to third floor		45
third to fourth floor		33
fourth floor to roof		80
Internal block partitions:		
to first floor level		174
first to second floor		196
second to third floor		196
third to fourth floor		166
fourth floor to roof		90
Roof screed		72
First fixings:		
carpenter		325
plumber		410
electrician		450
Plastering:		
to walls		590
to ceilings		375
floor screeds		350
Wall tiling		847
Second fixings and finishings:		
carpenter		480 + 150 = 630
plumber		320
electrician		220

Operation	*Plant-hours*	*Man-hours (productive trades)*
Painting and decorating:		
external		68
internal		840

Fig. 1.6 Schedule of sub-contractors

Operation	*Time required (weeks)*	*Notice required (weeks)*
Erect tower crane	1½	1
Asphalt roof covering	1	8
Curtain walling and glazing	6	12
Heat engineer:		
first fixing	6	16
second fixing	3	
Lift installation	8	20
Suspended ceilings	4	16
Cubicles to W.C.'s and showers	3	11
Timber wall panelling	2	7
Terrazzo floor finish	6	12
Wood block floor finish	2½	18
Mosaic to columns	1	10
Connect and lay services	½	2
Tarmac paving	1	8

1.3.12 Key operations

Key operations are those which determine the time required for the contract, e.g. sub-structure, frame erection, roofed-in stage, etc.

It is essential that key operations are planned realistically. If these are correct, then the programme will be practical.

1.3.13 Flexibility

To allow for wet time, absenteeism, etc., a 35-hour week has been used in the calculations and flexibility has also been provided by not overlapping operations the maximum time possible. Overtime working could be used to gain time if necessary.

1.3.14 General notes on overall programme (Fig. 1.8)

1. For operations in first six weeks, see notes with six-weekly programme.

2. Setting out

Setting out will be carried out during first week.

3. Frames

Carpenters on formwork will determine the time required for erection of the frame. The steelfixing gang will be balanced with the carpenter gang so that the time taken for steelfixing is approximately the same as that for formwork.

The staircases and lift walls are cast as work proceeds.

The floors are used to store bricks at each level before the next floor formwork is erected.

4. Brickwork and blockwork

Start of brickwork and blockwork is delayed until erection of frame is well advanced. This will also give continuity of work for bricklayers.

5. First and second fixing

Gang sizes are selected to provide continuity of work from first to second fixing. In the case of carpentry, it may be the firm's policy to have carpenters specializing in first and second fixing.

6. Plastering and tiling

Sufficient plasterers and tilers are used to overlap first fixing but to allow continuity of work for trades moving onto second fixing.

7. Suspended ceilings

Follow plastering to upper floors.

8. W.C. cubicles

Fixed to plastered or tiled walls and floor.

9. Wall panelling

Starts after commencement of wood blocks.

10. Floor finishes

Wood block, starts well after screeds to allow them to dry out.

11. Mosaic finishes to columns

Delayed to end of project to avoid damage.

12. Painting and decorating

External painting done first to allow scaffold to be stripped.

1.3.15 Crane

The crane will be released from site as soon as heavy lifting is complete. Whilst on site, the programme should provide maximum use, and to do this

Fig. 1.7 Calculations sheet

Operation no.	*Operation*	*Plant-hours*	
1	**Set up site**		
	Erect crane		
2	**Excavation**		
	Reduce level dig	7	35
	Excavate trenches, bases and internal drains	28	
3	**Formwork in frames bases, and slab**		
	Rough formwork to bases and edge of floor slab		
	In R.C. frame		
4	**Reinforcement in frames and bases**		
	Reinforcement in bases, foundations and floor slab		
	In R.C. frame		
5	**Internal drainage**		
6	**Concrete in frames, bases and trenches**		
	Concrete in bases and trenches	22	235
	In R.C. frame	213	
7	**Ground floor construction**		
	Hardcore fill and blinding	17	
	D.P.M. and concrete ground floor slab		
8	**Brickwork cladding**		
	to 1st floor level		
	1st to 2nd floor		
	2nd to 3rd floor		
	3rd to 4th floor		
	4th floor to roof		
9	**Curtain walling and glazing**		

Productive man-hours		*Labour and plant*	*Time required (weeks)*		*Remarks*
		2 carpenters 2/3 general labourers Plumber and mate Electrician and mate	2		Setting out for dig also undertaken in this period. G.P.O. to connect up for phones
		Crane erection gang (sub-contractor)	$1\frac{1}{2}$		
21 84	105	J.C.B. 4 and operator 3 labourers 2 lorries and drivers	1		Reduce level dig to loose fill can be moved with J.C.B. front bucket. Labourers dig widenings and bottom-up
196 4071	4267	4 carpenters for two weeks, then 8 carpenters Tower crane	16		
98 1402	1500	3 steel fixers Tower crane	15		
96		2 plumbers and mates	$1\frac{1}{4}$		
		300/200 litre mixer set up Tower crane 5 labourers, then 4 labourers	7		Work spread over $15\frac{1}{2}$ weeks. Floors loaded with bricks
152		7 labourers 300/200 litre mixer set up 5 labourers	$\frac{3}{4}$ $\frac{1}{2}$	$1\frac{1}{4}$	Work spread over a period of 2 weeks and integrated with bases and trenches
185 264 258 264 258	1229	6 bricklayers 4 labourers 150/100 litre mixer	6		1 labourer on 150/100 litre mixer
		Sub-contractor	6		12 weeks notice required

Fig. 1.7 *continued*

Operation no.	*Operation*	*Plant-hours*
10	**Roof screed**	
11	**Asphalt**	
12	**Internal blockwork to all levels**	
	to 1st floor level	
	1st to 2nd floor	
	2nd to 3rd floor	
	3rd to 4th floor	
	4th floor to roof	
13	**Internal block partitions**	
	to 1st floor level	
	1st to 2nd floor	
	2nd to 3rd floor	
	3rd to 4th floor	
	4th floor to roof	
14	**Carpentry**	
	1st fixing	
	2nd fixing	
15	**Plumbing**	
	1st fixing	
	2nd fixing	
16	**Electrical work**	
	1st fixing	
	2nd fixing	
17	**Heating**	
	1st fixing	
	2nd fixing	
18	**Plastering and floor screeds**	
	Plastering to walls	
	Plastering to ceilings	
	Floor screeds	
19	**Wall tiling**	
20	**Lift installation**	

Productive man-hours		*Labour and plant*	*Time required (weeks)*	*Remarks*
72		2 plasterers	1	
		2 labourers		
		150/100 litre mixer		
		Sub-contractor	1	8 weeks notice required
78		2 bricklayers	4	
45		1 labourer		
45	281	150/100 litre mixer		
33				
80				
174		4 bricklayers	6	
196		2 labourers		
196	822	150/100 litre mixer		
166				
90				
325		2 carpenters	5	
630		2 carpenters	9	
410		2 plumbers and mates	6	
320		2 plumbers and mates	4½	
450		1 electrician and mate	4	4 weeks total time for conduits laid on form-work before concreting by 1 electrician and mate
		2 electricians and mates	4½	
220		1 electrician and mate	6	
		Sub-contractor	6	16 weeks notice required
		Sub-contractor	3	
590		6 plasterers	6	
375	1255	4 labourers		
300		150/100 litre mixer		
847		6 tilers	4	
		3 labourers		
		Sub-contractor	8	20 weeks notice required

Fig. 1.7 Continued

Operation no.	*Operation*	*Plant-hours*
21	**Suspended ceilings**	
22	**Cubicles to W.C.'s and showers**	
23	**Wall panelling**	
24	**Floor finishes**	
	Terrazzo	
25	**Wood blocks**	
26	**Painting and decorating**	
	External	
	Internal	
27	**Mosaic finishes**	
EXTERNAL WORKS		
28	**Excavation**	
	Reduce level dig	14 } 38
	Excavate for drains and manholes	10
	Excavate for services	7
	Excavate for foundations to boundary walls	7
29	**Concrete foundations, beds to drains, ducts**	28
30	**Brickwork to manholes and boundary walls**	
	Brickwork to manholes and ducts, lay drains	
	Brickwork to boundary walls	
31	**Connect and lay services**	
32	**Backfill hardcore and blinding**	
33	**Tarmac paving**	
34	**Clear huts from site (includes hoardings)**	
35	**Generally clean and tidy-up**	

Productive man-hours	*Labour and plant*	*Time required (weeks)*	*Remarks*
	Sub-contractor	4	16 weeks notice required
	Sub-contractor	3	11 weeks notice required
	Sub-contractor	2	7 weeks notice required
	Sub-contractor	5	Toilet areas executed first to allow sanitary fiittings to be fixed; 12 weeks notice required
	Sub-contractor	$2\frac{1}{2}$	18 weeks notice required
65 840	2 painters 6 painters	1 4	
	Sub-contractor	1	10 weeks notice required
42 30 21 21 } 114	J.C.B. 4 and operator 3 labourers 2 lorries and drivers	1	Labourers dig branches and generally bottom-up
84	150/100 litre mixer 3 labourers	$\frac{3}{4}$	
21 85 } 106	2 bricklayers 1 labourer 150/100 litre mixer	$1\frac{1}{2}$	
	Sub-contractor	$\frac{1}{2}$	2 weeks notice required; laid by personnel from chemical works
205	3 labourers, vibrating roller for compacting	2	
	Sub-contractor	1	8 weeks notice required
68	2 carpenters	1	
208	3 labourers	2	

OVERALL BUILDING PROGRAMME FOR AMENITY CENTRE & OFFICE BLOCK — CONTRACT NO.

OP NO	OPERATION	
1	SET UP SITE	2/3 GENERAL LABS. 2 CARPS. PLUMB & MATE. ELECT. & MATE, CRANE ERECTION GANG IN 3RD & 4TH WEEKS. SUB CONTRACT
2	EXCAVATION	J.C.B. 4 & OPERATOR. 2 LORRIES & DRIVERS. 3 GENERAL LABS. INTEGRATED WITH OP NO.1
3	FORMWORK IN FRAMES BASES & SLAB	4/8 CARPS. TOWER CRANE & GANG
4	REINF. IN FRAMES & BASES	2/3 STEELFIXERS. TOWER CRANE & GANG
5	INTERNAL DRAINAGE	2 PLUMB & MATE (1 FROM OP NO.1)
6	CONC IN FRAMES BASES & TRENCHES	300/200 MIXER & 5 LABS, THEN 4 LABS (CRANE GANG) 500kg TOWER CRANE
7	GROUND FLOOR CONSTRUCTION	300/200 MIXER & 7 LABS INTEGRATED WITH OP NO 6 – LORRIES SPREAD HC AS DELIVERED
8	BRICKWORK CLADDING	6 BRICKLAYERS, 4 LABS 150/100 MIXER. HOIST AFTER HOLIDAY
9	CURTAIN WALLING	SUB-CONTRACT
10	ROOF SCREED	2 PLAST. 2 LABS 150/100 MIXER, HOIST
11	ASPHALT	SUB-CONTRACT. (SCAFFOLD REQ'D)
12	INTERNAL BLOCKWORK TO CILL LEVEL	2 BRICKLAYERS 1 LAB 150/100 MIXER. HOIST AFTER HOLIDAY
13	INTERNAL BLOCK PARTITIONS	4 B'LAYERS, 2 LABS, 150/100 MIXER, HOIST (2+1 FROM OP. NO.12)
14	CARPENTRY	1ST FIX, 2ND FIX. 2 CARPS. HOIST
15	PLUMBING	1ST FIX, 2ND FIX. 2 PLUMBERS & MATES HOIST
16	ELECTRICAL WORK	1ST FIX – CONDUITS ON FORMWORK. 2ND FIX. 2/ELECTICIANS & MATES
17	HEATING	SUB-CONTRACT
18	PLASTERING & FLOOR SCREEDS	6 PLASTERERS 4 LABS, 150/100 MIXER HOIST
19	WALL TILING	6 TILERS, 3 LABS
20	LIFT INSTALLATION	SUB-CONTRACT
21	SUSPENDED CEILINGS	SUB-CONTRACT
22	CUBICLES TO WC's & SHOWERS	SUB-CONTRACT
23	WALL PANELLING	SUB-CONTRACT
24	FLOOR FINISHES – TERRAZZO	SUB-CONTRACT.
25	WOOD BLOCK	SUB-CONTRACT
26	PAINTING & DECORATING	2 PAINTERS; (INTERNAL) 6 PAINTERS
27	MOSAIC FINISHES	SUB-CONTRACT
	EXTERNAL WORKS.	
28	EXCAVATION	JCB & OP 2 LORRIES & DRIVERS 3 LABS
29	CONC FDNS & BEDS TO DRAINS ETC	3 LABS 150/100 MIXER
30	BRICKWORK TO MH's & BOUNDARY WALLS	2 BRICKLAYERS 1 LAB 150/100 MIXER
31	CONNECT & LAY SERVICES	SUB CONTRACT
32	BACKFILL, HARDCORE & BLINDING	3 LABS VIBRATING ROLLER
33	TARMAC PAVING	SUB CONTRACT
34	CLEAR HUTS FROM SITE	2 CARPS
35	GENERALLY CLEAN & TIDY UP	3 LABS

Other chart notes: SCAFFOLD REQ'D; WALLS WATERTIGHT; ROOF WATER-TIGHT; START STRIPPING SCAFFOLD; BANK HOLIDAY; ANNUAL HOLIDAY; CONTRACT PERIOD; CALLING UP DATES SHOWN THUS ▷

Fig. 1.8

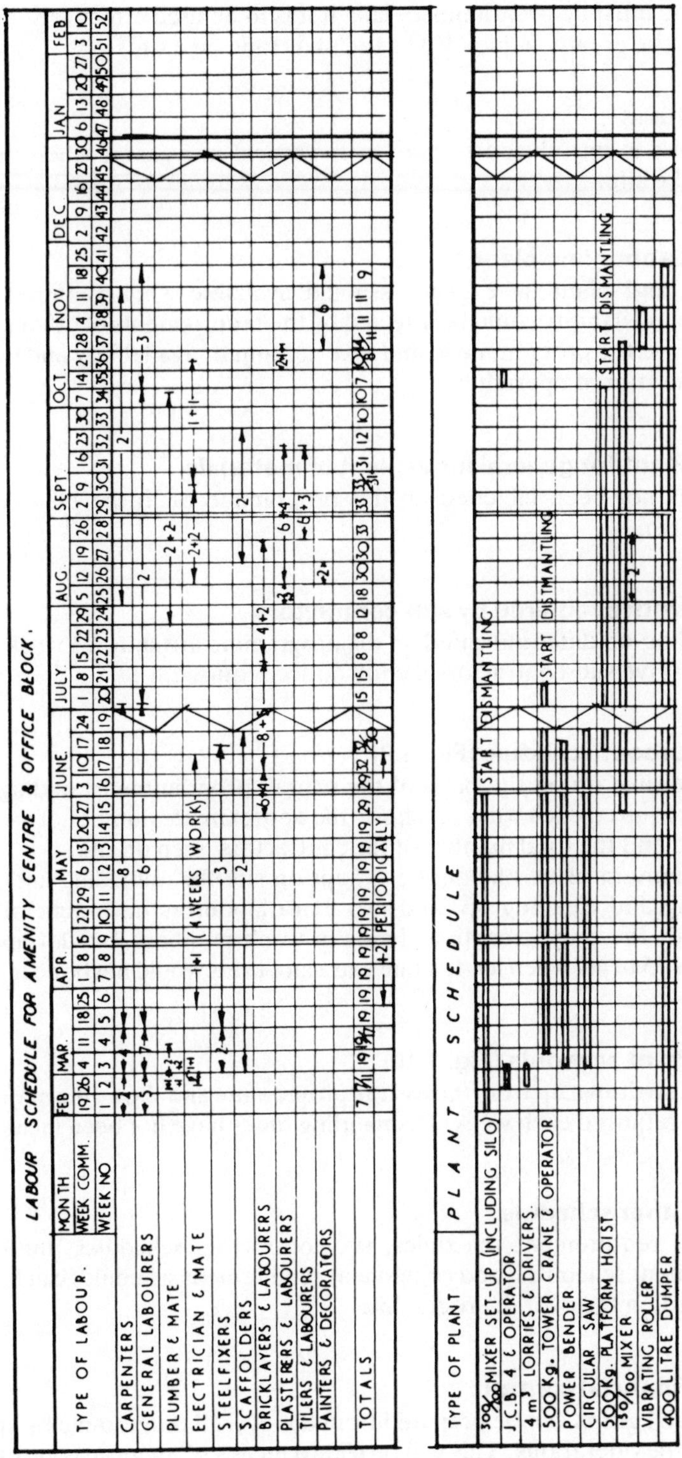

LABOUR SCHEDULE FOR AMENITY CENTRE & OFFICE BLOCK.

TYPE OF LABOUR.

MONTH: FEB, MAR., APR., MAY, JUNE., JULY., AUG., SEPT., OCT., NOV, DEC., JAN, FEB

WEEK COMM.: 19 26 4 11 18 25 1 8 15 22 29 6 13 20 27 3 10 17 24 1 8 15 22 29 5 12 19 26 2 9 16 23 30 7 14 21 28 4 11 18 25 2 9 16 23 30 6 13 20 27 3 10

WEEK NO: 1 2 3 4 5 6 7 8 9 10 11 12 13 14 15 16 17 18 19 20 21 22 23 24 25 26 27 28 29 30 31 32 33 34 35 36 37 38 39 40 41 42 43 44 45 46 47 48 49 50 51 52

CARPENTERS
GENERAL LABOURERS
PLUMBER & MATE
ELECTRICIAN & MATE (4 WEEKS WORK)
STEELFIXERS
SCAFFOLDERS
BRICKLAYERS & LABOURERS
PLASTERERS & LABOURERS
TILERS & LABOURERS
PAINTERS & DECORATORS

TOTALS: 7 7/11 19 19/17 19 19 19 19 19 19 19 19 19 19 29 29 32 32/30 15 15 8 8 12 18 30 30 33 33 33/31 31 12 10 10 7 10/8 14/11 11 11 9

+2 PERIODICALLY

TYPE OF PLANT — PLANT SCHEDULE

30/200 MIXER SET-UP (INCLUDING SILO) — START DISMANTLING
J.C.B. 4 & OPERATOR.
4 m³ LORRIES & DRIVERS.
500 Kg. TOWER CRANE & OPERATOR — START DISMANTLING
POWER BENDER
CIRCULAR SAW
500 Kg. PLATFORM HOIST — START DISMANTLING
150/100 MIXER
VIBRATING ROLLER
400 LITRE DUMPER

Fig. 1.9 (*top*) and **Fig. 1.10** (*lower*)

the crane must be programmed daily if it is to be used effectively. The gang sizes can be greatly influenced by use of a crane on a site.

1.3.16 Hoist

The hoist is introduced to take over vertical movement when the crane leaves the site.

1.3.17 Labour and plant

Labour and plant have been assumed available when required, but in practice availability could be affected by the requirements of other projects. In this case the programme would take account of availability and this could delay the start of operations.

1.3.18 Plant for general movement of materials

This has not been included on this programme or in the plant schedule (Fig. 1.10).

1.3.19 Notice required by sub-contractors

The calling-up dates indicated on the programme are shown 1 week earlier; this is to give sub-contractors the full notice required.

1.3.20 Labour schedule (Fig. 1.9)

This is drawn up on the basis of the overall programme by adding up the labour required each week. It shows the approximate number of each trade required and the total number of the contractors' men on site.

A minimum of two labourers is kept on site for unloading and general sweeping and cleaning up, and two extra labourers have generally been provided, the only exceptions being in week numbers 3 and 4 and week numbers 35 to 40, when it is felt that extra labourers would not be economical.

1.3.21 Plant schedule (Fig. 1.10)

This is also drawn up from the overall programme and shows the major items of plant required each week. Dismantling times have not been included.

1.3.22 Other schedules

Material requirement schedules, sub-contractors schedules, information requirement schedules, and component requirement schedules can be drawn up, based on the overall programme.

1.3.23 Financial control

A financial graph can be prepared from the overall programme by applying costs to the operations. This will provide a means of financial control.

1.4 SEQUENCE STUDIES

1.4.1 Introduction

It is necessary to compile sequence studies when closely related operations have to be coordinated, e.g. the construction of the columns, beams and slabs in a reinforced concrete framed building or the finishing trades in a building such as a multi-story office block which would have much repetitive work on the various floors. (The line of balance technique described in section 1.10.21 could also be used.)

1.4.2 Achieving continuity

To achieve continuity of work throughout these repetitive cycles, it is necessary to balance the gangs of workmen.

1.4.3 Procedure for preparing a sequence study

Assume the frame of a twelve storey reinforced concrete building is to be erected and the work content per floor is as follows:

Columns and Walls

Reinforcement	128 man hours
Concreting	20 man hours
Formwork	512 man hours

Floors (including beams and slab)

Reinforcement	160 man hours
Concreting	80 man hours
Formwork	640 man hours

An 8 hour day is used in the calculations. The hardening period for columns and walls is 2 days and for floors and beams 7 days, the props being left in. The preparation of the sequence study is then carried out as follows:

1. Calculate the work content of steelfixers and carpenters

	Reinforcement	*Formwork*
Columns and Walls	128	512
Floors	160	640
Total	288	1152

2. Balance the gangs

Ratio of carpenters to steelfixers $\frac{1152}{288} = 4{:}1$

∴ Use 8 carpenters and 2 steelfixers (could have used other combinations giving a 4:1 ratio).

NOTE: Concrete gang will be used intermittently as required and will be working on other operations when not required on the frame.

The sequence will be arranged to re-use formwork as much as possible. In this example sufficient formwork for one complete floor will be required.

3. Calculate operations times
 Time required for columns and walls

 Reinforcement $\dfrac{128}{2 \times 8} = 8$ days (half floor = 4 days)

Formwork $\dfrac{512}{8 \times 8} = 8$ days – assume 6 erect and 2 strip
 (half floor assume 3 erect and 1 strip)

Time required for floors and beams

Reinforcement $\dfrac{160}{2 \times 8} = 10$ days

 Formwork $\dfrac{640}{8 \times 8} = 10$ days – assume $7\frac{1}{2}$ erect $2\frac{1}{2}$ strip
 (half floor assume $3\frac{3}{4}$ erect $1\frac{1}{4}$ strip)

4. Prepare the sequence study providing continuity on the repetitive work cycles on each floor (see Fig. 1.11). Of the three trades involved the carpenters have the least flexibility because all their work takes place on the building itself. Steelfixers have a reasonable amount of flexibility because they can fabricate columns and beams when they are not actually fixing steel in position. When sufficient column cages have been erected it is possible to start the formwork and care must be taken to allow the required hardening periods before the various formwork elements are stripped. The sequence of formwork erection and stripping is shown in Fig. 1.11.

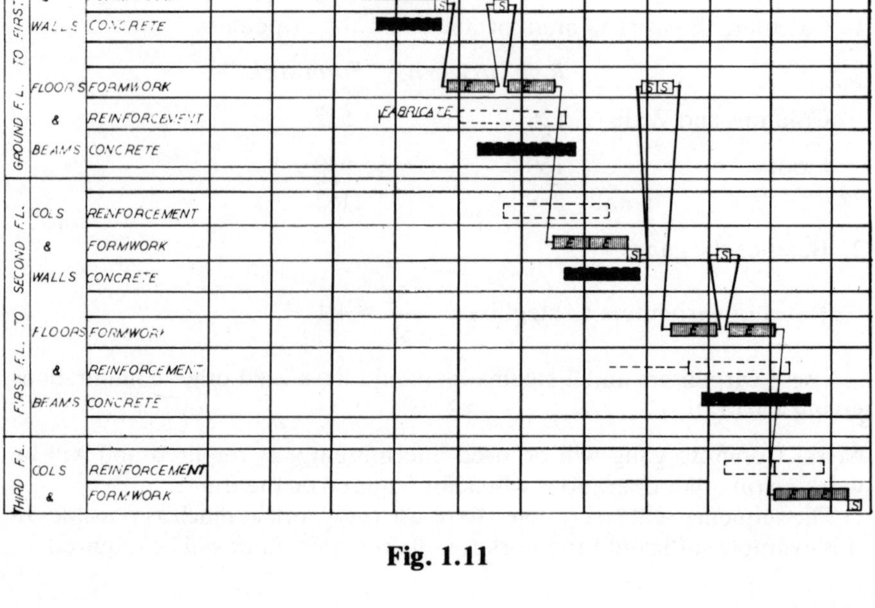

Fig. 1.11

As the formwork and steelfixing gangs have been balanced it is not necessary to show the fixing of steel in great detail but merely to indicate the period over which it would be carried out. As stated previously, the concrete would be placed intermittently as required.

If more detailed planning is to be carried out this would take place at the short term planning stage (see section 1.6).

1.4.4 Sequence study for the frame of the Amenity Centre

1. Formwork stripping times

The stripping times for formwork stated in the B.O.Q. are as follows:

Suspended floor and beam soffites	7 days (props left in)
Columns	1 day
Staircases	7 days

To provide continuity of work for carpenters it will be necessary to have sufficient formwork for half the columns on one floor, one complete floor soffite and all the stairs to one floor.

2. Preparing the sequence study
   (a) Using the quantities involved, calculate the work content for carpenters and steelfixers as follows:

Carpenters	4071 hours
Steelfixers	1402 hours

   (b) Balance the gangs to avoid non-productive time. It is difficult to provide continuity for the concrete gang as shown in the previous example:

Carpenters	4071 hours – with 8 carpenters = 509 gang hours
Steelfixers	1402 hours – with 3 steelfixers = 467 gang hours

   (c) Using the gang sizes calculated above calculate the work content for each operation (see sequence study chart Fig. 1.12).
   (d) Programme the work providing continuity on the repetitive work cycles on each floor.

3. General notes
   (a) As operatives become experienced in the repetitive operations the output would increase so that on the upper floors, working time would be reduced. This has not been allowed for in the programme.
   (b) Scaffolding has not been shown on the programme but this would be programmed in sequence with the frame.
   (c) The reinforcement cages for the columns and some of the beams are to be prefabricated. The tower crane will be used to lift the fabricated columns and beams into position and to move formwork from floor to floor.
   (d) Weeks only have been used on Fig. 1.12 for clarity in publication but days would also be included in practice.

	OPERATION		PLANT HRS.	PROD MAN HRS.	LAB REQD	TIME REQD. (DAYS)	WEEK NO. 3	4	5	6	7	8
GD. TO 1ST. FLOOR	COLUMNS	REINF		78	2/3 S	4						
		FWK		240	8 C	4.25						
		CONC	14			2						
	1ST FLOOR	FWK		584	8C	10.25						
		REINF		219	3S	10.25						
		CONC	31			4.5						
	STAIRS	FWK		178	8C	3.25						
		REINF		36	3S	1.75						
		CONC	9			1.25						
1ST. TO 2ND. FLOOR	COLUMNS	REINF		75	3S	3.5						
		FWK		202.5	8 C	3.5						
		CONC	9.5			1.25						
	2ND FLOOR	FWK		543	8C	9.75						
		REINF		194	3S	9.25						
		CONC	27			4						
	STAIRS	FWK		178	8C	3.25						
		REINF		36	3S	1.75						
		CONC	9			1.25						
2ND. TO 3RD. FLOOR	COLUMNS	REINF		75	3S	3.5						
		FWK		202.5	8C	3.5						
		CONC	9.5			1.25						
	3RD FLOOR	FWK		543	8C	9.75						
		REINF		194	3S	9.25						
		CONC	27			4						
	STAIRS	FWK		178	8C	3.25						
		REINF		36	3S	1.75						
		CONC	9			1.25						
3RD. TO 4TH. FLOOR	COLUMNS	REINF		75	3S	3.5						
		FWK		202.5	8C	3.5						
		CONC	9.5			1.25						
	4TH FLOOR	FWK		543	8C	9.75						
		REINF		194	3S	9.25						
		CONC	27			4						
	STAIRS	FWK		98	8C	1.75						
		REINF		18	3S	1						
		CONC	5			0.75						
4TH. FLOOR TO ROOF	COLUMNS	REINF		48	3S	2.25						
		FWK		125	8C	2.25						
		CONC	9.5			1.25						
	ROOF	FWK		322	8C	5.75						
		REINF		120	3S	5.75						
		CONC	14			2						
	STAIRS	FWK		32	8C	0.5						
		REINF		6	3S	0.25						
		CONC	3			0.5						

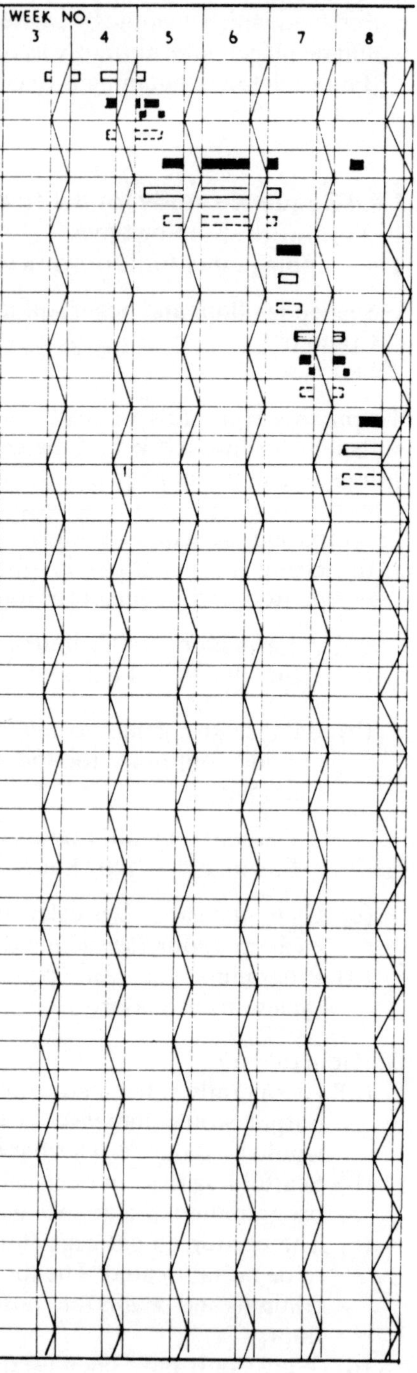

Fig. 1.12 Sequence study for amenity centre R.C. frame.

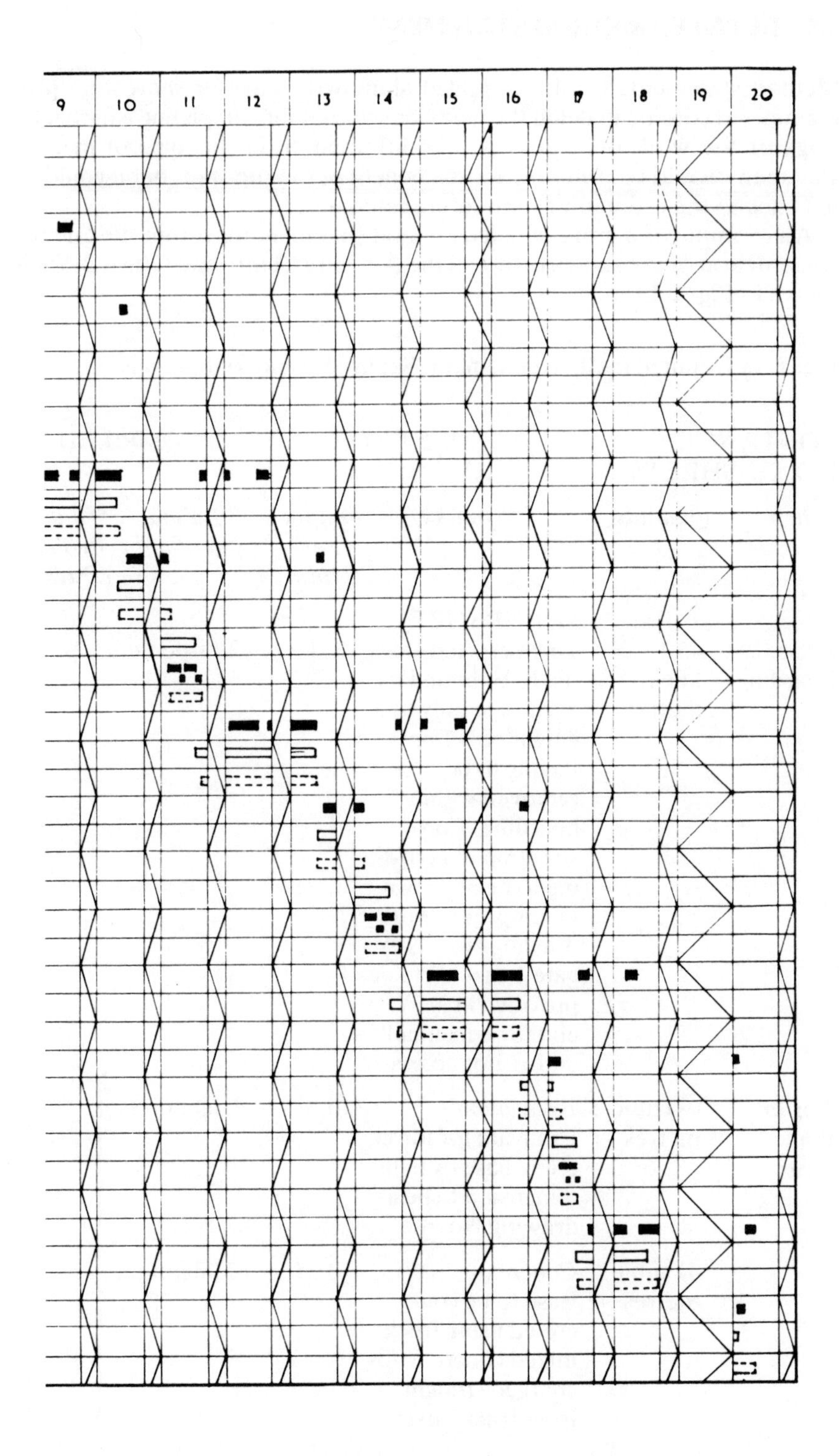
9
10
11
12
13
14
15
16
17
18
19
20

1.5 DETAILED METHOD STATEMENT

Method statements can be compiled along with stage or short term programmes recording in detail the methods selected for completing a particular stage of the work on a project. The information in this type of method statement would be primarily for the benefit of the site staff, but it could be fed back to the estimator for his future reference.

An example of a more detailed method statement for set-up site and the construction of the sub-structure of the Amenity Centre and Office Block is shown in Fig. 1.13.

Fig. 1.13 Detailed method statement. Set up site and foundations

CONTRACT SHEET NO. COMPILED BY
NO. OF SHEETS DATE:

Item	*Quantity*	*Remarks*	*Output Per Hour*	*Labour & Plant*	*Time Req'd (Days)*
		SET UP SITE			
Erect hoarding	45 linear metres	To be erected along the N.E. boundary of the site only to a height of 2·2 m. Paving to be removed within hoarding. Construction of hoarding to be as shown in temporary works drawing No. . A gate 1 m wide to be provided at N.W. end for personnel and visitor access.	1.4	2 carpenters	2
Tower crane base	$6\frac{1}{2}$ cubic metres	*Excavation* Excavate for tower crane base by hand, dimension to be as drawing No.	0.5	2 labourers	1 (Excavation and Concreting together)
	$4\frac{1}{2}$ cubic metres	*Concreting* Base to be concreted using truck mixed concrete discharged straight from truck mixer	2.25	2 labourers	

Fig. 1.13 *continued*

Items	*Quantity*	*Remarks*	*Output Per Hour*	*Labour & Plant*	*Time Req'd (Days)*
		standing on works service road. Mild steel angles cast into base to receive bottom section of tower crane mast.			
	–	*Disposal of spoil* Spoil to be left alongside excavations and removed later with reduced level dig.	–	2 lorries + drivers J.C.B. 4 + operator	
Hard-standings	–	Hard-standing to be concreted using readymixed concrete delivered via works service road. Concrete for mixer hard standing delivered via truck shute, conc. for other hard-standings delivered by wheelbarrow. Excavate spoil to be placed alongside hard-standing area for removal with reduced level dig.	–	3 labourers	2
Set-up Mixer	–	Mixer and silo delivered to site on low loader off loaded and located as shown on site layout. Mobile crane hired from client for off-loading and erection of silo. Walls between aggregate to be	–		

Fig. 1.13 *continued*

Items	*Quantity*	*Remarks*	*Output Per Hour*	*Labour & Plant*	*Time Req'd (Days)*
		constructed from railway sleepers and rolled steel I section stakes.			}
Site Accom-modation	–	To be located in areas shown on site layout plan. Sectional timber construction based on 1 m module. All provided with insulation, heating and lighting. All units to be sup-ported clear of the ground on railway sleepers.		2 carpenters	2
All services	21 cubic metres	*Excavation* Excavation for all temporary services to be carried out by hand, any surplus material being deposited alongside the excavation for removal at same time as reduced level dig.	0.3	3 labourers	3
	–	*Water service & W.C.* Service provided to areas as indicated on site layout plan – water pipes to be P.V.C. and placed 300 mm below surface. W.C.'s provided in mobile amenity unit located over and connected to exist-	–	Plumber & mate	3

Fig. 1.13 *continued*

Items	*Quantity*	*Remarks*	*Output Per Hour*	*Labour & Plant*	*Time Req'd (Days)*
		ing m.h. – special cover (see drawing No. 365) used to make gas-tight.			
	–	*Electrical service* Electrical service provided in form of 415 V three phase supply to all electrically operated plant and 220 V single phase to accommodation for heating and lighting. Location of supply to be as shown on site layout drawing. Supply to mess-room from dist. board in clerk of works office to be by overhead cable. All other electrical service routes to be underground in trench 300 mm deep – protected by pitch fibre pipe.	–	Electrician & mate	2
Erect Tower Crane	–	Erection of crane to be carried out by fitters from plant department.		Crane erection gang	7
Foundations	105 cubic metres	*Reduced Level dig* Reduced level dig will be carried out over area of building by J.C.B. 4 using front bucket, as going is light – all spoil will be loaded into lorries and	15	J.C.B. 4 & operator 3 labourers	1

Fig. 1.13 *continued*

Items	*Quantity*	*Remarks*	*Output Per Hour*	*Labour & Plant*	*Time Req'd (Days)*
		taken to tip within works, total haul distance approx. 400 m.			
	140 cubic metres	*Excavate Trenches and bases* Tie beam trenches excavated using rear back-acter bucket of J.C.B. 4 surplus spoil loaded into lorries and removed to tip. Bottoming up and trimming at isolated foundation to be completed by hand.	5	J.C.B. 4 & operator 3 labourers	4
	See Bar Schedule for quantities	*Reinforcement* Reinforcement for foundations prepared in cutting and bending areas and carried to foundation by hand.	–	2 steel fixers	7
	See Formwork Schedule for quantities	*Rough formwork* Formwork to foundation and edge of slab to consist of sawn boards and bearers, prefabricated in site shop and transported to fixing position by hand *x* square metres being used, stripped and refixed as required.	–	4 carpenters	7
Internal Drainage	See Drain-	All internal drainage is in cast iron –	–	Plumber & mate	7

Fig. 1.13 *continued*

Items	*Quantity*	*Remarks*	*Output Per Hour*	*Labour & Plant*	*Time Req'd (Days)*
	age Schedule for quantities	trench excavation carried out by J.C.B. along with excavations for tie beams – transported from stores area by hand, surrounded with granular fill when fixed.			
Concrete in bases & trenches	66 cubic metres	Concrete produced in site batching plant and delivered to bases etc. by dumper.	3	300/200 litre mixer + operator & 7 labourers	3
Hardcore fill	100 cubic metres	Limestone hardcore delivered to site in 10 tonne lorries and tipped into place. Sleepered area provided to bridge tie beams thus allowing lorries to pass into area within tie beams. Limestone spread from back of lorries as it is tipped, final levelling being carried out by hand.	1.0	7 labourers	$3\frac{1}{2}$ (Hardcore fill and Blinding together)
Blinding	14 cubic metres	Concrete for blinding delivered from batching plant by dumper tipped into position and spread and levelled by hand.	1	300/200 litre mixer + operator & 7 labourers	

Fig. 1.13 *continued*

Items	*Quantity*	*Remarks*	*Output Per Hour*	*Labour & Plant*	*Time Req'd (Days)*
Concrete Ground Floor	51 cubic metres	*Mixing* Mixed in site batching plant. *Transportation* From batching plant to forms by dumper. *Reinforcement* Mesh reinforcement placed in centre of slab. *Placing* Compaction carried out using vibrating tamping beam. *Construction* Hit and miss construction – stop ends placed as shown on sketch below (Fig. 1.14).	3	300/200 litre mixer + operator & 7 labourers	$2\frac{1}{2}$

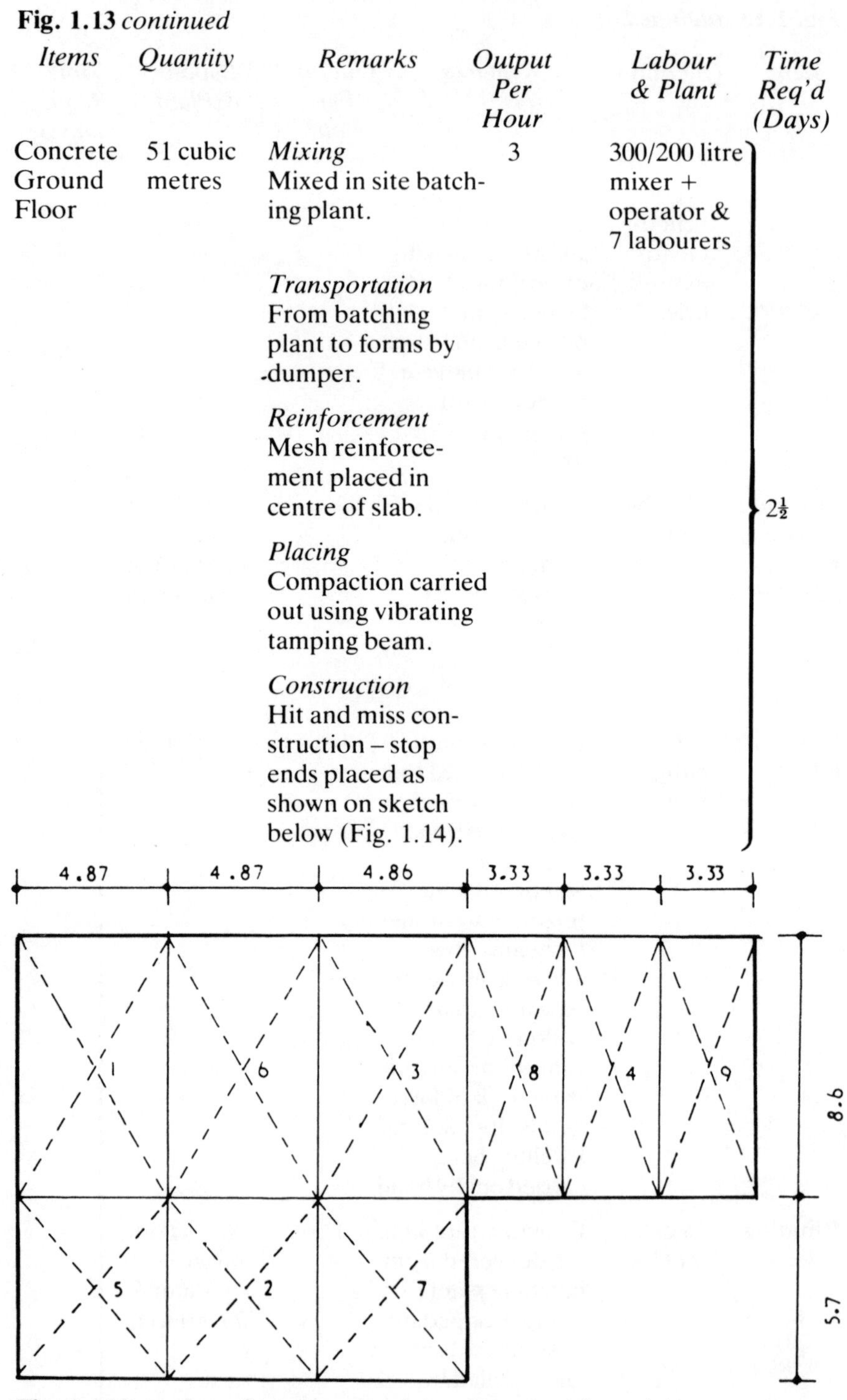

Fig. 1.14 Location of construction joints in ground floor slab

1.6 SHORT TERM PLANNING

1.6.1 Period of time covered

Short term programming can cover a period of four to six weeks or may cover a stage of the work. The programme considered here will cover a period of six weeks. A new programme is drawn up every fourth week, thus giving a two week overlap and allowing a review to be made of the work outstanding from the previous month; any work behind schedule can then be included in the current programme.

1.6.2 Purpose

The purpose of short term programming is to ensure that work proceeds in accordance with the overall programme. The overall programme is thus converted into a working schedule and must be updated regularly, whilst a review is simultaneously made of all requirements by checking schedules.

1.6.3 Time of preparation

The most suitable time for preparation of the programme is after the monthly planning meeting, which is in turn best held immediately after the architect's monthly meeting. In this way account can be taken of any alterations or variations which are necessary.

1.6.4 Originator of programme

The programme should be originated by the site manager, if necessary with the assistance of a projects manager and a planning engineer. The co-operation of foremen and sub-contractors should be obtained by getting them involved and keeping them informed. They would be present at the monthly site meeting to give their point of view and agree the general plan of action.

1.6.5 Degree of detail necessary

The operations in the overall programme will have to be broken down to show each element of work to be completed, e.g. the reinforced concrete frame – which was one operation on the overall programme – will be broken down into the elements of work on each floor:

Reinforcement in columns
Formwork to columns
Concrete to columns
Formwork to beams and slab
Reinforcement to beams and slab
Concrete to beams and slab.

1.6.6 Basis of calculations

The basis of calculations at this stage must be the operational records from previous contracts. There should be an integrated cost control/bonusing/planning and surveying system which will provide the basic information for all departments. The expected outputs for bonus operations should be used in calculating the time required.

The bill of quantities cannot be used in calculating time periods for these detailed operations as the outputs used to price the bill are averages and do not take into account the conditions of work in individual operations.

1.6.7 The six-weekly programme (Fig. 1.15)
The period covered is the first six weeks of the contract.

1.6.8 Procedure for programming
1. Break down the operations in the overall programme into more detailed operations as shown in the six-weekly programme (Fig. 1.15), taking care to include all operations. Any preparatory work necessary must be included.
2. Fill in plant-hours and/or productive man-hours for the detailed operations.
3. Decide on the number of operatives and plant necessary and calculate the time required, plotting each operation on the programme as it is considered. The aim is to provide continuity of work and to meet the targets set in the overall programme.

1.6.9 Labour requirements
These can now be obtained from the six-weekly programme by adding up the labour required each day (Fig. 1.15). Extra labourers necessary for off-loading, etc. are not included here.

1.6.10 Plant requirements
These can also be obtained from the six-weekly programme.

1.6.11 Other requirements
Requirements for details, nominations, materials, etc. can also be obtained from this programme and a check made with the schedules originated at the overall programming stage to ensure that all requirements will be available when required.

1.6.12 General notes on six-weekly programme

1.6.12.1 Water for works and drinking
During the first week this can be obtained from a tap on the works near to the site. After the first week, it will be laid on.

1.6.12.2 Tower crane base and all hardstandings
Ready mixed concrete will be used for pouring the tower crane base and hardstandings.

1.6.12.3 Cutting, bending and fixing steel
The steelfixers' time includes cutting and bending. Fixing of steel proceeds after rough formwork in bases and slab, with steel in bases and trenches being fixed first to allow concreting to proceed.

CONTRACT – AMENITY CENTRE & OFFICE BLOCK.
CONTRACT NO.

SIX WEEKLY PROGRAMME

PERIOD – FEB 19TH TO MAR 29TH

OP. NO	OPERATION	PLANT HRS or PROD M.HRS	LABOUR &/OR PLANT	TIME REQ DAYS
	SET UP SITE			
1	ERECT HOARDING	28	2 CARPS	2
2	TOWER CRANE BASE	13	2 LABS.	1
3	SET UP MIXER & ALL H'D'ST'G'S	40	3 LABS.	2
4	SITE ACCOMMODATION	110	2 CARPS	8
5	EXCAVATE FOR ALL SERVICES	63	3 LABS.	3
6	WATER SERVICE & W.C.	20	PL. & M'TE	3
7	ELECTRIC SERVICE	13	EL. & M'TE	2
8	ERECT TOWER CRANE	ERECT'N GANG		7
	FOUNDATIONS			
9	REDUCE LEVEL DIG	7	JCB & OP, 2 LORRY	1
10	EXC. TRENCHES & BASES	28	3 LABS	4
11	REINF. IN FOUNDS. ETC.	98	2 ST'F'X'S	7
12	RO. FWK. IN BASES & SLAB	196	4 CARPS.	7
13	INTERNAL DRAINAGE	96	2 PL. & M'TE	7
14	CONC. IN BASES & TRENCH'S	22	MXR. & 7 L.	3
15	H.C. FILL & BLINDING	152	7 LABS	3½
16	D.P.M. & G.F. CONC.	17	MXR + 7 L.	2½
	COLS TO U/S. F.F. BEAMS			
17	REINF.	78	2/3 ST'F'X'S	4
18	FORMWORK	240	8 CARPS	4½
19	CONCRETE	14	MIXER & CRANE + 4	2
	1ST. FLOOR INC. CANOPY			
20	FORMWORK	584	8 CARPS	10½
21	REINF.	253	3 ST'F'X'S	12
22	CONCRETE	31	MIXER & CRANE + 4	4½
23	CONDUITS (IN FL. SLAB)	27	1 EL. & M'TE	4
24	SCAFFOLDING			

Bar chart columns: FEBRUARY 19, 26; MARCH 4, 11, 18, 25 — days M T W T F S S. Op 8: CONC. MATURING. Op 14: INTEGRATE WITH OP 15. Op 15: INTEGRATE WITH OP 14.

REMARKS — LABOUR REQUIREMENTS

CARPS.	2, 2, 4, 4, 8, 8, 8
GENERAL LABS.	2, 3, 3, 3, 7, 7, 4, 4
PLUMBER & MATE	1 1, 2 2, 2 2
ELECT. & MATE	1 1, 1 1
STEELFIXERS	2, 2, 3, 3, 3
CR. ERECT'N GANG	SUB-CONTRACT
SCAFFOLDERS	2, 2, 2, 2

REMARKS — PLANT REQUIREMENTS

- 300/200 MIXER
- J.C.B. 4 & OPERT'R
- 4 m³ LORR. & DR. (2)
- 500 Kg. TOWER CR. & OP
- POWER BENDER
- CIRCULAR SAW
- 400 litre DUMPER

Fig. 1.15

1.6.12.4 Rough formwork in bases and slab
Formwork to bases has priority so that concreting can proceed.

1.6.12.5 Plant-hours
It has been assumed that the plant fixes the speed of operations for excavating and concreting and that sufficient labour is used to attend the plant.

1.6.12.6 Formwork
The two-day period between 'rough formwork in bases and slab' and 'formwork to columns' can be used for adjusting the form panels, fixing stops, etc.

1.6.12.7 Use of crane
When the crane is not used on concreting operations it will be employed for moving formwork, placing steel, distributing materials, etc., in accordance with the daily programme (Fig. 1.18).

1.6.12.8 Stripping times for columns
It is assumed that the stripping time for columns is 24 hours. The hours given includes erect, strip, clean, and re-oil.

1.6.12.9 Concreting to first floor, including canopies
Concreting to first floor includes the canopies and is placed in two sections. The first section is placed when half the formwork and steel is fixed, thus allowing half the formwork to be stripped earlier and re-used on the next floor.

1.6.13 Sub-contract programmes

If required these can be extracted from the detail programmes for handing to sub-contractors and show how the sub-contractor's work ties in with the main contractor's work.

1.7 WEEKLY PLANNING

1.7.1 Purpose

The purpose of weekly programming is to ensure that the six-weekly programme is effectively carried out. It is used as a basis for operational instructions, for communicating the plan to the trades foremen, gangers, and operatives, and to help in co-ordinating the requirements of different sites.

1.7.2 Time for preparation

It should be prepared after a weekly site meeting at which gangers and foremen are present, together with others who will be involved in the following week's work.

1.7.3 Originator of programme

The programme should be originated by the site manager with the assistance of a planning engineer if necessary. The work should be discussed with foremen and sub-contractors when they are affected.

1.7.4 Degree of detail necessary

Considerable detail is necessary here, so the programme will show days and sometimes even hours.

1.7.5 Basis of calculations

As with the six-weekly programme, the basis of calculations should be operational records. Since weekly programming is tied up very much with the bonus system, the expected outputs for bonus purposes are probably the best basis and can be evolved from work study synthetics from the work study department. Information from bonus feed-back may be used to update this programme.

1.7.6 Procedure for programming (Fig. 1.16)

1. List all the operations to be carried out during the next week in the order of starting time.

2. Insert the quantity of men and plant required and (taking into account the work content) enter the time required on the programme, providing continuity of work wherever possible.

CONTRACT: AMENITY CENTRE & OFFICE BLOCK — WEEKLY PROGRAMME — WEEK NO. 5
CONTRACT NO. — WEEK COMM. MAR 18.

OP NO	OPERATION	LABOUR & PLANT	DAY / DATE	MONDAY	TUESDAY	WEDN'DAY	THURSDAY	FRIDAY	SATURDAY	SUNDAY
				MAR. 18 TH	19TH	20TH	21ST	22ND	23RD	24 TH
1	ERECT REINF. TO COLS.	3 STEELFIXERS								
2	ERECT FWK. TO COLS	8 CARPS								
3	POUR CONC TO COLS	4 LABS. 300/200 MIXER								
4	STRIP FWK TO COLS.	8 CARPS								
5	PREPARE REINF TO BEAMS & FLOOR	3 STEELFIXERS								
6	ERECT FORMWORK TO BEAMS & FLOOR SLAB	8 CARPS.								
7	FIX REINFORCEMENT TO BEAMS & FLOOR	3 STEELFIXERS								
8	ERECT SCAFFOLDING	2 SCAFFOLDERS								
	LABOUR									
	CARPENTERS	PLANNED		8	8	8	8	8		
		ACTUAL								
	STEELFIXERS	PLANNED		3	3	3	3	3		
		ACTUAL								
	LABOURERS	PLANNED		4	4	4	4	4		
		ACTUAL								
	SCAFFOLDERS	PLANNED		2	2	2	2	2		
		ACTUAL								

Fig. 1.16

1.7.7 Labour requirements

The labour requirements are entered below the weekly programme and this will assist in co-ordinating the requirements of different sites.

1.7.8 Plant and other requirements

Plant and other requirements should now be finally checked to ensure that they will be available when required.

1.7.9 General notes on weekly programme

1.7.9.1 Erect reinforcement to columns

This operation is the completion of the fixing of the column cages which were started the previous week. The crane would assist as necessary in lifting cages into position.

1.7.9.2 Erect formwork to columns

The carpenters follow the steelfixers to complete the column formwork started the previous week.

1.7.9.3 Pour concrete to columns

The crane gang follows the carpenters erecting column forms. Crane and gang leave this operation for short periods to lift column cages.

1.7.9.4 Strip formwork to columns

Stripping of the columns which were poured first can start as soon as the last column forms are erected. This will leave plenty of time for concrete to harden.

1.7.9.5 Prepare reinforcement to beams and floors

The steelfixers cut, bend, and fabricate steel for floors after reinforcement to columns has been erected.

1.7.9.6 Erect formwork to beams and floor slab

The carpenters move onto this after stipping columns; crane will assist by lifting formwork panels into position. The floor will be cast in two equal sections to allow earlier release of some of the formwork.

1.7.9.7 Fix reinforcement to beams and floors

The steelfixers follow carpenters on floors, fixing steel. The crane is used to lift beam cages as necessary in accordance with the programme for the crane.

1.7.9.8 Erect scaffolding

Scaffolders erect scaffold to the outside of the building as work proceeds.

1.7.9.9 Crane and operator

The tower crane and operator have been omitted from the weekly programme as a separate programme for the crane will be drawn up each day.

1.7.10 Communicating the programme

There are many methods of communicating the programme to trades foremen, gangers and operatives, including pictorial diagrams, sequence studies, simplified bar charts, and work lists; the method chosen will depend upon the type of work being undertaken.

The operation and location must be clearly defined. On small projects the work of several gangs can be shown on one sheet or diagram, but on large projects a separate sheet or diagram will be necessary for each trade or gang.

The weekly site meetings are of great assistance in communicating a programme.

1.7.11 Sequence studies

The sequence study should be adhered to when drawing up programmes on site.

1.7.12 Simplified bar chart (Fig. 1.17)

A simplified bar chart will be used to communicate the weekly programme for the amenity centre and office block.

The work of each trade or gang is set down on one bar line, and the chart is read in conjunction with an outline layout drawing which can be used for each floor. The plan is referenced so that the extent of work to be carried out in the week can be clearly shown, foremen and gangers being given a copy of both the chart and the diagrams.

CONTRACT: AMENITY CENTRE & OFFICE BLOCK						WEEK NO. 5 WEEK COMM MAR 18	
CONTRACT NO.	MONDAY	TUESDAY	WEDNESDAY	THURSDAY	FRIDAY	SATURDAY	SUNDAY
CARPENTERS (8)	ERECT FORMWORK TO COLUMNS.	STRIP FORMWORK TO COLUMNS		ERECT FORWORK TO BEAMS & FLOOR SLAB			
STEELFIXERS (3)	ERECT REINF. TO REST OF COLS.	PREPARE REINFORCEMENT TO BEAMS & FLOORS			FIX REINF TO BEAMS & FLOORS		
LABOURERS (4)	POUR CONCRETE TO ALL COLUMNS		WORK AS DAILY PROGRAMME FOR CRANE				
SCAFFOLDERS (2)	ERECT SCAFFOLD OUTSIDE BUILDING TO ALLOW POURING OF CONCRETE						

Fig. 1.17

1.8 DAILY PROGRAMMING FOR THE TOWER CRANE

1.8.1 Planning in detail

The work of the tower crane must be planned in detail if it is to be used effectively.

If the tower crane is to be employed on one operation all day it may be necessary for the crane and gang to work overtime the previous day, preloading the scaffold with materials for other trades.

1.8.2 The work of the crane gang

When the crane gang (four labourers) are not all working on operations with the crane, any surplus men can be pre-loading pallets, preparing loads, etc.

1.8.3 The daily programme (Fig. 1.18)

The programme is for Thursday, 21st March. It is assumed that progress is as programmed on the weekly site plan.

CONTRACT AMENITY CENTRE & OFFICE BLOCK	DAILY PROGRAMME FOR TOWER CRANE											DAY : THURSDAY DATE : 21 ST MARCH.	
OPERATION	GANG	7.00–8.00	8.00–9.00	9.00–10.00	10.00–11.00	11.00–12.00	12.00–1.00	1.00–2.00	2.00–3.00	3.00–4.00	4.00–5.00	5.00–6.00	6.00–7.00
STRIP COLUMN FORMWORK	CARPENTERS		X	X									
ERECT FORMWORK TO FIRST FLOOR	CARPENTERS				X	X				X	X		
LOAD UP GROUND FLOOR	4 LABOURERS							X	X			X	

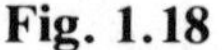

Fig. 1.18

1.8.4 General notes

1. Strip column formwork

The crane is working with carpenters, mainly moving formwork to be cleaned and re-oiled for its next application.

2. Erect formwork to first floor

Again working with the carpenters, the crane is engaged in lifting beam and column formwork into position ready for fixing. Carpenters are then left to fix the formwork lifted whilst the crane moves to the next operation, subsequently returning to assist the carpenters.

3. Load up ground floor

Bricks are loaded on the ground floor while it is still accessible to the crane. The loading areas will be pre-determined to avoid obstructing formwork supports.

The crane is programmed to service specific gangs or operatives throughout the day, thus avoiding a situation where everyone requires the crane at the same time. On the day shown on the programme, the operations listed are the only ones which require the use of the crane: it will carry out any lifting required by the gangs involved during their time allocation.

1.9 CONTROL OF PROGRESS ON PROJECTS

1.9.1 Introduction

Control is complementary to planning, therefore progressing is complementary to programming.

Control involves comparing, at regular intervals, the actual achievement with the plans and then taking any necessary corrective action to bring things back on schedule.

1.9.2 Use of meetings

The monthly and weekly meetings mentioned earlier in this chapter are invaluable in helping to control progress. Work which is not proceeding as planned will receive particular attention, and explanations will be required where sufficient progress is not being achieved. The action necessary for correcting under-production will be considered and the best solution will then be incorporated into the programme for the next period.

1.9.3 Effective control

If a programme is to be really effective as a control document, it must represent time and quantity of work done. This is particularly important on programmes where operations take several weeks or months to perform, since time alone will not clearly indicate whether work in progress is behind plan or not.

1.9.4 Methods of recording progress

Progress can be recorded on pictorial diagrams by colouring plans and elevations when certain sections of work are completed. This method is often used on housing projects and gives a quick visual impression of overall progress, but other ways are also available depending on the type of building and the amount of detail necessary.

In this instance progress will be shown on the planning charts, which clearly indicates what is happening and where corrective action will have to be taken.

1.9.5 Progressing the weekly programme (Fig. 1.19)

The example given shows the progress of operations at 5 p.m. on Wednesday, 20th March.

CONTRACT : AMENITY CENTRE & OFFICE BLOCK — WEEKLY PROGRAMME — WEEK NO 5
CONTRACT NO. — WEEK COMM. MAR. 18

OP. NO	OPERATION	LABOUR & PLANT	MONDAY MAR. 18TH	TUESDAY 19TH	WEDN'DAY 20TH	THURSDAY 21ST	FRIDAY 22ND	SATURDAY 23RD	SUNDAY 24TH
1	ERECT REINF TO COLS	3 STEELFIXERS							
2	ERECT FWK TO COLS.	8 CARPS.							
3	POUR CONC. TO COLS	4 LABS 300/200 MIXER							
4	STRIP FWK TO COLS.	8 CARPS							
5	PREPARE REINF TO BEAMS & FLOOR.	3 STEELFIXERS.							
6	ERECT FORMWORK TO BEAMS & FLOOR SLAB	8 CARPS.							
7	FIX REINFORCEMENT TO BEAMS & FLOORS	3 STEELFIXERS							
8	ERECT SCAFFOLDING	2 SCAFFOLDERS							

CURSOR

KEY: PROGRAMME; WK. COMPLTD; TIME LINE

LABOUR		MON	TUE	WED	THU	FRI	SAT	SUN
CARPENTERS	PLANNED	8	8	8	8	8		
	ACTUAL	8	8	8				
STEELFIXERS	PLANNED	3	3	3	3	3		
	ACTUAL	2	2	2				
LABOURERS	PLANNED	4	4	4	4	4		
	ACTUAL	4	4	4				
SCAFFOLDERS	PLANNED	2	2	2	2	2		
	ACTUAL	2	2	2				

Fig. 1.19

1.9.6 Use of cursor

A cursor is used to show the position reached on the time scale.

1.9.7 The programme line

The programme line is used to represent time and quantity of work.

1.9.8 The time line

The time line is used to indicate the time at which an operation starts and finishes.

1.9.9 Analysis of progress

1. Erect reinforcement to columns

This operation is now complete but took half a day longer than planned because only two steelfixers were available instead of the three planned.

2. Erect formwork to columns

This operation is now complete, but also took half a day longer than planned due to previous operation finishing late.

3. Pour concrete to columns

This operation is now complete. It took slightly longer than planned because of the formwork finishing late.

4. Strip formwork to columns

This operation started late because the carpenters were delayed on operation number 2. The production of the carpenters is slightly higher than expected as they have done 25 per cent more work than planned.

5. Prepare reinforcement to beams and floor

This operation started late because of the delay in operation number 1. There are still only two steelfixers available and work is falling behind plan.

1.9.10 Taking corrective action

It is obvious from the information given that unless another steelfixer is obtained the work on the frame will be delayed and the carpenters will not be able to work to capacity. The site manager would therefore contact the projects manager to get a steelfixer as soon as possible. It may in fact be necessary to get two steelfixers for a short period to catch up on work behind plan.

By progressing the plan, corrective action can be taken before serious delays occur.

1.9.11 Up-dating the six-weekly programme

The six-weekly programme can be up-dated at weekly intervals from the weekly site plan and a review made of the operations in the coming week to see if they are affected.

1.9.12 Up-dating the overall programme

This can also be up-dated weekly from the weekly site plan, and will show the overall position of progress to date.

1.9.13 Keeping head office informed

This can be done by means of a shuttle programme which is up-dated by the site manager each week.

Alternatively a progress report can be sent to head office giving the percentage completion of all operations, with reasons for delay and any other relevant information.

1.10 PROGRAMMING FOR REPETITIVE CONSTRUCTION

1.10.1 Introduction

When a project or part of a project consists of operations which are repeated many times, the programme must (as in bar charts for non-repetitive projects) attempt to provide continuity of work for the operatives and plant. The most obvious example of this situation is in housing projects, where the units have more or less the same work content whether they are two-storey houses, maisonettes, or flats. Another example is formwork, reinforcement, and concrete in multi-storey reinforced concrete buildings.

1.10.2 Balancing of gangs

In repetitive construction, the work of the different trades or gangs must take approximately the same period of time to complete, otherwise unnecessary non-productive time results.

1.10.3 Example of the use of balanced gangs

A number of identical buildings are to be built, the operations involved, together with their work content, being as follows:

Excavation	4 man-days
Concrete foundations	16 man-days
Brickwork	32 man-days
Roof construction	12 man-days
Roof finishings	8 man-days

If, in manning the operations, the need to provide the same period of time for each trade is ignored, the result may be as follows:

Operation	*Work content (days)*	*Number of men to be used*	*Period of time taken (days)*
Excavation	4	2	2
Concrete foundations	16	2	8
Brickwork	32	6	$5\frac{1}{3}$
Roof construction	12	2	6
Roof finishings	8	4	2

This would result in the programme shown in Fig. 1.20, which shows that some trades have considerable non-productive time as follows:

Bricklayers are non-productive for $2\frac{2}{3}$ days in 8 days
Carpenters are non-productive for 2 days in 8 days
Roofers are non-productive for 6 days in 8 days

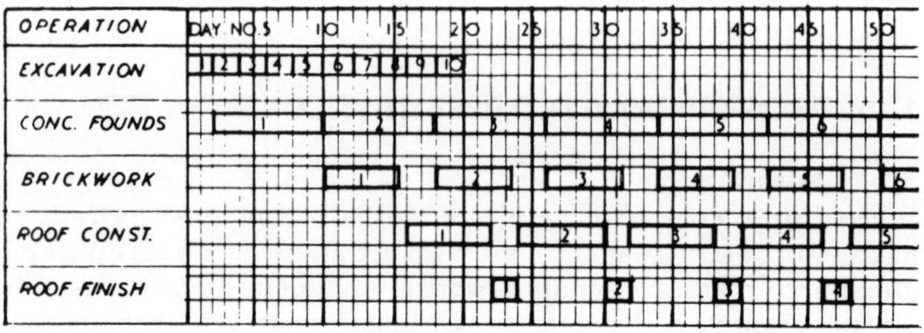

Fig. 1.20

Whilst it may be possible to employ these men on other sites, the gangs could equally well have been balanced to give continuity of work for all the operatives:

Operation	*Work content (days)*	*Number of men to be used*	*Period of time taken (days)*
Excavation	4	2	2
Concrete foundations	16	4	4
Brickwork	32	8	4
Roof construction	12	3	4
Roof finishings	8	2	4

This would result in the programme shown in Fig. 1.21.

Excavation is planned on a faster cycle but since it is the first operation this does not result in non-productive time and the men engaged on it can leave the site on completion. Obviously some flexibility is essential and this can be provided by delaying the start of operations on housing which follows.

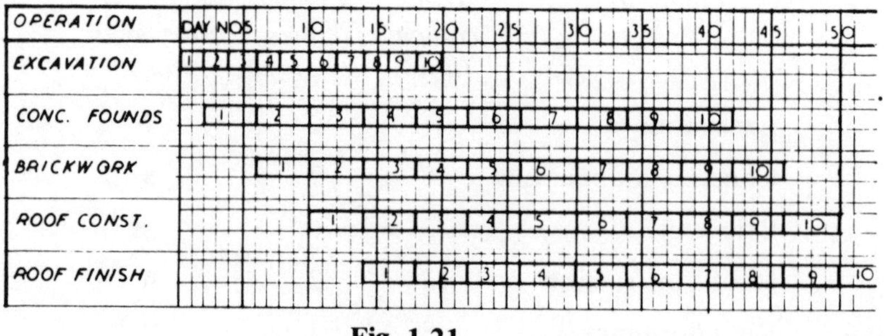

Fig. 1.21

1.10.4 Description of a housing project

An example will now be considered on a housing project consisting of an estate of 30 pairs of semi-detached houses. The site layout and working drawings are shown in Figs 1.22 and 1.23.

1.10.5 Construction of houses

1.10.5.1 Foundations

Concrete strip foundations, brick cavity walls up to damp-proof course level.

1.10.5.2 Ground floor

Hardcore filling, blinded with ashes. Damp-proof membrane, site concrete (ground floor slab).

1.10.5.3 Superstructure

External walls: cavity constructed with brick outer leaf and clinker block inner leaf.
Internal garage wall: brickwork.
All other internal walls: insulating blocks.
First floor tongued and grooved boarding on timber joists.
External features: cedar boarding on battens on building paper on insulating blocks.

1.10.5.4 Roof construction

Traditional construction with T.D.A. roof trusses.
Eaves finish: fascia board and soffit board.
Verges: barge boards.

1.10.5.5 Roof coverings and rainwater goods

Main roof: concrete slates on battens on felt.
Garage projections: nuralite on boarding on firrings on joists.
All rainwater goods: cast iron.

1.10.5.6 Internal finishes and services

Bathroom and kitchen walls: partial tiling.
Other walls: two coats of plaster.
Ceilings: plaster board and skim.
Heating: gas fired boiler.
Fireplace: stonework built on site.
Staircase: traditional softwood.
Ground floor finishes: wood blocks on screed to dining room and hall.
P.V.C. tiles on screed to kitchen.
Joinery work including fittings: softwood.
Soil pipes: cast iron.
Painting: knot prime stop and three coats of oil paint on all softwood, three coats of emulsion paint on all walls.

1.10.5.7 External works
Paving to drive and around houses.
Rustic fencing.

1.10.6 The site

The site is located four miles from the contractor's office, and is easily accessible. The ground is level and the sub-soil is sandy clay.

When the site was acquired by the contractor, the roadworks, together with drains and mains, had been completed.

1.10.7 Working week

A 40-hour, 5-day week has been used throughout, and working days only have been shown on the programme.

1.10.8 Productive hours

Plant-hours and man-hours (for productive operations only) have been established for all operations (Fig. 1.24).

1.10.9 Sub-contract work

The sub-contractors on this project have worked for the contractor previously and have agreed periods on site for each operation. They have also agreed to work to the programme produced providing sufficient notice is given.

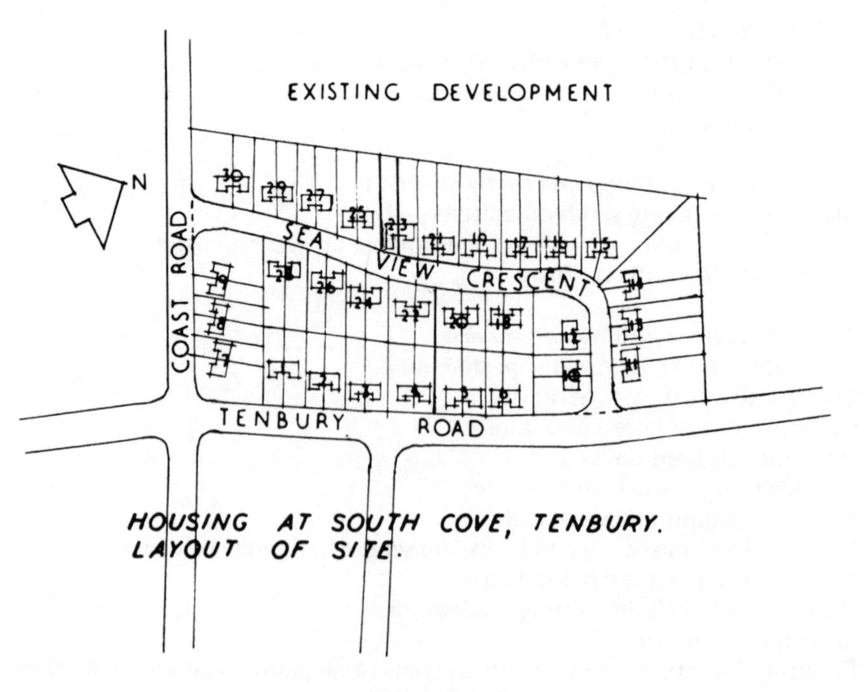

Fig. 1.22

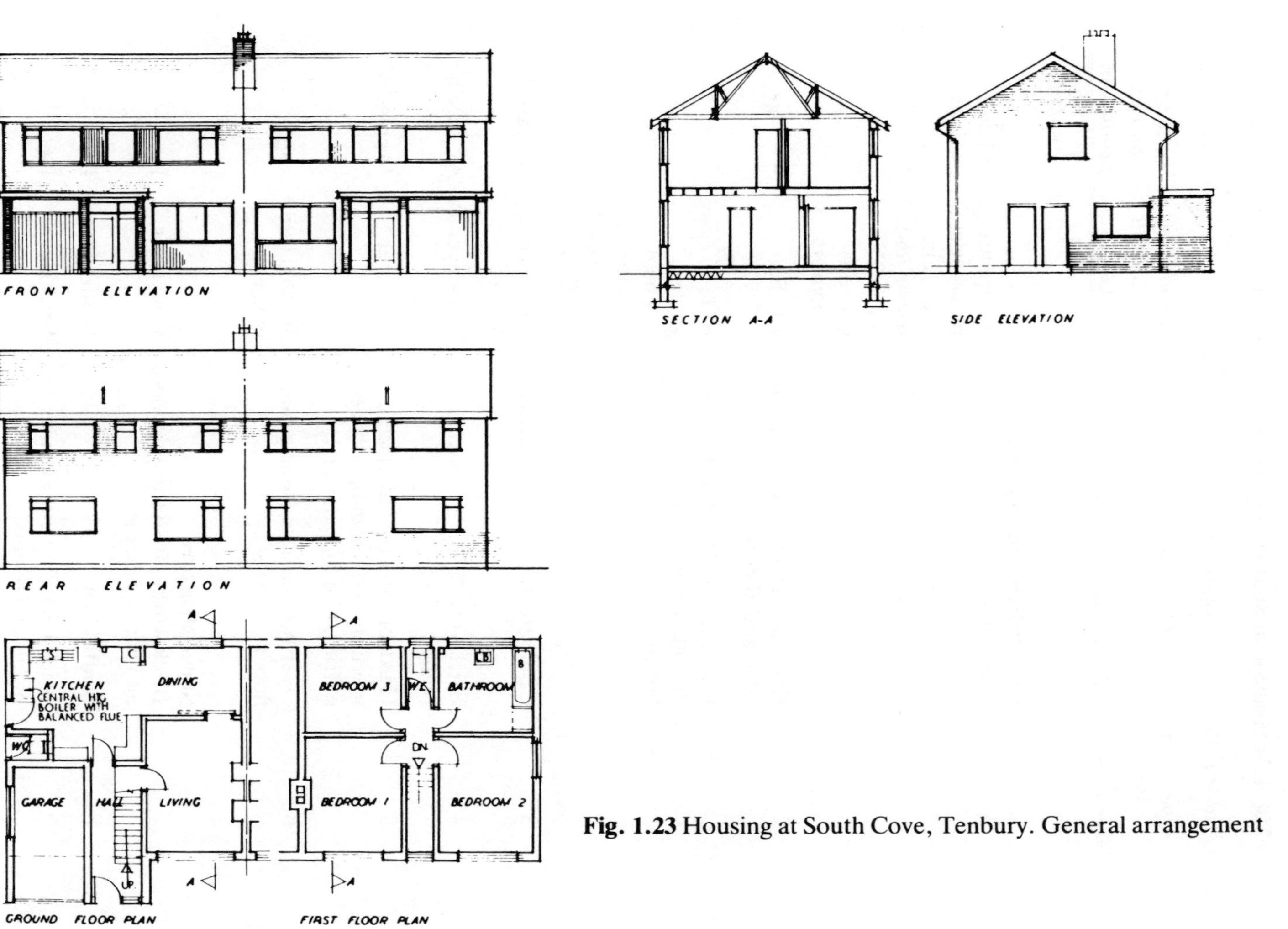

Fig. 1.23 Housing at South Cove, Tenbury. General arrangement

Fig. 1.24 Housing at South Cove, Tenbury

Table 1: Schedule of operations

Operation no.	*Operation*	*Productive trade/skill or plant*	*Hours per pair for productive trade/skill or plant*
	Sub-structure up to D.P.C.		
1	Preliminaries	2 carpenters, plumber and mate, electrician and mate, 2 general labourers	35 hours total for site
2	Excavate over site, remove spoil 30 m and deposit	J.C.B.	Average 14
	Excavate trenches for foundations and level, and ram bottoms	NOTE: easy going assumed, therefore use front bucket of J.C.B.	
3	Concrete in foundation trenches	300/200 litre mixer	Average 7
4	Brickwork to D.P.C. (includes filling cavity)	Bricklayer	Average 80
5	Hardcore and backfill (includes levelling and ramming service entries, blinding, d.p.m.	Labourer	Average 66
6	Over site concrete (solid ground floor)	300/200 litre mixer	6
	Super-structure		
7	Brick work: 1st lift (includes door and window frames)	Bricklayer	166
8	Scaffolding	Scaffolder	105
9	Brickwork: 2nd lift (includes lintels and beams)	Bricklayer	166
10	First floor joists and roof to garage	Carpenter	40

Fig. 1.24 *continued*

Operation no.	*Operation*	*Productive trade/skill or plant*	*Hours per pair for productive trade/skill or plant*
11	Brickwork: 3rd lift (includes window frames)	Bricklayer	110
12	Brickwork: 4th lift (includes lintels, chimney, making good putlog holes and around pipes)	Bricklayer	64
13	Roof carcass (includes barge boards, fascia, soffit, etc.)	Carpenter	71
14	Brickwork top-out (includes beam filling and raking cutting etc.)	Bricklayer	48
15	Roof coverings (includes battens and felt)	Roofer (sub-contractor)	40
16	Flashings, gutters, downpipes and soilstack (includes garage projection with nuralite roof covering)	Plumber	29
17	Glazing	Plumber	25
	External works		
18	Excavate for drains and services (includes assisting with backfill)	J.C.B. and labourers	Average 27
19	Build manholes, lay drains and test; build stopcock boxes	Bricklayer	Average 48
20	Service mains		
21	Backfilling	Labourer	Average 30
22	Paving, fences, generally clean and tidy site	Labourer	Average 103
	Internal trades and finishings		
23	Plumber: 1st fixing (pipe carcass etc.; includes gas pipes)	Plumber	28
24	Heating engineer: 1st fixing (pipe carcass etc.)	Heating engineer (sub-contractor)	60

Fig. 1.24 *continued*

Operation no.	*Operation*	*Productive trade/skill or plant*	*Hours per pair for productive trade/skill or plant*
25	Electrician: 1st fixing (conduit carcassing)	Electrician	21
26	Fix stone fireplace built *in situ*	Bricklayer	14
27	Floor boards and staircase	Carpenter	40
28	Plasterboard to ceilings (includes 12 mm plasterboard soffit to garage)	Plasterer	45
29	First floor partitions (includes external cills)	Bricklayer	95
30	Carpenter: 1st fixing (linings, grounds etc.)	Carpenter	36
31	Electrical drops and switchboxes	Electrician	7
32	Plastering to walls and ceilings, screed floors	Plasterer	272
33	Wood block floors	Floorlayer (sub-contractor)	140
34	P.V.C. tile floor	Floorlayer (sub-contractor)	20
35	Plumber: 2nd fixing (sanitary fittings etc.)	Plumber	14
36	Tiling to bathroom and kitchen	Plasterer	53
37	Carpenter: 2nd fixing (includes doors, architraves, skirtings, cupboards etc.)	Carpenter	112
38	Electrician: 2nd fixing (wiring and boards)	Electrician	16
39	Heating boiler, valves to radiators, and radiator brackets	Heating engineer (sub-contractor)	48
40	Make good generally	Plasterer	7

Fig. 1.24 *continued*

Operation no.	*Operation*	*Productive trade/skill or plant*	*Hours per pair for productive trade/skill or plant*
41	Painting (internal and external)	Painter	260
42	Final fixings: electric-light holders	Electrician	3½
43	Plumbing (chains and seats)	Plumber	3½
44	Radiators	Heating engineer	3
45	Furniture	Carpenter	30
46	Clean floors and scrub out	Cleaners	28
47	Final inspection	Site manager and client	

1.10.10 Procedure for programming

1. Study the drawings and other information available before starting the programme.
2. List operations in the order of starting time in each stage of the project.
3. Calculate total number of productive man-hours for each trade or skill in each stage of the work (see first two columns in Fig. 1.25). This is done because operations within each stage are dependent on each other, as they were in the previous example (section 1.10.3).
4. Using optimum gang sizes, or man-power available if this is limited, calculate the period of time required by each trade or skill.

The trade or skill requiring the longest period of time will determine the pace for the whole project, and in this example it is bricklaying, which requires 8 days. In housing the key trade is usually bricklaying, plastering or carpentry.

5. By adjusting gang sizes, balance the gangs so that the period of time required by each is approximately equal to that required by the key trade (wherever possible) and fill-in labourers required to assist tradesmen (Fig. 1.25).

A two-day cycle has been used for the work in sub-structure up to D.P.C., but this will not affect the other stages as this is the first stage to be carried out. A four-day cycle has been used for external works, as this is to some extent independent of the construction of the houses.

6. Programme the work, providing as much continuity of work as possible and providing flexibility (Figs. 1.26–1.30).

Fig. 1.25 Housing at South Cove, Tenbury

Table 2: Productive hours in each stage after balancing gangs

Operation	*Hours per pair for productive trade/ skill or plant*	
Sub-structure up to D.P.C.		
Excavate oversite and trenches for foundations	Average 14	
Concrete in foundation trenches	7	13
Oversite concrete	6	
Brickwork to D.P.C.	80	
Hardcore fill and backfill	66	
Superstructure		
Bricklayer		
1st lift	166	554
2nd lift	166	
3rd lift	110	
4th lift	64	
top-out	48	
Scaffolding	105	
Carpentry		
First floor joists and roof to garage	40	111
Roof carcass	71	
Roof coverings	40	
Plumbing, glazing, flashings, soil stack and rainwater goods	29	54
Glazing	25	
External works		
Excavate for drains and services	Average 27	
Build manholes, lay drains and test, build stopcock boxes	Average 48	

Labour and plant required	*Time required (days)*	*Notes*
J.C.B. and driver, 1 labourer	2	
300/200 litre mixer, 2 labourers at mixer, 400 litre dumper and driver, 1 labourer placing concrete	2	
6 bricklayers 3 labourers 200/150 litre mixer	2	
5 labourers	2	
Gang A: 3 bricklayers Gang B: 3 bricklayers Gang C: 2 bricklayers Gang D: 2 bricklayers Gang D: 2 bricklayers All gangs assisted by 5 labourers and 200/150 litre mixer	55 hours = 8 55 hours = 8 55 hours = 8 32 hours } 56 hours = 8 24 hours }	Total of 10 bricklayers take 8 days, each gang working on lifts shown and taking 8 days
2 scaffolders	8	
Gang A: 2 carpenters	8	
Sub-contractor, 2 roofers, and 1 labourer	3	
Gang A: 1 plumber and mate	8	
J.C.B. and driver, 5 labourers, 150/100 litre mixer	4	
2 bricklayers, 1 labourer, 150/100 litre mixer	3½	

Fig. 1.25 *continued*

Operation	*Hours per pair for productive trade/skill or plant*	
Service mains		
Backfilling	Average 30	133
Paving, fences and generally clean and tidy site	Average 103	
Internal trades and finishings		
Plumber		
1st fixing	28	45½
2nd fixing	14	
Final fixing	3½	
Heating engineer		
1st fixing	21	72
Heating boiler, valves to radiators, radiator brackets	48	
Radiators	3	
Electrician		
1st fixing	21	45
Drops and boxes	7	
2nd fixing	14	
Light holders etc.	3½	
Bricklayer		
Stone fireplace	14	109
1st floor partitions	95	
Carpenter and joiner		
Floor boards and staircase	40	218
1st fixing	36	
2nd fixing	112	
Door furniture	30	
Plasterer		
Plasterboard	45	377
Plaster walls, screed floors	272	
Making-good	7	
Tiling to bathroom and kitchen	53	
Flooring		
Wood block	140	160
P.V.C. tile	20	
Painting	260	
Cleaners	28	

Labour and plant required	*Time required (days)*	*Notes*
5 labourers	4	
Gang B: 1 plumber and mate	6½	
Sub-contractor, 2 heating engineers, 4 labourers	5	
1 electrician and mate	6½	
Gang E: 2 bricklayers, 1 labourer	8	
4 joiners	8	Gang C on 2nd fixing, gang B on remainder
7 plasterers, 4 labourers 150/100 litre mixer	8	Gang B on plastering and screed floors, gang A on remainder
Sub-contractor, 3 floorlayers, 1 labourer	8	
5 painters	7½	
2 cleaners	2	

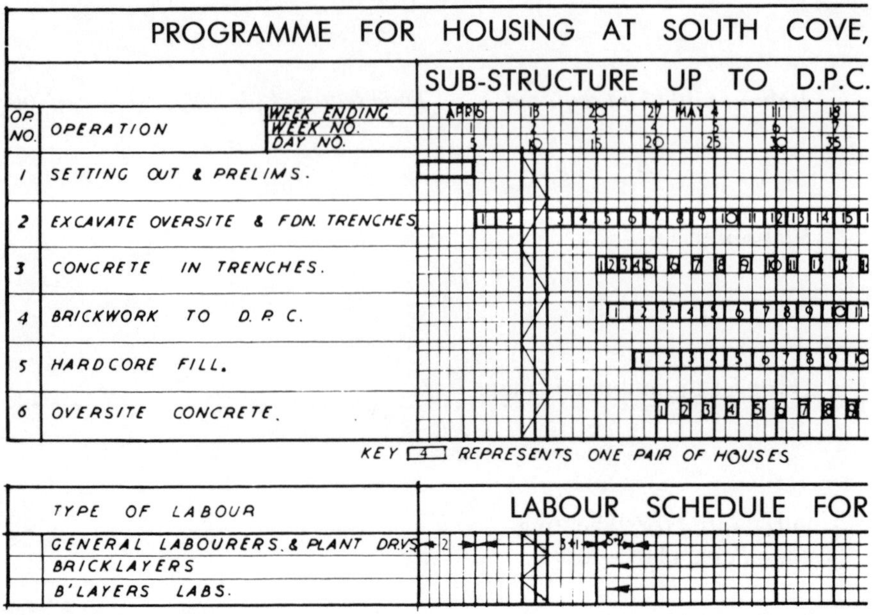

PROGRAMME FOR HOUSING AT SOUTH COVE,

SUB-STRUCTURE UP TO D.P.C.

OP. NO.	OPERATION	WEEK ENDING / WEEK NO. / DAY NO.
		APR 6 / 1 / 5; 13 / 2 / 10; 20 / 3 / 15; 27 / 4 / 20; MAY 4 / 5 / 25; 11 / 6 / 30; 18 / 7 / 35
1	SETTING OUT & PRELIMS.	
2	EXCAVATE OVERSITE & FDN. TRENCHES	1 2 3 4 5 6 7 8 9 10 11 12 13 14 15
3	CONCRETE IN TRENCHES.	1 2 3 4 5 6 7 8 9 10 11 12 13
4	BRICKWORK TO D. P. C.	1 2 3 4 5 6 7 8 9 10 11
5	HARDCORE FILL.	1 2 3 4 5 6 7 8 9 10
6	OVERSITE CONCRETE.	1 2 3 4 5 6 7 8 9

KEY [4] REPRESENTS ONE PAIR OF HOUSES

TYPE OF LABOUR	LABOUR SCHEDULE FOR
GENERAL LABOURERS & PLANT DRVS	2 — 3+1 — 5+2
BRICKLAYERS	
B'LAYERS LABS.	

Fig. 1.26

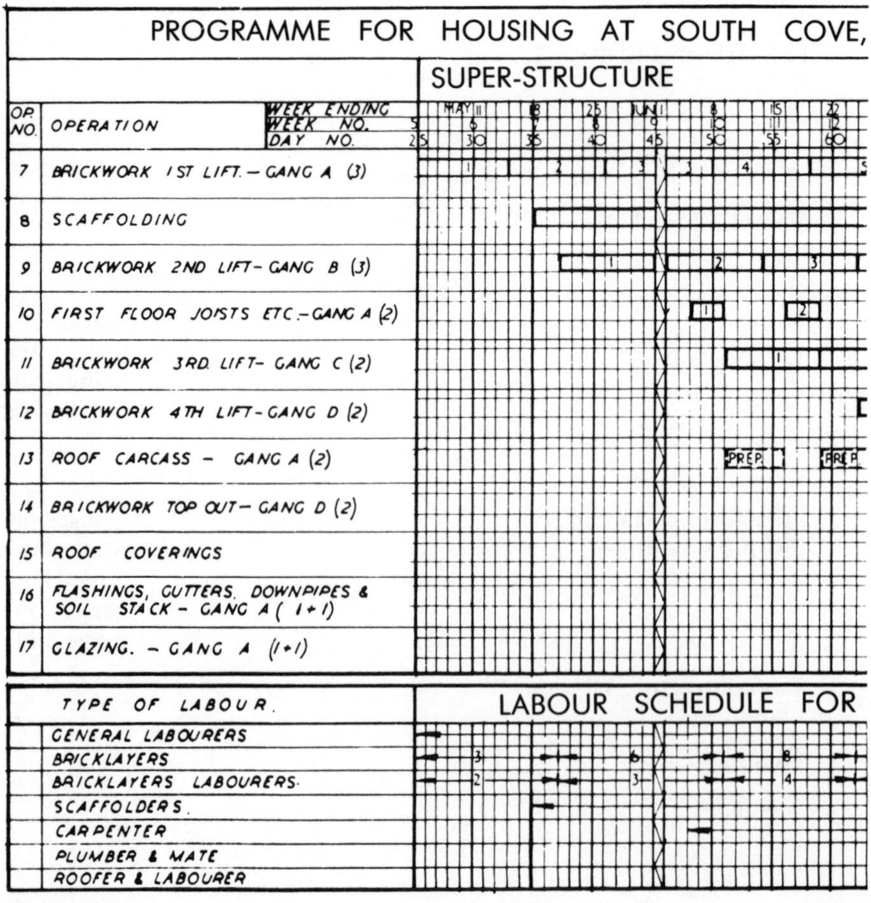

PROGRAMME FOR HOUSING AT SOUTH COVE,

SUPER-STRUCTURE

OP. NO.	OPERATION	WEEK ENDING / WEEK NO. / DAY NO.
		5 / 25; MAY 11 / 6 / 30; 18 / 7 / 35; 25 / 8 / 40; JUN 1 / 9 / 45; 8 / 10 / 50; 15 / 11 / 55; 22 / 12 / 60
7	BRICKWORK 1ST LIFT. – GANG A (3)	1 2 3 3 4 5
8	SCAFFOLDING	
9	BRICKWORK 2ND LIFT – GANG B (3)	1 2 3
10	FIRST FLOOR JOISTS ETC. – GANG A (2)	1 2
11	BRICKWORK 3RD. LIFT – GANG C (2)	1
12	BRICKWORK 4TH LIFT – GANG D (2)	
13	ROOF CARCASS – GANG A (2)	PREP. PREP.
14	BRICKWORK TOP OUT – GANG D (2)	
15	ROOF COVERINGS	
16	FLASHINGS, GUTTERS, DOWNPIPES & SOIL STACK – GANG A (1+1)	
17	GLAZING. – GANG A (1+1)	

TYPE OF LABOUR.	LABOUR SCHEDULE FOR
GENERAL LABOURERS	
BRICKLAYERS	3 — 6 — 8
BRICKLAYERS LABOURERS	2 — 3 — 4
SCAFFOLDERS.	
CARPENTER	
PLUMBER & MATE	
ROOFER & LABOURER	

Fig. 1.27

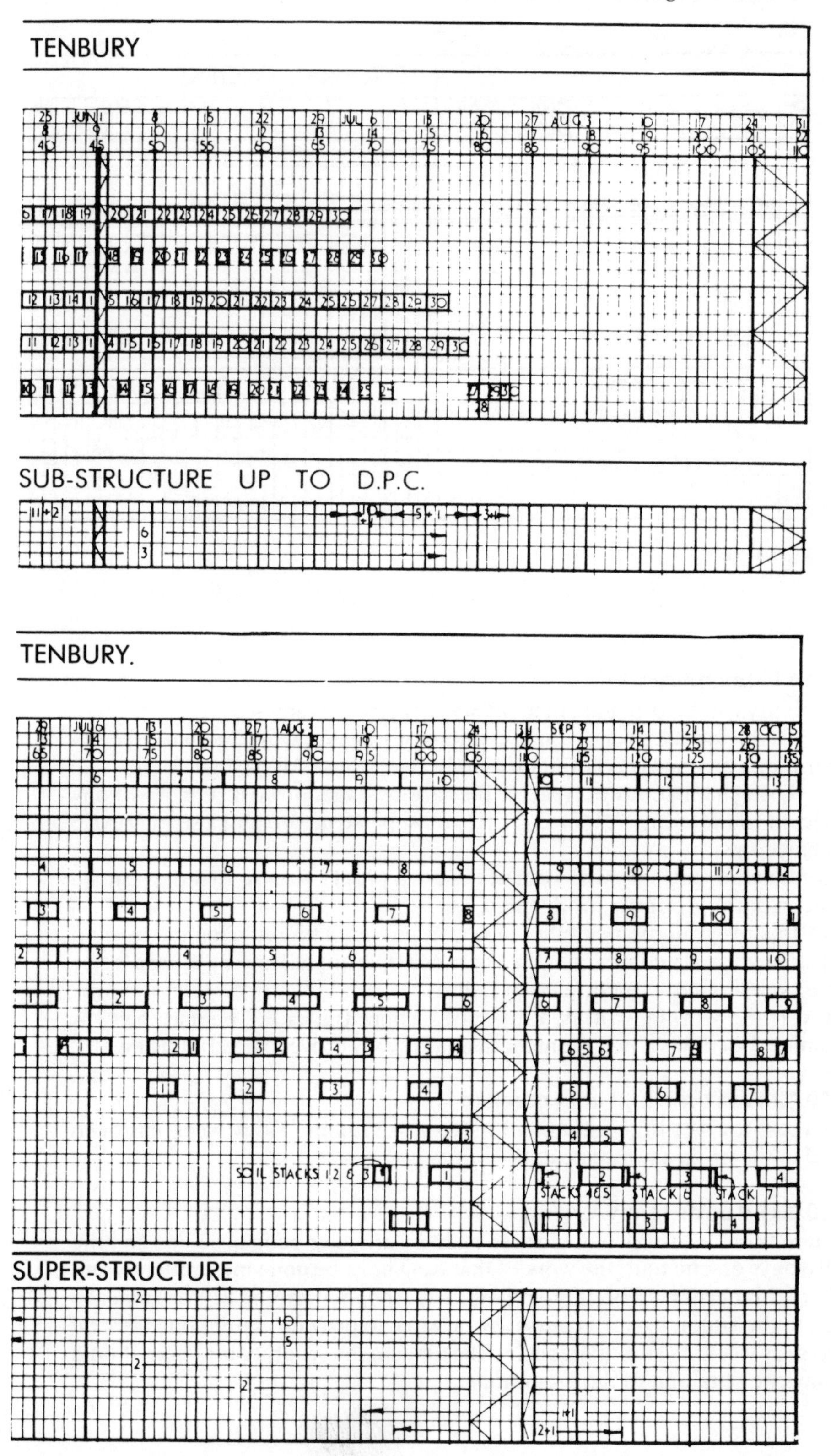
TENBURY
JUN
JUL
AUG
SUB-STRUCTURE UP TO D.P.C.
TENBURY.
JUL
AUG
SEP
OCT
SOIL STACKS 1 2 & 3
STACKS 4 & 5
STACK 6
STACK 7
SUPER-STRUCTURE

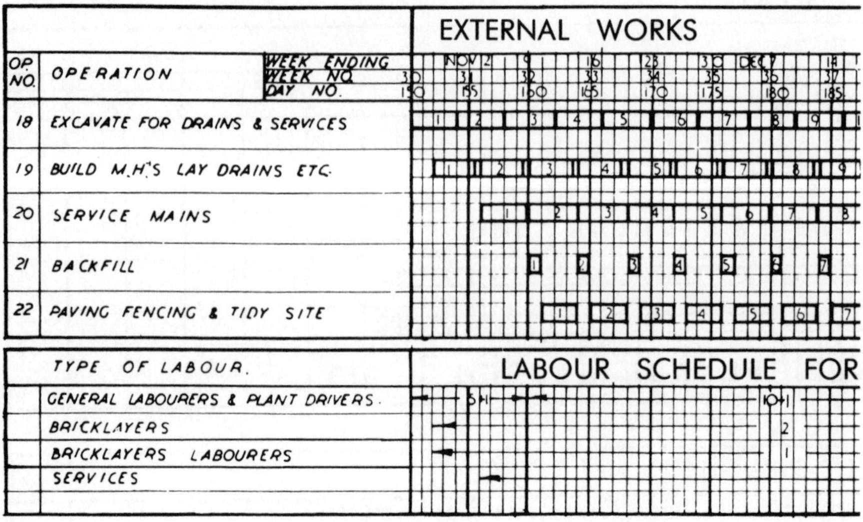

EXTERNAL WORKS

OP. NO.	OPERATION
18	EXCAVATE FOR DRAINS & SERVICES
19	BUILD M.H.'S LAY DRAINS ETC.
20	SERVICE MAINS
21	BACKFILL
22	PAVING FENCING & TIDY SITE

LABOUR SCHEDULE FOR

TYPE OF LABOUR.
GENERAL LABOURERS & PLANT DRIVERS.
BRICKLAYERS
BRICKLAYERS LABOURERS
SERVICES

Fig. 1.28

1.10.11 Flexibility

Flexibility can be provided in many ways. Typical examples are:

1. Reducing the overlap on operations or stages of work, so that delays in operations or stages do not immediately affect those following.
2. Using a longer time cycle for the various stages of work, e.g. a two-day cycle on foundations, a four-day cycle on superstructure and a six-day cycle on finishings.
3. Leaving any site works out of the main programme so that they can be done in periods of over production.

In this programme flexibility has been provided by reducing the overlap between stages (and in some cases between operations) and by using a faster time cycle for sub-structure up to D.P.C.

Overtime working will be introduced where trades fall behind the programme, but if this is persistent the gang sizes may prove to need adjustment.

1.10.12 Wet time and absenteeism

To allow for wet time, absenteeism, etc., a 35-hour, 5-day week has been used in the programme calculations.

1.10.13 Intermittent working

Where there is insufficient work to keep a particular gang occupied and the full 8 days on one unit, the work of that gang must be done intermittently and where this occurs it is better to provide a reasonable amount of work for each visit to the site; it may, of course, be possible to provide continuous work on two sites working simultaneously. The agreement of sub-contractors should always be obtained to the programme.

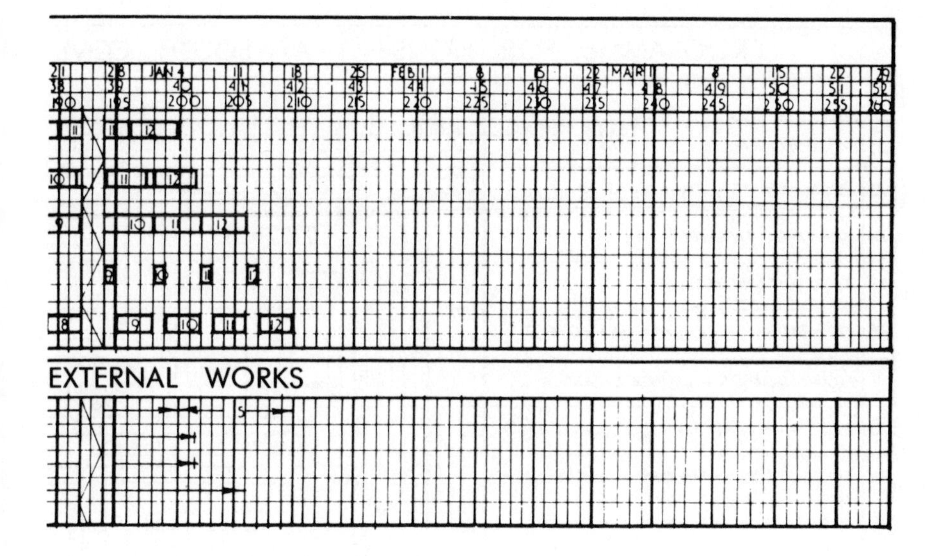

1.10.14 Rate of selling

The rate at which the houses are expected to sell is another factor which must be considered when deciding on the time cycle of production. In this case it is assumed that selling the houses presents no problem.

1.10.15 Alterations to standard design

Alterations to the design must be limited as excessive alterations will disrupt the programme. This would have the effect of unbalancing the gangs, causing non-productive time, and therefore an increase in cost.

1.10.16 Non-productive time

Where the work content for a particular gang does not take the full time cycle for the stage of work, some non-productive time is probable.

1.10.17 Accuracy of outputs

A detailed programme like the one shown is only possible where the operation times can be established with a reasonable degree of accuracy. Feedback information should be available from previous work and can be used as a basis, whilst work study synthetic time can be extremely useful in building up accurate output rates.

1.10.18 Availability of gangs

Continuity of work from site to site will be aimed for in the overall programming policy, so the availability of gangs from previous projects would have an influence on the programme in practice. It has been assumed that gangs can start on site when required.

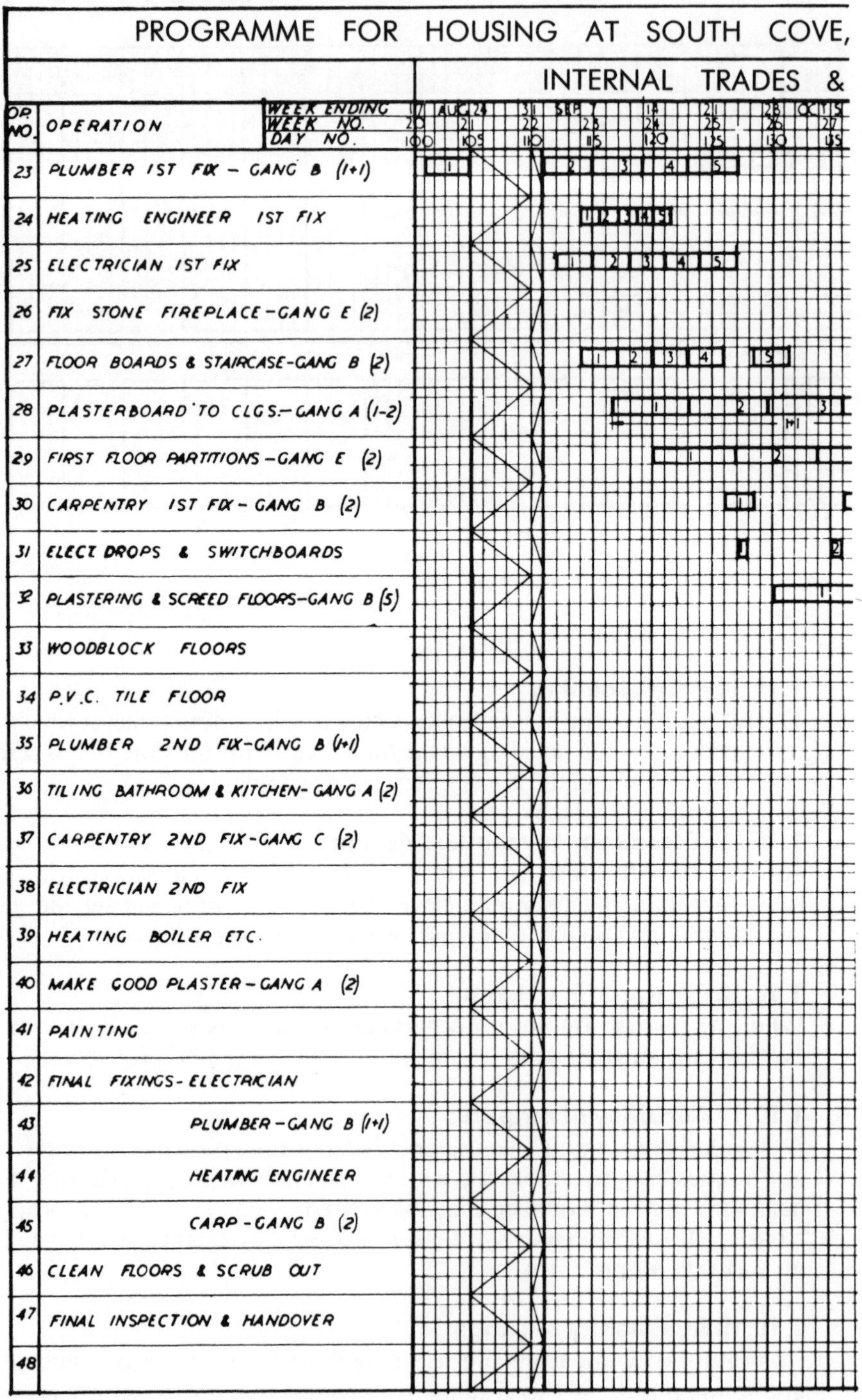

PROGRAMME FOR HOUSING AT SOUTH COVE,

INTERNAL TRADES &

WEEK ENDING	17 AUG	24	31	SEP 7	14	21	28	OCT 5
WEEK NO.	20	21	22	23	24	25	26	27
DAY NO.	100	105	110	115	120	125	130	135

OP. NO.	OPERATION
23	PLUMBER 1ST FIX – GANG B (1+1)
24	HEATING ENGINEER 1ST FIX
25	ELECTRICIAN 1ST FIX
26	FIX STONE FIREPLACE – GANG E (2)
27	FLOOR BOARDS & STAIRCASE – GANG B (2)
28	PLASTERBOARD TO CLGS. – GANG A (1-2)
29	FIRST FLOOR PARTITIONS – GANG E (2)
30	CARPENTRY 1ST FIX – GANG B (2)
31	ELECT. DROPS & SWITCHBOARDS
32	PLASTERING & SCREED FLOORS – GANG B (5)
33	WOODBLOCK FLOORS
34	P.V.C. TILE FLOOR
35	PLUMBER 2ND FIX – GANG B (1+1)
36	TILING BATHROOM & KITCHEN – GANG A (2)
37	CARPENTRY 2ND FIX – GANG C (2)
38	ELECTRICIAN 2ND FIX
39	HEATING BOILER ETC.
40	MAKE GOOD PLASTER – GANG A (2)
41	PAINTING
42	FINAL FIXINGS – ELECTRICIAN
43	PLUMBER – GANG B (1+1)
44	HEATING ENGINEER
45	CARP – GANG B (2)
46	CLEAN FLOORS & SCRUB OUT
47	FINAL INSPECTION & HANDOVER
48	

Fig. 1.29

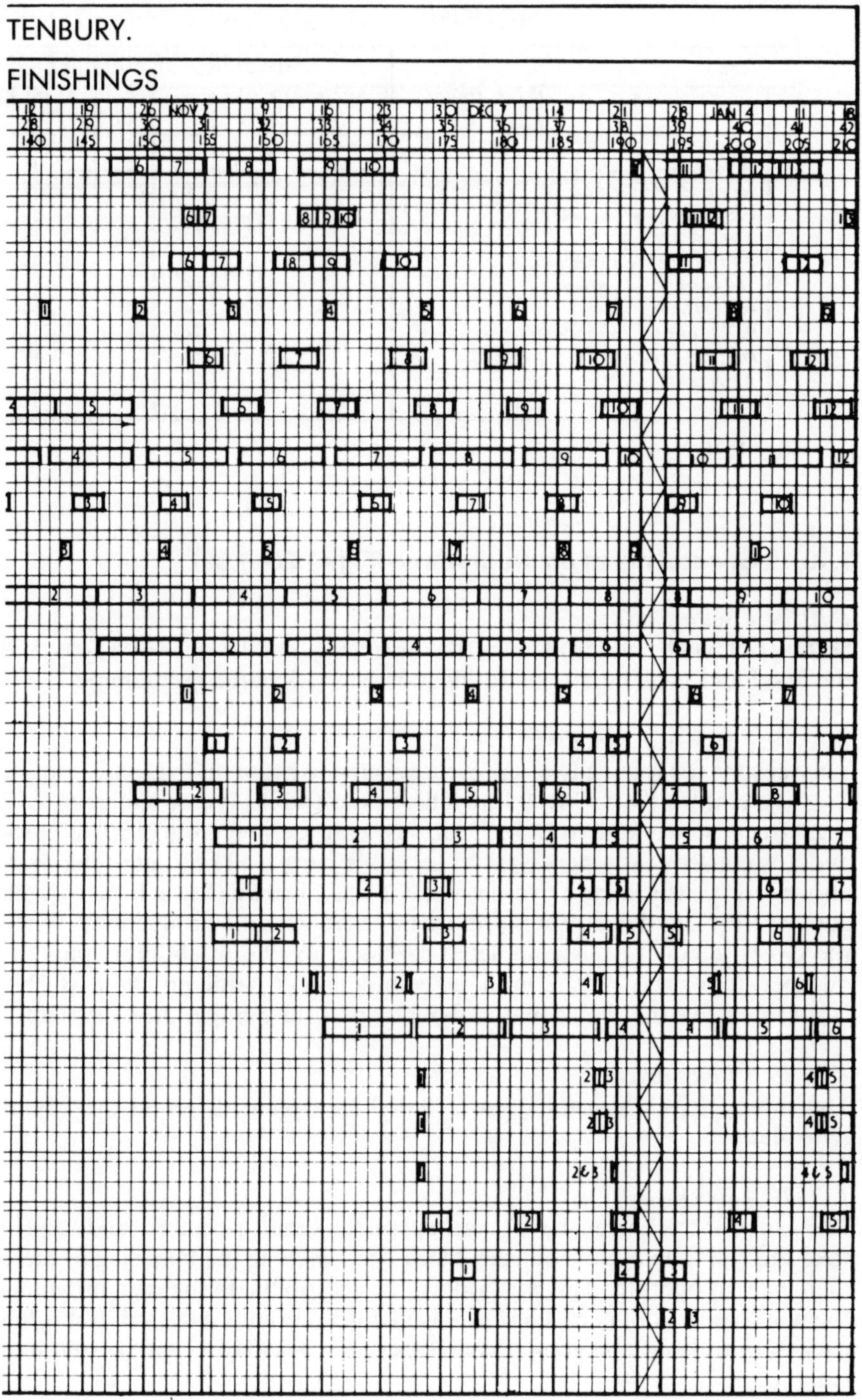

TENBURY.

FINISHINGS

12 19 26 NOV 2 9 16 23 30 DEC 7 14 21 28 JAN 4 11 18
28 29 30 31 32 33 34 35 36 37 38 39 40 41 42
140 145 150 155 160 165 170 175 180 185 190 195 200 205 210

6 7 8 9 10 11 11 12 13
6 7 8 9 10 11 12 13
6 7 8 9 10 11 12
1 2 3 4 5 6 7 8 9
6 7 8 9 10 11 12
4 5 6 7 8 9 10 11 12
4 5 6 7 8 9 10 10 11 12
1 3 4 5 6 7 8 9 10
3 4 5 6 7 8 9 10
2 3 4 5 6 7 8 8 9 10
1 2 3 4 5 6 6 7 8
1 2 3 4 5 6 7
1 2 3 4 5 6 7
1 2 3 4 5 6 7 8
1 2 3 4 5 5 6 7
1 2 3 4 5 6 7
1 2 3 4 5 5 6 7
1 2 3 4 5 6
1 2 3 4 4 5 6
1 2 3 4 5
1 2 3 4 5
1 2&3 4&5
1 2 3 4 5
1 2 3
1 2 3

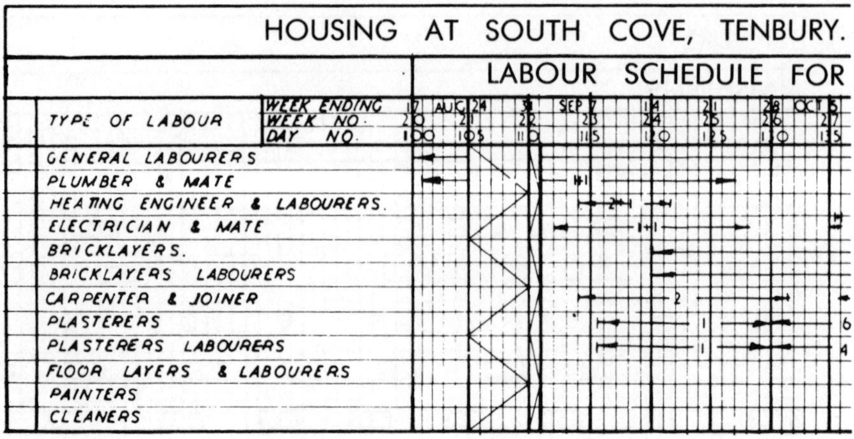

HOUSING AT SOUTH COVE, TENBURY.

LABOUR SCHEDULE FOR

TYPE OF LABOUR	WEEK ENDING / WEEK NO. / DAY NO.
GENERAL LABOURERS	
PLUMBER & MATE	
HEATING ENGINEER & LABOURERS.	
ELECTRICIAN & MATE	
BRICKLAYERS.	
BRICKLAYERS LABOURERS	
CARPENTER & JOINER	
PLASTERERS	
PLASTERERS LABOURERS	
FLOOR LAYERS & LABOURERS	
PAINTERS	
CLEANERS	

Fig. 1.30

1.10.19 General notes on programme (Figs 1.26–1.30)

1.10.19.1 Setting out

Setting out will be undertaken by the site manager as the work proceeds, and the first blocks will be set out during the first week of the programme.

1.10.19.2 Sub-structure up to D.P.C.

The reasons for using a two-day cycle on sub-structure up to D.P.C. are:

1. To employ the J.C.B. to the fullest extent whilst it is on site, and
2. Because work up to D.P.C. can vary considerably from block to block and it is consequently necessary to keep this fairly well in front of the work above D.P.C. (by using a two-day cycle it will move quickly in front of work above D.P.C. and does not therefore interfere with the production). When the oversite concrete is completed to block no. 26, 5 labourers plus the dumper driver stay on site until the completion of hardcore, after which 2 labourers leave the site so that 3 labourers plus the dumper driver remain.

1.10.19.3 Plant drivers

The mixer driver has not been included as a plant driver in the labour schedule.

1.10.19.4 Brickwork

The gangs of bricklayers have been kept on the same operation – and sized to take the full time cycle – on each block, thus making the bonusing of these men easy to operate. This method will be acceptable to the men providing the work content of each stage is reasonably accurately assessed. If it is felt necessary to keep one gang on one block, the bricklayers will be split into gangs of 5.

The brickwork in the ground floor partitions is to be built with first and second lift brickwork, whilst the work in first floor partitions is to be

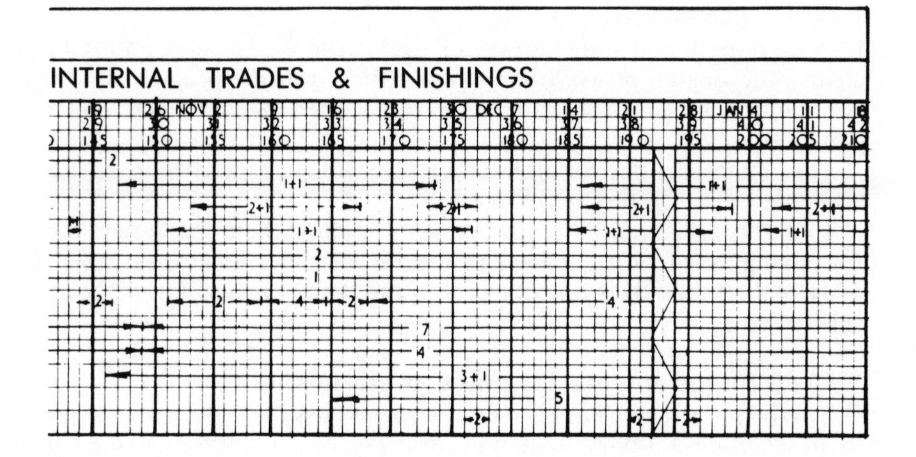

undertaken after the floor boards have been laid. Making good the brickwork after the scaffold is removed and around pipes is to be carried out by bricklayers gang D.

1.10.19.5 Roof carcass
The fixing of barge boards, fascia boards etc. has been delayed until after the bricklayers top out.

1.10.19.6 Gangs working on more than one stage
It is of course possible for a gang in a particular trade to work on two stages of the work if there is insufficient work within one stage to take up the full time cycle, e.g. if one gang of bricklayers did not have the full 8 days work on the carcass they could also work on partitions on the first floor. However, it is sometimes desirable to separate trades into specialisms e.g. carpenter and joiner specialising in either carcass, first fixing, second fixing or finishings. This depends mainly on the size of the firm and the amount of work carried out.

It is assumed here that one gang of carpenters will work on floor boards, staircase, first fixing and finishings.

1.10.19.7 External works
External works will be carried out in sections to give continuity of work to the gangs involved during the periods of site. The excavator is followed by 5 labourers carring out any necessary hand digging to branches, and concreting beds to drains and manholes.

1.10.19.8 Internal trades and finishings
The bricklayers working on partitions also build the stone fireplace if they are capable of work of this type; if they are not, the fireplaces would have to be built intermittently by a specialist in stonework.

1.10.19.9 Labour schedules

Labour requirements are shown for each stage of the work and indicate clearly the degree of work continuity achieved in the programme. More continuity will be available for some trades when the first block is nearing completion.

The labour required for preliminaries has not been included in the labour schedules, apart from general labourers.

Two general labourers have been kept on throughout the project period for tidying the site, off-loading and general assistance. If necessary they will assist with any hand digging of trenches and bottoming up of trenches in the sub-structure.

1.10.20 Requirements for plant, materials, etc.

When the programme has been drawn up it is possible to determine requirement schedules for plant, materials, drawings etc., in addition to the labour requirements shown on the charts.

1.10.21 The line of balance method (elemental trend analysis)

1.10.21.1 The advantages of the method

One of the main advantages of this method is that it gives a better indication of the dependence of one activity on another. it is very useful when progressing because it is immediately obvious when corrective action needs to be taken.

1.10.21.2 Example of line of balance method

Normally the line of balance method is used only when large numbers of units are to be built, but in order to illustrate the method a smaller project will be used.

Assume ten identical units are to be constructed and the work content and sequence for some of the internal activities is shown in Fig. 1.31. No overlap is to be used between units. A five day week is to be worked.

Fig. 1.31

Activity	*Time required*
Plumbing carcass	4 gang days
Electrical carcass	3 gang days
Carpenter 1st fix	2 gang days
Plastering	4 gang days
Plumbing sanitary fittings	3 gang days
Electrical fittings	1 gang day
Carpenter 2nd fix	5 gang days

In all types of repetitive construction, it is necessary to balance the gangs of operatives if non-productive time and project duration are to be kept to a minimum.

To illustrate tnis point, assume a line of balance schedule is prepared using one gang on each activity. Each gang is to be continuously occupied once they start on site and no buffer is to be used between the activities. The line of balance schedule would be as shown in Fig. 1.32. It is clear that the project duration is excessive, which results in high overhead costs.

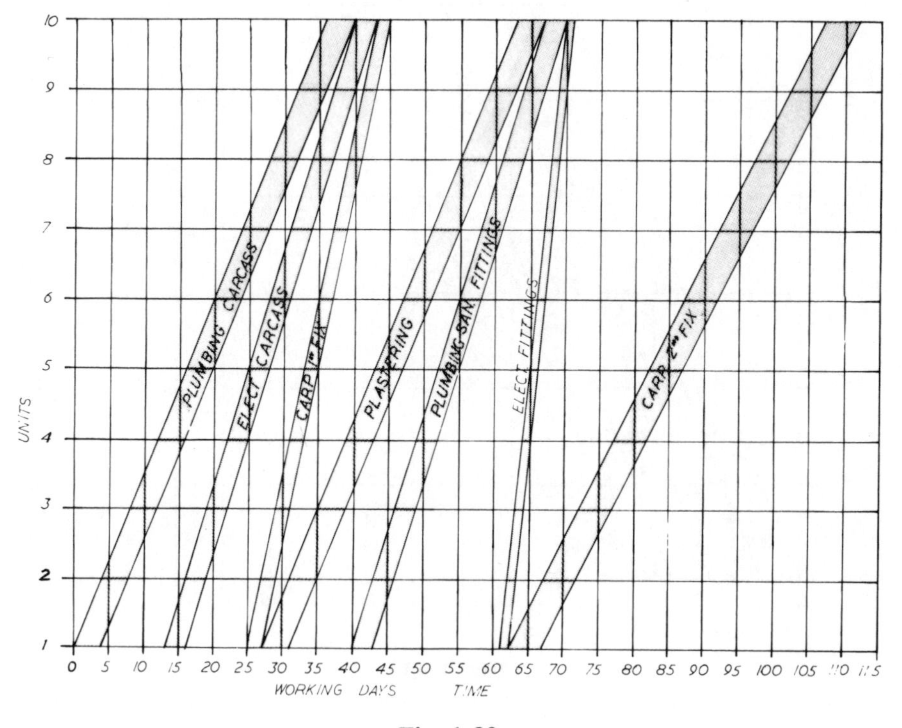

Fig. 1.32

If a schedule is produced using one gang on each activity, as above, but starting each activity immediately the previous one is completed on each unit, the line of balance schedule would be as shown in Fig. 1.33. This would reduce the project duration but would result in excessive non-productive time which in turn would result in high direct costs.

In order to prepare a schedule to give continuity of work and to complete the project in reasonable time more gangs would be required. To clarify the method it is assumed that the completion rate for each unit is to be one per day. The number of gangs required to achieve this is shown in Fig. 1.34.

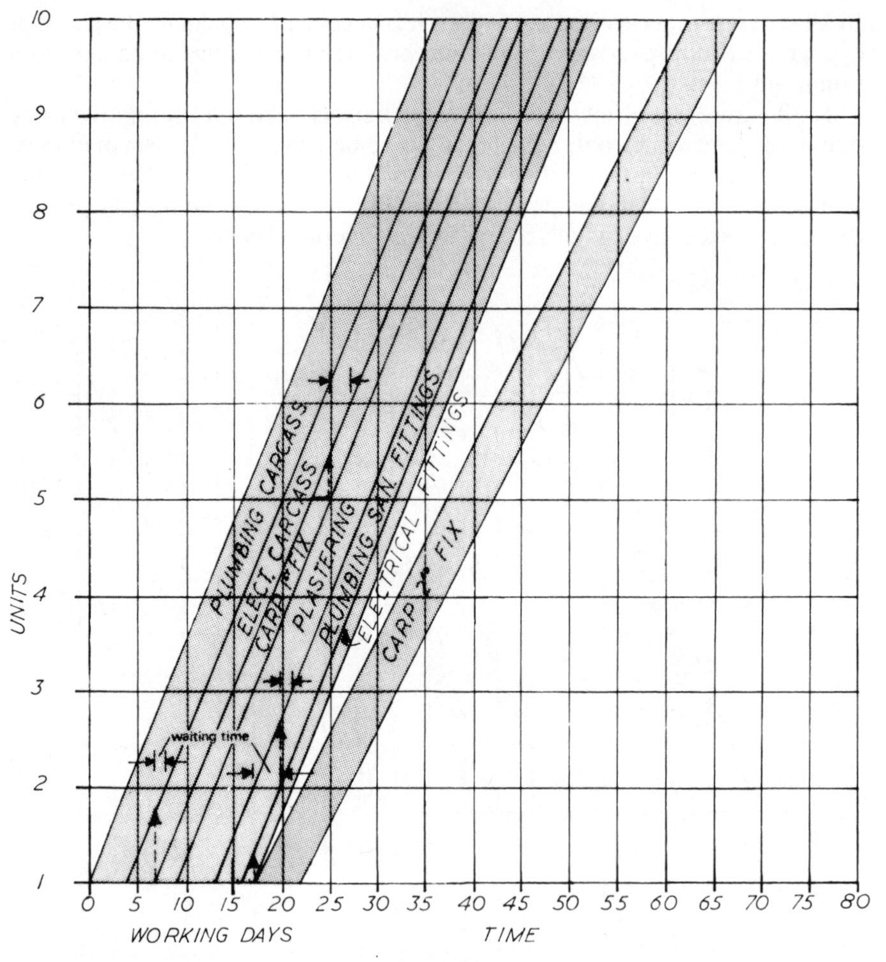

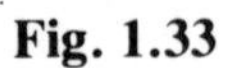
Fig. 1.33

Fig. 1.34

Activity	*Time required*	*Number of gangs*
Plumbing carcass	4 gang days	4
Electrical carcass	3 gang days	3
Carpenter 1st fix	2 gang days	2
Plastering	4 gang days	4
Plumbing sanitary fittings	3 gang days	3
Electrical fittings	1 gang day	1
Carpenter 2nd fix	5 gang days	5

The work could be carried out using one gang in each unit (i.e. if two gangs are required, each gang would work on alternate units), or by using more than one gang in each unit providing this does not result in overcrowding.

A line of balance schedule using one gang in each unit is shown in Fig. 1.35, which also shows the movement of some of the gangs. A two day buffer has been included to provide some flexibility.

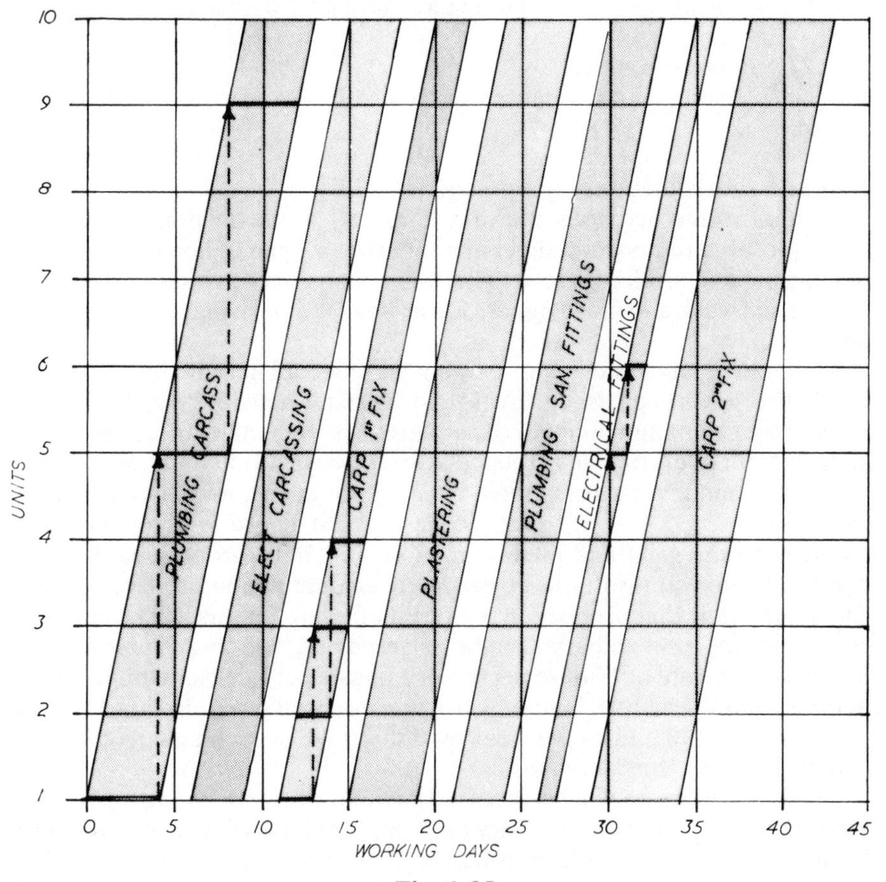

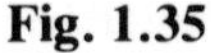
Fig. 1.35

1.10.21.3 Line of balance method on the housing project

The housing project described in paragraph 1.10.4 will be used to illustrate the Line of Balance method on a whole project. In this example it is assumed that the project is a Local Authority development and the handover rate required is four houses per week. External works and drainage have been omitted for clarity.

1.10.21.4 Avoiding cyclic queues

In repetitive construction it is best to arrange the schedule so that operatives work on one activity or a group of successive activities on each unit. This is not always possible for small firms producing houses at the rate shown in the previous programme, i.e. one house every four days. However, the problem can sometimes be overcome by grouping activities together even when slow rates of production are essential.

When operatives work on separate activities, on the same unit, cyclic queues are formed as shown in Fig. 6.40. This results in an increase in non-productive time.

In the following example cyclic queues have been avoided.

1.10.21.5 Balancing gangs

In practice, gangs do not balance exactly and they have to be rounded off as shown.

1.10.21.6 Procedure for programming

1. Group the activities shown in Fig. 1.24 wherever practicable. The hours for sub-structure and superstructure are for a pair of houses, as it is not practicable to consider these activities separately for each house. The hours for internal work and finishings are for single houses (see Fig. 1.36 columns 1 and 2).

2. Draw a network diagram showing the sequence of activities. A straight line network has been used to avoid overlap of the activities (see Fig. 1.37).

3. Using minimum number of operatives, or items of plant, calculate the gang or plant hours for each grouped activity (see Fig. 1.36 columns 3 and 4).

4. The handover rate required would result in one house being handed over every $35 \div 4 = 8\frac{3}{4}$ hours, once the first house is completed. By considering the gang and plant hours in Fig. 1.36 it can be seen that this handover rate will result in a considerable amount of non-productive time. The hours show that production of one house every 7 hours will be far more economical as most activities can be balanced based on this. The result will be a handover rate of 1 house per day or 5 houses per week and this has been adopted in the schedule. The sub-structure has been scheduled at a rate of 1 pair per day or 10 houses per week as at this stage the work content in each activity can vary considerably.

5. Some activities will be out of balance and adjustments to gang sizes or amount of plant may be necessary to bring them into balance, e.g. the time for first floor joists was originally based on a gang of two carpenters, this has been adjusted to two carpenters and one labourer.

6. Calculate the number of gangs or items of plant required to give the handover rate required, i.e. $\dfrac{\text{gang or plant hours}}{7\text{ (hours per unit)}}$.

This will balance the gangs as nearly as possible (see Fig. 1.36, column 5).

7. Using the number of gangs or items of plant, calculated in 6 above, calculate the number of completions per week for each activity, i.e. $\dfrac{35\,g}{G}$, where calculations are based on a 35 hour week, g = number of gangs or items of plant and G = gang or plant hours (see Fig. 1.36, column 6).

8. Decide on the number of gangs or items of plant to be used on each unit (P) (see Fig. 1.36, column 7).

9. Calculate the time required to complete each unit, i.e. $\frac{G}{7P}$, (see Fig. 1.36, column 8).

10. Calculate the overall time from the start of a particular activity on the first unit to the start of that activity on the last unit, i.e. $\frac{5(N-1)}{R}$, where N = total number of units (see Fig. 1.36, column 9).

11. Plot the line of balance schedule based on the above information, allowing stage and/or activity buffers as necessary to allow flexibility in the schedule.

NOTE: When the activity preceding the one being considered has a smaller overall time, the buffer will be plotted at unit 1. If the overall time is greater, the buffer will be plotted at the last unit (see Fig. 1.38).

1.10.21.7 Overlapping activities

In practice, some activities can proceed concurrently as shown in Fig. 1.39.

The line of Balance Programme for this network is shown in Fig. 1.40.

It can be seen that the programme is much more difficult to read and is more difficult to progress. A good method when using this type of network is to draw concurrent activities on separate transparent sheets and lay them over the line of balance programme or produce separate programmes for these activities.

1.10.21.8 An alternative method of allowing buffers between activities

Instead of showing buffers separately for each activity, the activity times can be increased to allow for the buffers. This results in a much simpler schedule, see Fig. 1.41. However, the period allowed for the buffer is not obvious using this method and this may be considered a disadvantage.

1.10.21.9 Factors governing handover rate

The handover rate for Local Authority work is usually determined by the Local Authority. If the required handover rate does not fit in with the natural rhythm of the activities, to be performed, either it must be adjusted as in this example, or non-productive time will result, thereby increasing direct costs.

For private development, the forecast rate of selling will obviously have an influence on the rate of production, but again adjustments would be made to allow the natural rhythm of the activities to be achieved.

1.10.21.10 Material schedules

Material schedules can be obtained easily from the Line of Balance Schedule. The amount of material required can be obtained by reading off the starts of activities at any particular time and allowing for such factors as call up times, size of loads, etc.

Fig. 1.36

1 *Operation*	2 *Productive operative or plant hours per pair*	3 *Min. number of productive operatives or items of plant*	4 *Time req'd. using min. gang or min. items of plant in hours*
Sub-structure			
(based on 1 pair every 7 hours)			*G*
Excavate oversite and trenches	14	1	14
Concrete trenches	7	1	7
Brickwork to D.P.C.	80	2	40
Hardcore fill and backfill	66	1	66
Site concrete	6	1	6
Super-structure			
(based on 1 pair every 14 hours)			*G*
Brickwork 1st and 2nd lift	332	2	166
First floor joists and roof to garage	40	2 carps. + 1 lab.	13·3
Brickwork to completion	222	2	111
Roof carcass	71	2 carps. + 1 lab.	27
Roof coverings	40	3	13·3
Flashings and glazing	54	1	54

Operation	*Productive operative or plant hours per house*	*Min. number of productive operatives or items of plant*	*Time req'd. using min. gang or min. items of plant in hours*
Internal work and Finishings			*G*
(based on 1 house ever 7 hours)			
Floor boards and staircase	20	2 carps. + 1 lab.	6·66
Partitions and stone fireplace	54·5	2	27·25
Carpenter 1 fix	18	2 carps. + 1 lab.	6
Electrician 1st fix	14	1	14
Plumber 1st fix	14	1	14
Heating Engineer 1st fix	30	1	30
Plastering and floor screed	158·5	2	79·25
Carpenter 2nd fix	71	2	35·5
Electrician 2nd fix	7	1	7
Plumber 2nd fix	7	1	7
Heating Engineer 2nd fix	25·5	1	25·5
Wall tiling	26·5	1	26·5
Floor finish	80	2	40
Painting and decorating	130	2	65
Snagging and final finish	14	2	7
Clean up	14	2	7

5 *Total number of gangs or items of plant req'd.*	6 *Number of pairs per week*	7 *Number of gangs or plant to be used on each pair*	8 *Time req'd. per pair in days*	9 *Total time from start of first pair to start of last pair in days*
$g = \frac{G}{t}$	$R = \frac{35g}{G}$	(P)	$T = \frac{G}{7P}$	$\frac{5(N-1)}{R}$
2	5	1	2	29
1	5	1	1	29
6	5·25	3	1·9	27·6
10	5·3	10	0·9	27·3
1	5·8	1	0·9	24·7
$g = \frac{G}{14}$	$R = \frac{35g}{G}$	(P)	$T = \frac{G}{7P}$	$\frac{5(N-1)}{R}$
12	2·5	3	7·9	58
1	2·63	1	1·9	55·3
8	2·52	2	7·9	57·5
2	2·59	1	3·9	55·9
1	2·63	1	1·9	55·2
4	2·59	2	3·9	55·9

Total number of gangs or items of plant req'd.	*Number of houses per week*	*Number of gangs or plant to be used on each house*	*Time req'd. per house in days*	*Total time from start of first house to start of last house in days*
$g = \frac{G}{t}$	$R = \frac{35g}{G}$	(P)	$T = \frac{G}{7P}$	$\frac{5(N-1)}{R}$
1	5·25	1	1·0	56·2
4	5·14	1	3·9	56·9
1	5·83	1	0·9	50·6
2	5·0	1	2·0	59·0
2	5·0	1	2·0	59·0
4	4·67	1	4·3	63·2
12	5·25	2	5·7	56·2
5	4·93	1	5·1	59·8
1	5·0	1	1·0	59·0
1	5·0	1	1·0	59·0
4	5·49	1	3·6	53·7
4	5·28	1	3·8	55·9
6	5·25	1	5·7	56·2
9	4·85	1	9·3	60·9
1	5·0	1	1·0	59·0
1	5·0	1	1·0	59·0

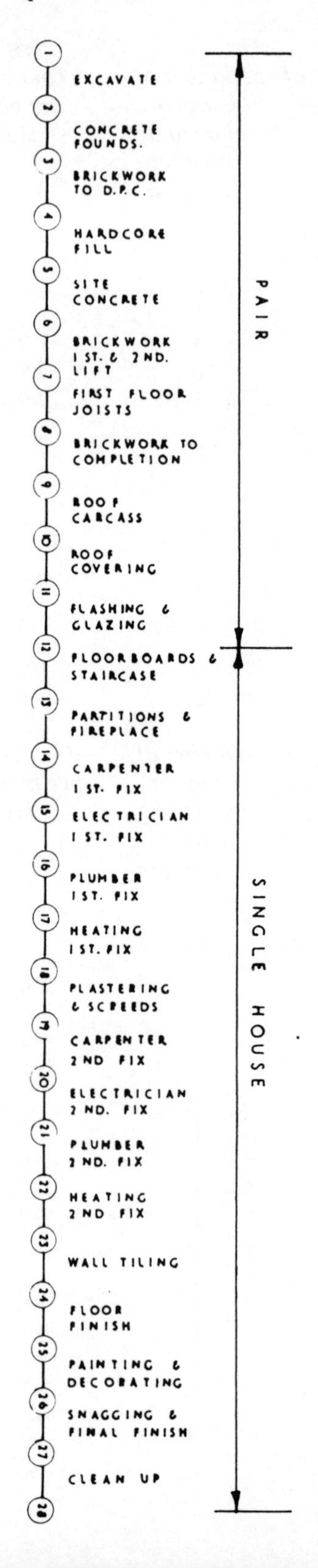
1
EXCAVATE
2
CONCRETE FOUNDS.
3
BRICKWORK TO D.P.C.
4
HARDCORE FILL
5
SITE CONCRETE
6
BRICKWORK 1 ST. & 2ND. LIFT
7
FIRST FLOOR JOISTS
8
BRICKWORK TO COMPLETION
9
ROOF CARCASS
10
ROOF COVERING
11
FLASHING & GLAZING
12
FLOORBOARDS & STAIRCASE
13
PARTITIONS & FIREPLACE
14
CARPENTER 1 ST. FIX
15
ELECTRICIAN 1 ST. FIX
16
PLUMBER 1 ST. FIX
17
HEATING 1 ST. FIX
18
PLASTERING & SCREEDS
19
CARPENTER 2 ND FIX
20
ELECTRICIAN 2 ND. FIX
21
PLUMBER 2 ND. FIX
22
HEATING 2 ND FIX
23
WALL TILING
24
FLOOR FINISH
25
PAINTING & DECORATING
26
SNAGGING & FINAL FINISH
27
CLEAN UP
28
PAIR
SINGLE HOUSE

Fig. 1.37

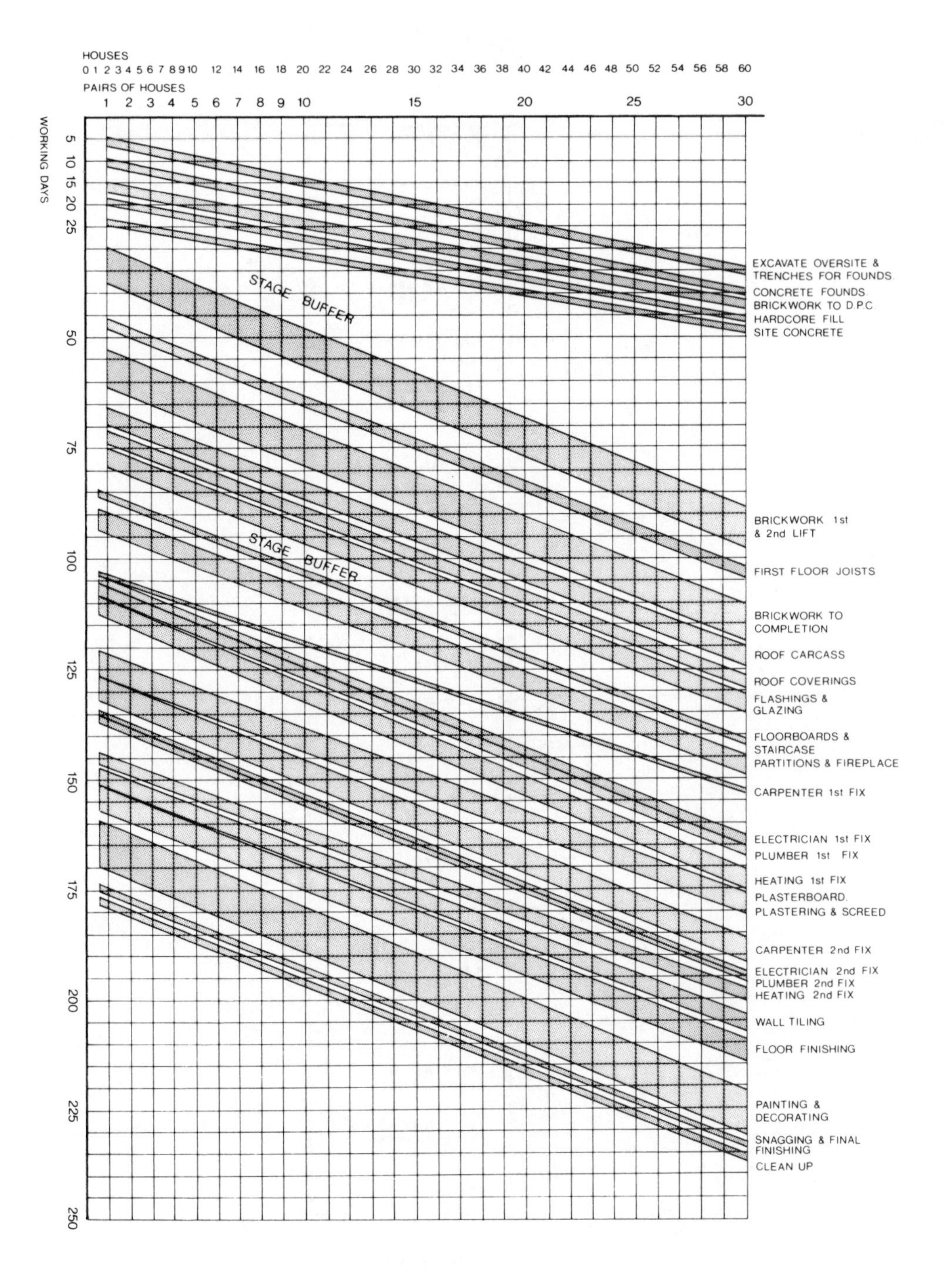
HOUSES
0 1 2 3 4 5 6 7 8 9 10 12 14 16 18 20 22 24 26 28 30 32 34 36 38 40 42 44 46 48 50 52 54 56 58 60
PAIRS OF HOUSES
1 2 3 4 5 6 7 8 9 10 15 20 25 30
WORKING DAYS
5 10 15 20 25 50 75 100 125 150 175 200 225 250
STAGE BUFFER
STAGE BUFFER
EXCAVATE OVERSITE & TRENCHES FOR FOUNDS.
CONCRETE FOUNDS.
BRICKWORK TO D.P.C.
HARDCORE FILL
SITE CONCRETE
BRICKWORK 1st & 2nd LIFT
FIRST FLOOR JOISTS
BRICKWORK TO COMPLETION
ROOF CARCASS
ROOF COVERINGS
FLASHINGS & GLAZING
FLOORBOARDS & STAIRCASE
PARTITIONS & FIREPLACE
CARPENTER 1st FIX
ELECTRICIAN 1st FIX
PLUMBER 1st FIX
HEATING 1st FIX
PLASTERBOARD.
PLASTERING & SCREED
CARPENTER 2nd FIX
ELECTRICIAN 2nd FIX
PLUMBER 2nd FIX
HEATING 2nd FIX
WALL TILING
FLOOR FINISHING
PAINTING & DECORATING
SNAGGING & FINAL FINISHING
CLEAN UP

Fig. 1.38

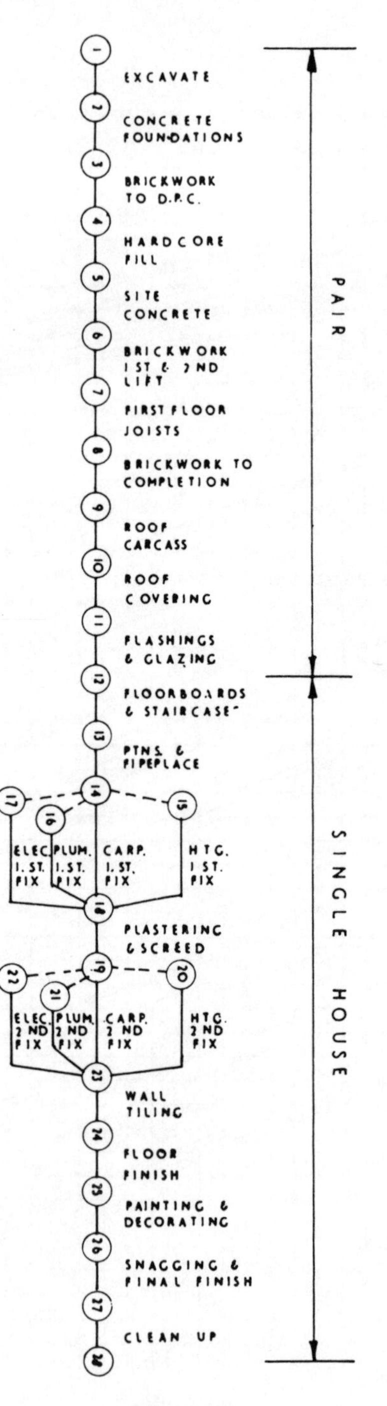
EXCAVATE
CONCRETE FOUNDATIONS
BRICKWORK TO D.P.C.
HARDCORE FILL
SITE CONCRETE
BRICKWORK 1ST & 2ND LIFT
FIRST FLOOR JOISTS
BRICKWORK TO COMPLETION
ROOF CARCASS
ROOF COVERING
FLASHINGS & GLAZING
FLOORBOARDS & STAIRCASE
PTNS. & FIREPLACE
ELEC. 1. ST. FIX
PLUM. 1. ST. FIX
CARP. 1. ST. FIX
HTG. 1 ST. FIX
PLASTERING & SCREED
ELEC. 2 ND FIX
PLUM. 2 ND FIX
CARP. 2 ND FIX
HTG. 2 ND FIX
WALL TILING
FLOOR FINISH
PAINTING & DECORATING
SNAGGING & FINAL FINISH
CLEAN UP
PAIR
SINGLE HOUSE

Fig. 1.39

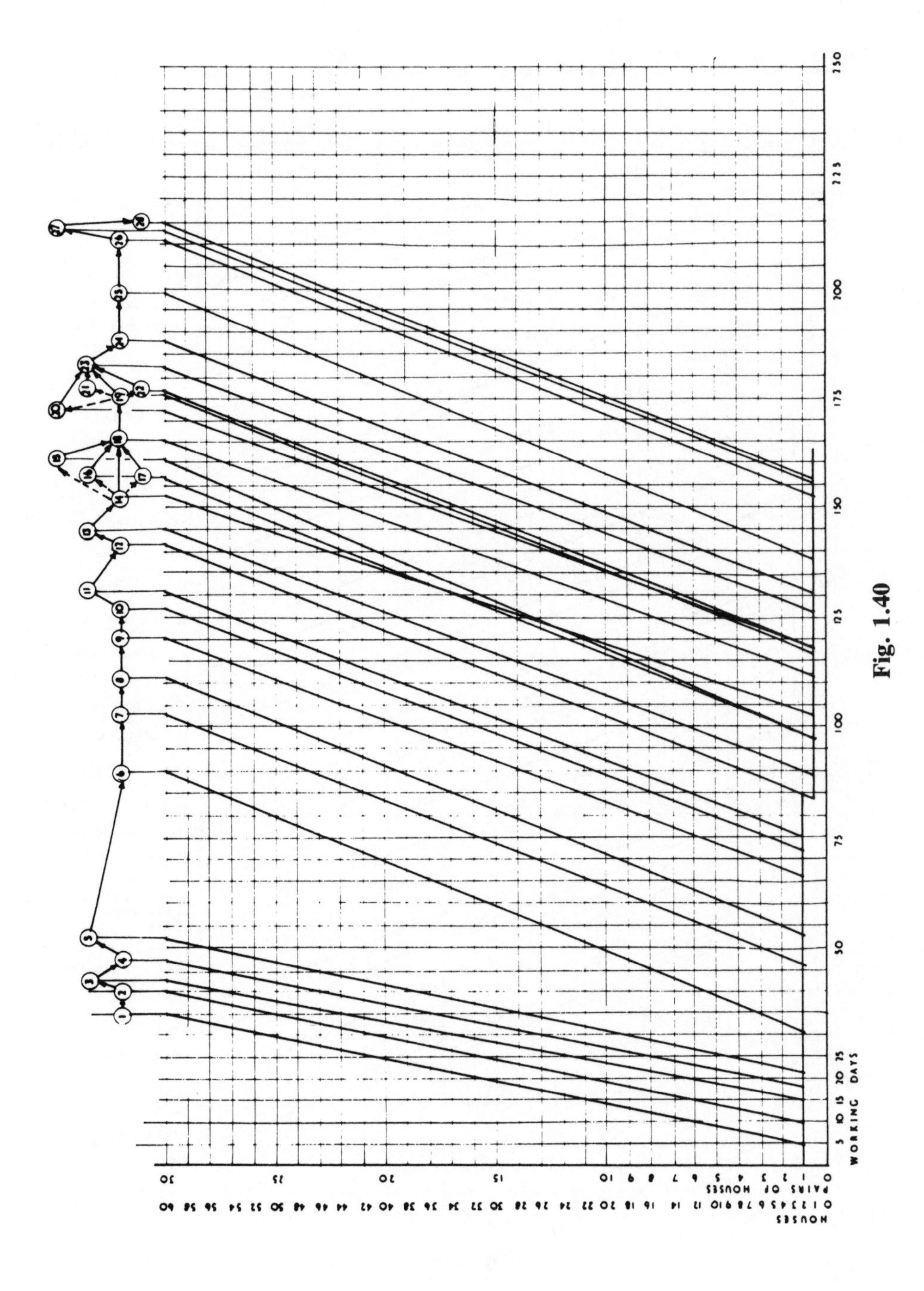

Fig. 1.40

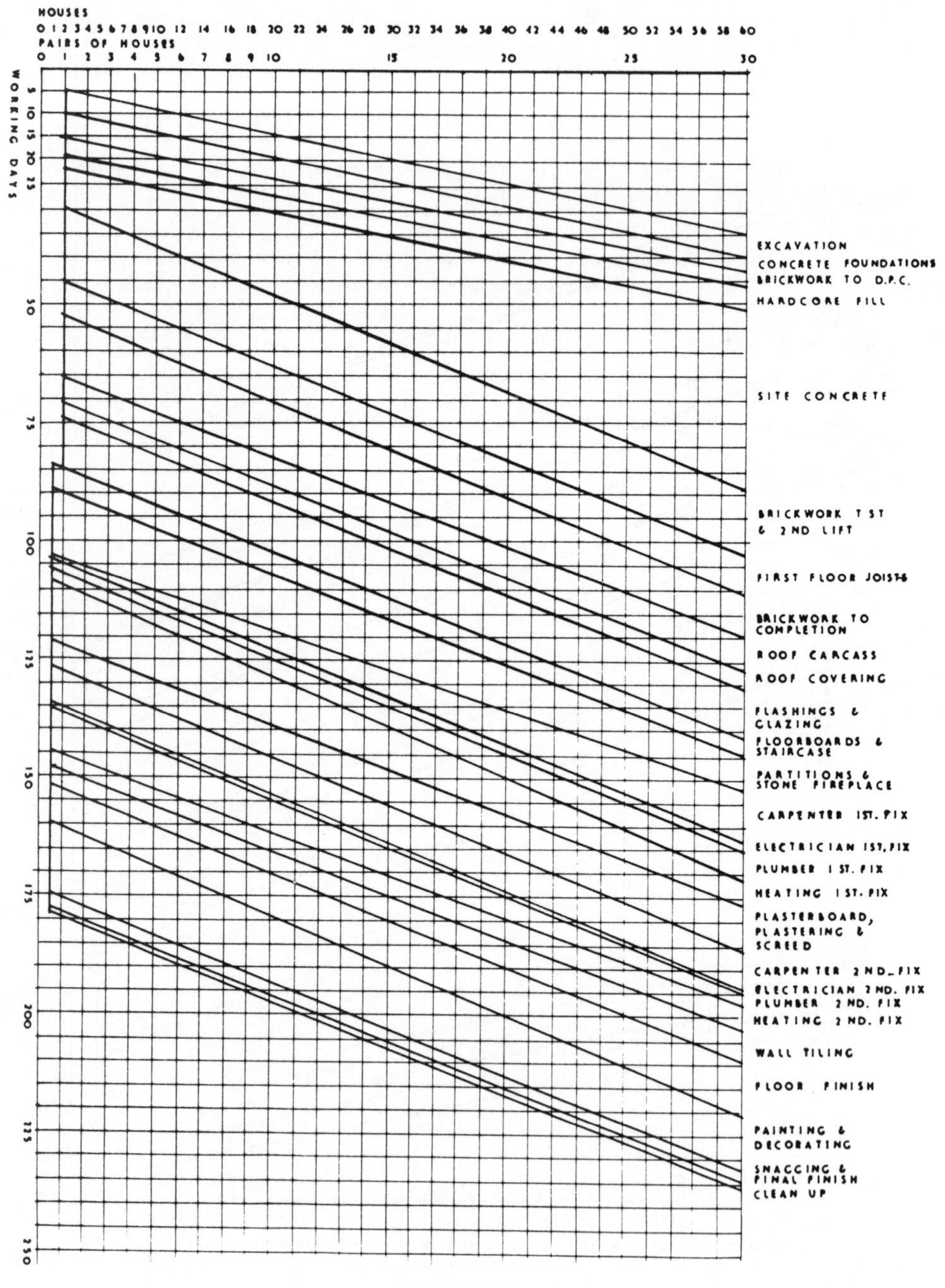
HOUSES
0 1 2 3 4 5 6 7 8 9 10 12 14 16 18 20 22 24 26 28 30 32 34 36 38 40 42 44 46 48 50 52 54 56 58 60
PAIRS OF HOUSES
0 1 2 3 4 5 6 7 8 9 10 15 20 25 30
WORKING DAYS
5 10 15 20 25 50 75 100 125 150 175 200 225 250
EXCAVATION
CONCRETE FOUNDATIONS
BRICKWORK TO D.P.C.
HARDCORE FILL
SITE CONCRETE
BRICKWORK 1ST & 2ND LIFT
FIRST FLOOR JOISTS
BRICKWORK TO COMPLETION
ROOF CARCASS
ROOF COVERING
FLASHINGS & GLAZING
FLOORBOARDS & STAIRCASE
PARTITIONS & STONE FIREPLACE
CARPENTER 1ST. FIX
ELECTRICIAN 1ST. FIX
PLUMBER 1 ST. FIX
HEATING 1 ST. FIX
PLASTERBOARD, PLASTERING & SCREED
CARPENTER 2ND. FIX
ELECTRICIAN 2ND. FIX
PLUMBER 2ND. FIX
HEATING 2 ND. FIX
WALL TILING
FLOOR FINISH
PAINTING & DECORATING
SNAGGING & FINAL FINISH
CLEAN UP

Fig. 1.41

1.10.21.11 Labour and plant schedules

Labour and plant schedules can be obtained in a similar way to that used on the previous example (Fig. 1.30) simply by accumulating requirements as the starts occur and dropping them off when the first unit is complete.

1.10.21.12 Cash flows

Cash flows can be obtained in a similar manner to that shown in Fig. 5.2.3.

1.10.21.13 Progressing the schedule

Progress can be shown on the Line of Balance Schedule, e.g. assume the progress for the first thirty days is as follows:

Excavation	completed up to and including pair 28
Concrete foundations	completed up to and including pair 19
Brickwork to D.P.C.	completed up to and including pair 13
Hardcore and backfill	completed up to and including pair 12
Site concrete	completed up to and including pair 9

Method 1

Shade in the activities as they are completed on each particular unit (see Fig. 1.42). An analysis of the progress to date shows that:

1. Excavation is ahead of schedule.
2. Concrete foundations are on schedule.
3. Brickwork to D.P.C. is falling well behind.
4. Hardcore is approximately on schedule.
5. Site concrete is slightly in front.

It can be seen that unless brickwork to D.P.C. proceeds as scheduled originally, i.e. 5·25 pairs per week, hardcore and backfill will be delayed resulting in non-productive time for the operatives on that activity.

Method 2

Plot the actual starts and completions of each unit as they occur, see Fig. 1.43 An analysis of the progress will be as in method 1 above.

For this method it is an advantage to use job cards for each activity which are issued at the start of each activity on each unit and handed in at the completion of the activity. The cards would also be used to calculate bonus earnings.

Method 3

Plot the cumulative starts and completions for each activity at say weekly intervals. The result will be similar to Fig. 1.43 (as method 2).

1.10.21.14 Other uses of the line of balance method

It rarely occurs that only one house type exists on a particular development. If differing types are included in a development the line of balance method can still be used by using a weighted average work content for each activity. This will not apply of course if the differences are substantial.

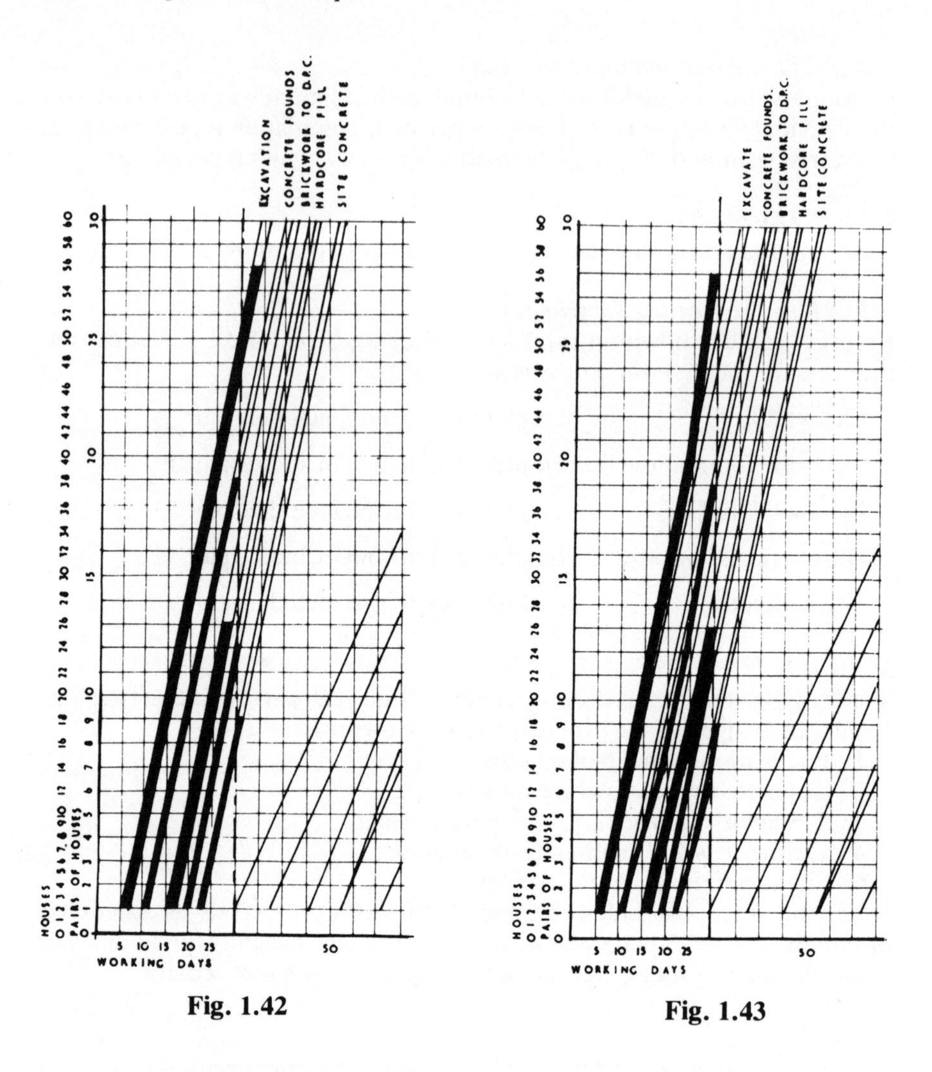

Fig. 1.42 **Fig. 1.43**

The method can also be used in other situations where repetitive work occurs, such as finishings in multi-storey buildings.

1.10.21.15 Learning curves

It has been established that as operatives become experienced on a particular activity, the time required for performing it is reduced. This reduction can be as much as fourteen per cent each time the number of units they complete is doubled. This has not been allowed for in the Line of Balance Schedule, but it could be incorporated.

1.10.21.16 Introducing costs to the line of balance method

The type of construction for repetitive projects can vary from traditional labour-intensive projects to system-built capital-intensive projects and this will have an influence on the type of line of balance schedule used.

Basically line of balance schedules can be produced by one of two methods:

Method 1
Parallel scheduling. In this method the lines of balance are drawn parallel to each other, progressing at the required rate of production, often with no buffer between them. To ensure that all activities should be able to progress at this rate, the number of gangs required is always rounded upwards. This method of scheduling would be used on capital-intensive projects, where site plant and site overheads are expensive. The cost of loss of production caused by overmanning is more than saved by the reduction in the overall planned duration of the project.

Method 2
Resource scheduling. In this method the schedule is determined by the progress of each activity and interference between activities is avoided at the planning stage. This results in a longer project duration and consequently greater project overheads. This method is used on labour-intensive projects. To reduce the project duration, some of the activities can be carried out in stages using a different number of gangs in each stage.

The method of scheduling is determined mainly on economic grounds.

Example
Part of a refurbishing project consists of the erection of the shells for 100 extensions. The completion rate is to be ten units per week and the working week is five days of eight hours per day. All activities are to start at the beginning of a day and a buffer of two days must be provided at all times.

The cost of labour is £48 per man per day and site overheads are to be charged at £240 per day.

The activity data are shown in Fig. 1.44. As the construction consists of small extensions, no overlapping of activities will be allowed.

Fig. 1.44

Act No.	*Activity*	*Total man hours per unit*	*Gang size*
1	Demolition of existing lean-to and load rubble	24	2
2	Construct foundation and make drain connection	16	2
3	Brickwork to D.P.C. level	15	3 (2+1)
4	Oversite concrete and hardcore	11	2
5	Brickwork to roof level	69	3 (2+1)

Method 1 – Parallel scheduling
The calculation sheet is shown in Fig. 1.45. This is prepared as described in 1.10.21.6, except that the number of gangs is rounded upwards. The line of balance schedule is shown in Fig. 1.46 and the cost calculations are shown in Fig. 1.47. The start time for brickwork to roof could be delayed, and the activity could still be completed by day 66. This would result in a saving of £1,781 as shown. This would not however be possible if the roof construction was to follow immediately after. The cost could of course be reduced by omitting the buffers, but in a project of this type, this would result in considerable interference between gangs which could outweigh any theoretical savings.

Method 2
(1) Resource scheduling with complete continuity for all activities
The additional calculations are shown in Fig. 1.48, and again they are prepared as described in 1.10.21.6, the number of gangs being rounded to the nearest number. The line of balance schedule is shown in Fig. 1.49 and the cost calculations in Fig. 1.50. It can be seen that the project duration is extensive due to the different rates of progress of the activities, even though the gangs have been balanced as near as possible. If the gangs had been rounded upwards, the times from the start of the first unit to the start of the last one would have been as shown in Fig. 1.45 for each activity. The project duration would have been reduced by two days to 83 days resulting in a saving of 2 × £240 = £480.

(2) Resource scheduling with some activities carried out in stages
A number of solutions are possible when adapting the schedule and the optimum solution will depend on a number of factors, e.g. site overheads, cost of labour, the extent to which the activities are out of parallel, the effects of lack of continuity etc. All the relevant factors must be taken into account when developing a solution, but due to all these factors determination of the optimum solution is extremely complex. In practical terms it is more appropriate to analyse each schedule logically and determine an economic solution based on this approach.

To simulate the effects of non-continuity of work a charge of £30 per man will be made each time resources are brought back to site. In this example one solution is to alternate the number of gangs on 'Brickwork to DPC' and on 'Oversite concrete'.

Considering the activity 'Brickwork to DPC', to achieve a 'start first' to 'start last' duration of 49.5 duration required for 10 units per week, it is necessary to alternate between two gangs and one gang. The number of units to be completed in each case can be calculated as follows:

Activity duration in each unit using one gang = 0.625 days.
Average activity duration in each unit using two gangs = 0.3125 days.

$$x \times 0.625 + (100 - x) \times 0.3125 = 50$$
$$0.625x - 0.3125x = 50 - 31.25 = 18.75$$
$$x = 18.75 \div 0.3125 = 60 \text{ units.}$$

Fig. 1.45

Act. No.	*Activity*	*Total man hours per unit*	*Min. gang size*	*Time reqd. by min. gang*	*Number of gangs*	*No. of units per week*	*No. of gangs per unit*	*Time per unit (days)*	*Time from start first to start last (days)*
1	Demolition	24	2	12	3	10	1	1.5	49.5
2	Foundations and drains	16	2	8	2	10	1	1	49.5
3	Brickwork to D.P.C.	15	3 (2+1)	5	2	16 (10)	1	0.625	30.94 (49.5)
4	Oversite concrete	11	2	5½	2	14.55 (10)	1	0.7	34.02 (49.5)
5	Brickwork to roof	69	3 (2+1)	23	6	10.43	1	2.075	47.46 (49.5)

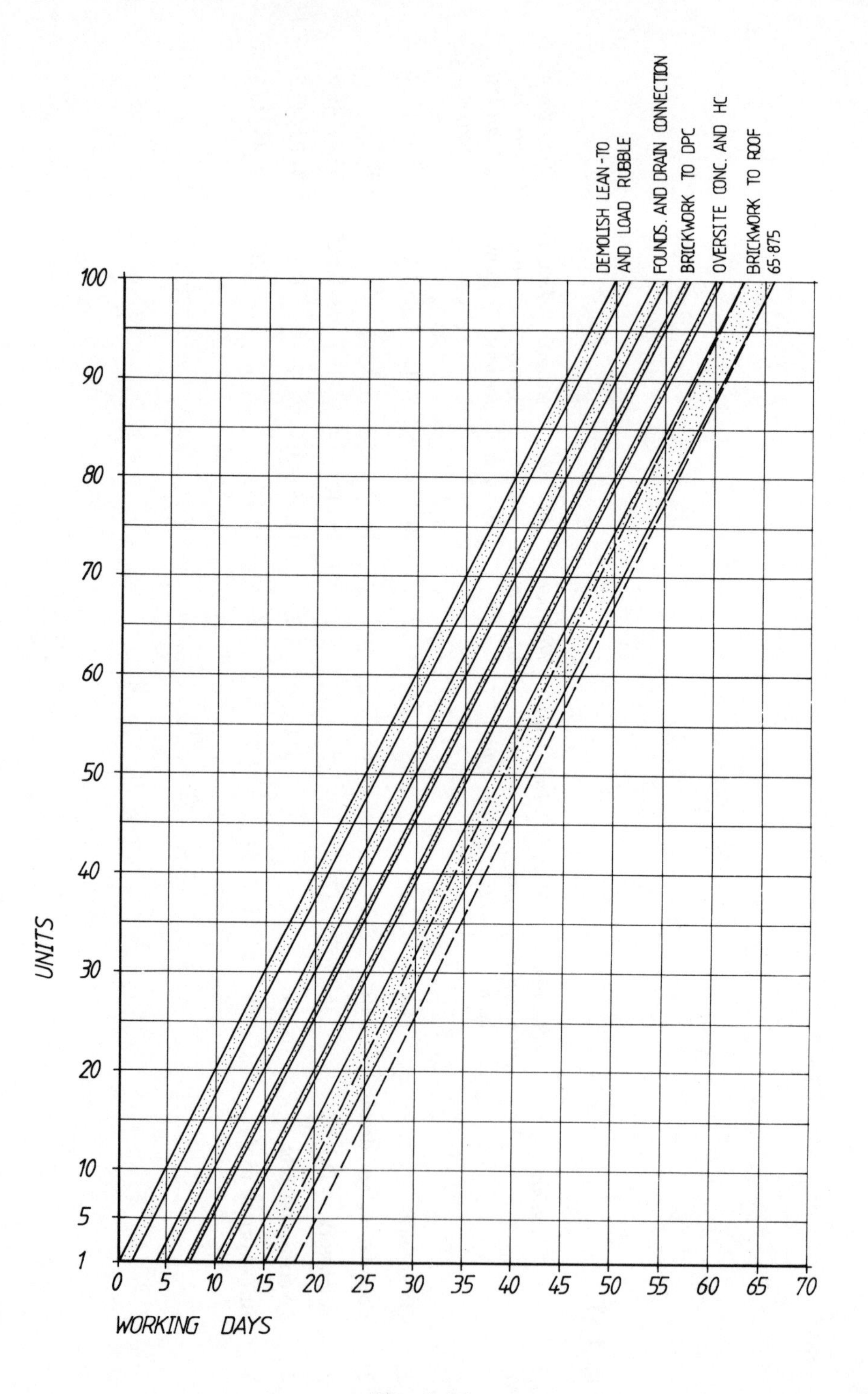
DEMOLISH LEAN-TO
AND LOAD RUBBLE
FOUNDS. AND DRAIN CONNECTION
BRICKWORK TO DPC
OVERSITE CONC. AND HC
BRICKWORK TO ROOF
65·875
100
90
80
70
60
50
40
30
20
10
5
1
UNITS
0
5
10
15
20
25
30
35
40
45
50
55
60
65
70
WORKING DAYS

Fig. 1.46

Fig. 1.47 Cost calculations for parallel scheduling

Demotion	6 × £48 = £288 per day	Cost = 500/10 × £288 = £ 14 400
Foundations	4 × £48 = £192 per day	Cost = 500/10 × £192 = £ 9600
Brickwork to D.P.C.	6 × £48 = £288 per day	Cost = 500/10 × £288 = £ 14 400
Oversite concrete	4 × £48 = £192 per day	Cost = 500/10 × £192 = £ 9600
Brickwork to roof	18 × £48 = £864 per day	Cost = 500/10 × £864 = £ 43 200
		£ 91 200
	Overheads 66 × £240	£ 15 840
		£107 040
	Delay start of brickwork to roof	Cost = 500/10.43 × £864 = £ 41 419
		Saving £ 1781

Fig. 1.48

Act. No.	*Activty*	*Total man hours per unit*	*Min. gang size*	*Time reqd. by min. gang*	*Number of gangs*	*No. of units per week*	*No. of gangs per unit*	*Time per unit (days)*	*Time from start first to start last*
3	Brickwork to D.P.C.	15	3 (2+1)	5	1	8	1	0.625	61.88
4	Oversite contrete	11	2	5½	1	7.27	1	0.7	68.09

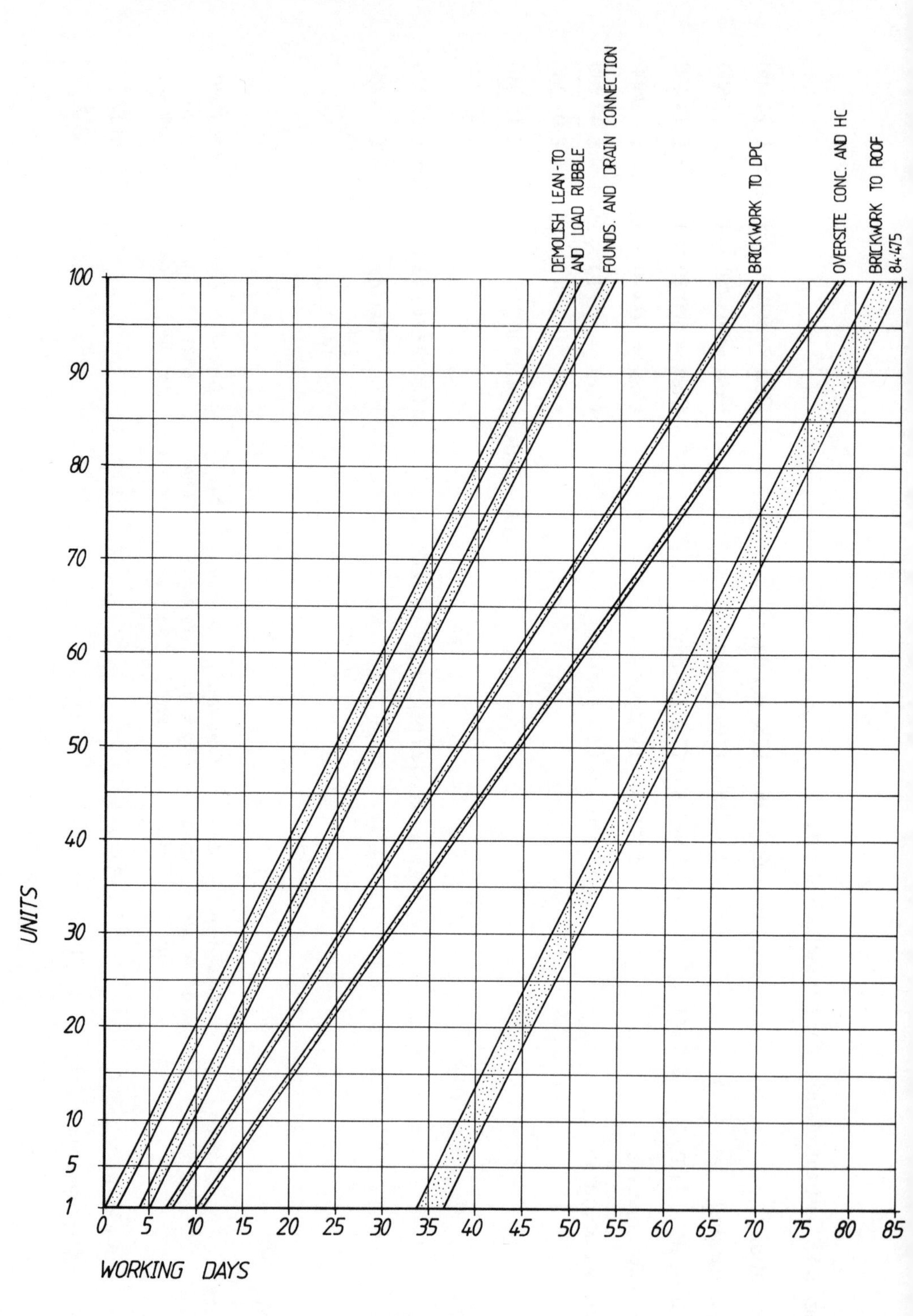
DEMOLISH LEAN-TO
AND LOAD RUBBLE
FOUNDS. AND DRAIN CONNECTION
BRICKWORK TO DPC
OVERSITE CONC. AND HC
BRICKWORK TO ROOF
84.475
100
90
80
70
60
50
40
30
20
10
5
1
UNITS
0 5 10 15 20 25 30 35 40 45 50 55 60 65 70 75 80 85
WORKING DAYS

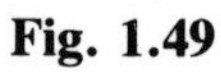

Fig. 1.49

Demolition	6 × £48 = £288 per day
Foundations	4 × £48 = £192 per day
Brickwork to D.P.C.	3 × £48 = £144 per day
Oversite concrete	2 × £48 = £ 96 per day
Brickwork to roof	18 × £48 = £864 per day

Cost = 500/10	× £288 =	£ 14 400
Cost = 500/10	× £192 =	£ 9600
Cost = 500/8	× £144 =	£ 9000
Cost = 500/7.27	× £ 96 =	£ 6603
Cost = 500/10.43	× £864 =	£ 41 419
		£ 81 022
Overheads 85 × £240		£ 20 400
		£101 422

Fig. 1.50 Cost calculations for resource scheduling

Therefore 60 units are to be built using one gang and 40 units using two gangs.

As units are to be built in stages, build 15 units with one gang and 10 units with two gangs in each stage.

Additional calculations are necessary to determine the time from 'start first' to 'start last' in each stage as follows:

Fifteen units using one gang $5(15-1) \div 8 = 8.75$

Ten units using two gangs $5(10-1) \div 16 = 2.81$

Similar calculations for oversite concrete would not produce an economic programme and it has therefore been decided to programme this activity in parallel with brickwork to D.P.C.

The line of balance schedule is shown in Fig. 1.51.

The adjustment to the price relative to the resource scheduling example shown in Fig. 1.49 is set out below:

Saving in cost
Overheads
(85 − 69) × £240 = £3840

Additional costs
Brickwork to D.P.C.
3 (men) × 3 (returns to site) × £30 = £ 270
Oversite concrete
2 × 3 × £30 = £ 180

£ 450 £450

Savings = £3390

1.11 SITE LAYOUT

The physical factors of the site often affect the method and sequence adopted in the construction programme, so the programme and the site layout are usually prepared together.

1.11.1 Factors affecting site layout

Before attempting to construct a site layout drawing, a list should be compiled of all accommodation, plant and material storage areas required. The areas on site which are available to the contractor should then be investigated to assist in determining the overall arrangement of the items mentioned above, i.e. where would the staff accommodation or the major items of plant be best situated? This should be considered in overall terms at an early stage, followed by a consideration of the more detailed aspects of the site layout.

For this work a detailed site drawing is necessary to a scale of at least 1:100, showing all foundations, drains, etc. The most suitable drawings are usually the site general arrangement, the ground floor plan and any other drawing that gives the full extent of the site and shows all obstructions. As changes will be made frequently in arriving at the site layout, two common methods may be used to facilitate alterations:

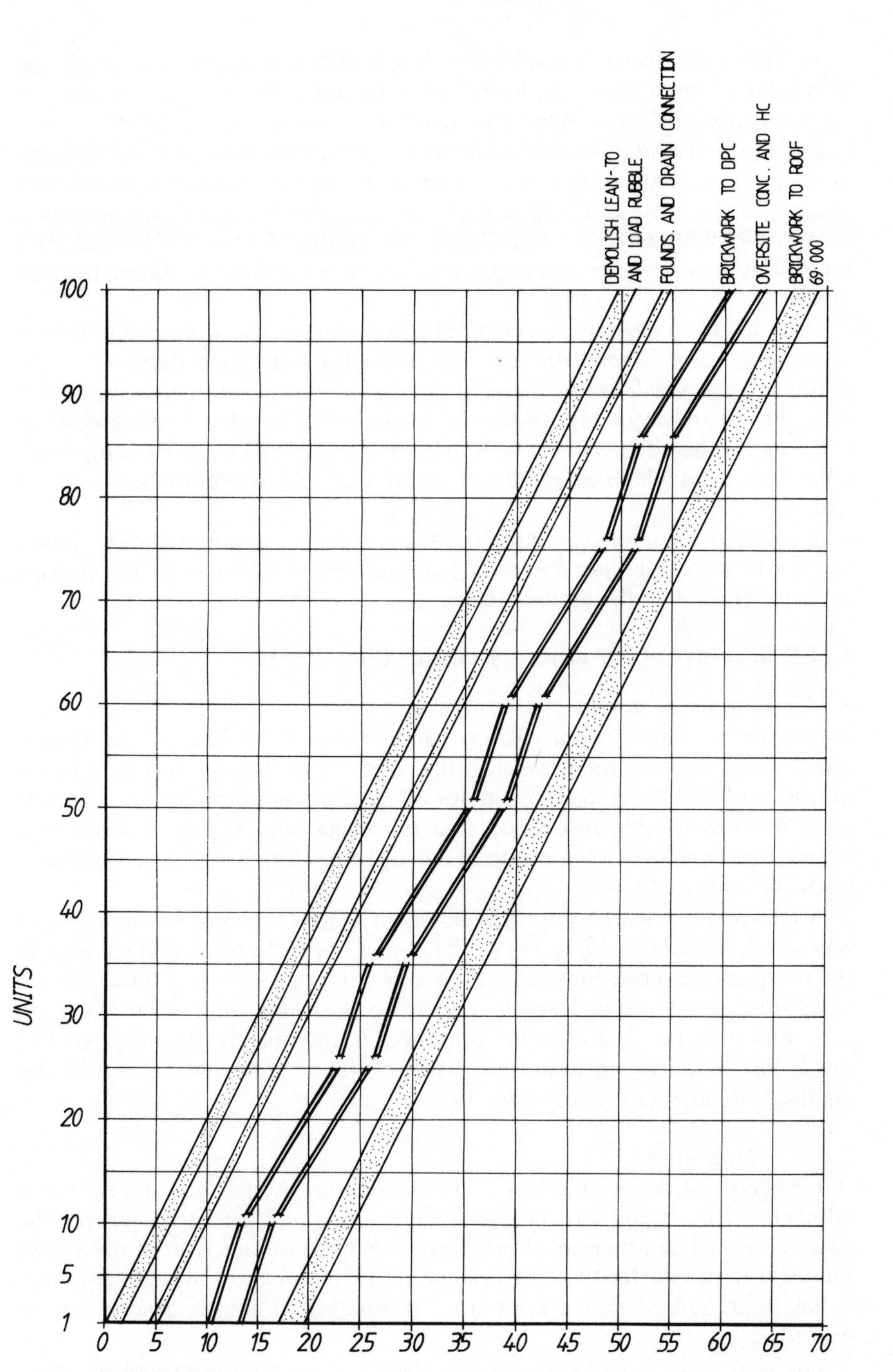
DEMOLISH LEAN-TO AND LOAD RUBBLE
FOUNDS. AND DRAIN CONNECTION
BRICKWORK TO DPC
OVERSITE CONC. AND HC
BRICKWORK TO ROOF 69·000
UNITS
100
90
80
70
60
50
40
30
20
10
5
1
0
5
10
15
20
25
30
35
40
45
50
55
60
65
70
WORKING DAYS

Fig. 1.51

1. The scale drawing can be overlayed with a sheet of clear plastic on which the planner will draw his tentative layout with a wax crayon that can easily be erased if parts of the layout are later found to be unsuitable.
2. Silhouettes or models of all items of accommodation, plant and storage areas are produced to the same scale as the layout drawing and and then placed on the scale drawing in the areas set aside for particular activities; they can now be moved around until the required layout is achieved. This method is a work study technique which is very useful in obtaining the best layout.

Temporary roads may or may not be necessary, depending upon the type of sub-soil experienced on site, but when they are used they should be planned to serve all major items of plant and material storage compounds; hardcore or railway sleepers may be employed as construction material. In some cases the site service roads may be completed before building work commences, in which case they could be used to provide access about the site.

A number of factors can influence the location of accommodation, plant, storage areas and temporary roads, and these are considered in the following example (based on the amenity centre).

1.11.2 Site layout for amenity centre (Fig. 1.52)

1.11.2.1 Access to the site

Ideally there should be an entrance and an exit to the site, as this has the effect of stimulating the flow of traffic. In this case the site is situated on a major road and the police would not allow the contractor access from that side: in view of this restriction and the shape and size of the site it is impracticable to provide more than one opening, which will have to serve as both entrance and exit.

A cross-over must be provided adjacent to this entrance as the footpath was not designed to take heavy vehicles, and the kerb stones and pavings on the footpath must be removed in the area of the cross-over and replaced by a temporary sleepered cross-over. As the access is not visible from the main road and is to be via the works entrance, it must be clearly signposted to direct lorries delivering materials. Visitors and personnel will enter the site via the gate adjacent to the wages office.

1.11.2.2 Hoarding

A licence must be obtained from the local authority to erect the hoarding adjacent to the main road and extending 0.6 m onto the footpath so that the boundary wall can be built. Permission has been obtained from the works manager to use the footpath at the rear of the site for extra storage space and it has been agreed that a temporary footpath and hoarding need not be provided.

The gate in the hoarding will open inwards so as not to cause obstruction in the road – a useful alternative would be the provision of sliding doors. As the main access to the site is via the works entrance (which is manned by the works police) it is considered that additional security is not necessary.

1.11.2.3 Temporary roads

Due to the shape of the site there will be only one temporary road, as shown in the plan. This will be constructed by hardcore laid at a level which is suitable for it to become part of the car park sub-base at a later date.

1.11.2.4 Plant

1. Tower crane

The crane selected has a lifting capacity of 500 kg at a maximum radius of 21 m and has been selected because with this jib length it will cover all the building and stacking grounds situated in the position indicated and will lift the required loads. As the crane is to remain static, provision must be made for anchoring the base of the mast by casting a concrete foundation block.

2. Concrete mixer

It has been decided to mix the concrete on the site because of delivery difficulties and varying quantity demand which ready mixed would have had difficulty in copying with. The size of the mixers selected is a 300/200 litre, which is electrically powered and provided with a cement silo of 20 tonnes capacity mounted directly above it, thus reducing the space required. The aggregate storage areas will be situated adjacent to the road for ease of access, the aggregates themselves will be kept apart by sleepered division walls. The mixer hopper is to be charged by means of a hand scraper drawn by an electrically operated winch situated on the mixer.

The concrete will be transported from the mixer by the tower crane using a roll-over skip.

3. Power bender

This will be situated in the area set aside for reinforcing. It will be located on a concrete hard standing and is covered over with a tarpaulin on a scaffold tube frame to provide a waterproof working area.

A hand operated cutter will also be provided in the covered area.

4. Circular saw

This will be situated in the area set aside for formwork, and requires a hard standing and weather sheeting in a similar manner to the power bender.

5. Hoist

A hoist will replace the tower crane when the frame is completed and the finishings are being carried out. It is to be situated in a central position, thus supplying all the building.

6. 150/100 litre mixer

The mixers are placed in the areas shown, paving first being removed to avoid damaging it. The mixers and their related material storage areas will be located adjacent to the hoist to facilitate the delivery of mixed material into the building.

7. Scaffolding

The scaffolding is of the independent type and will be used: (i) to give access to columns and beams when concreting, (ii) to give access to brickwork, and

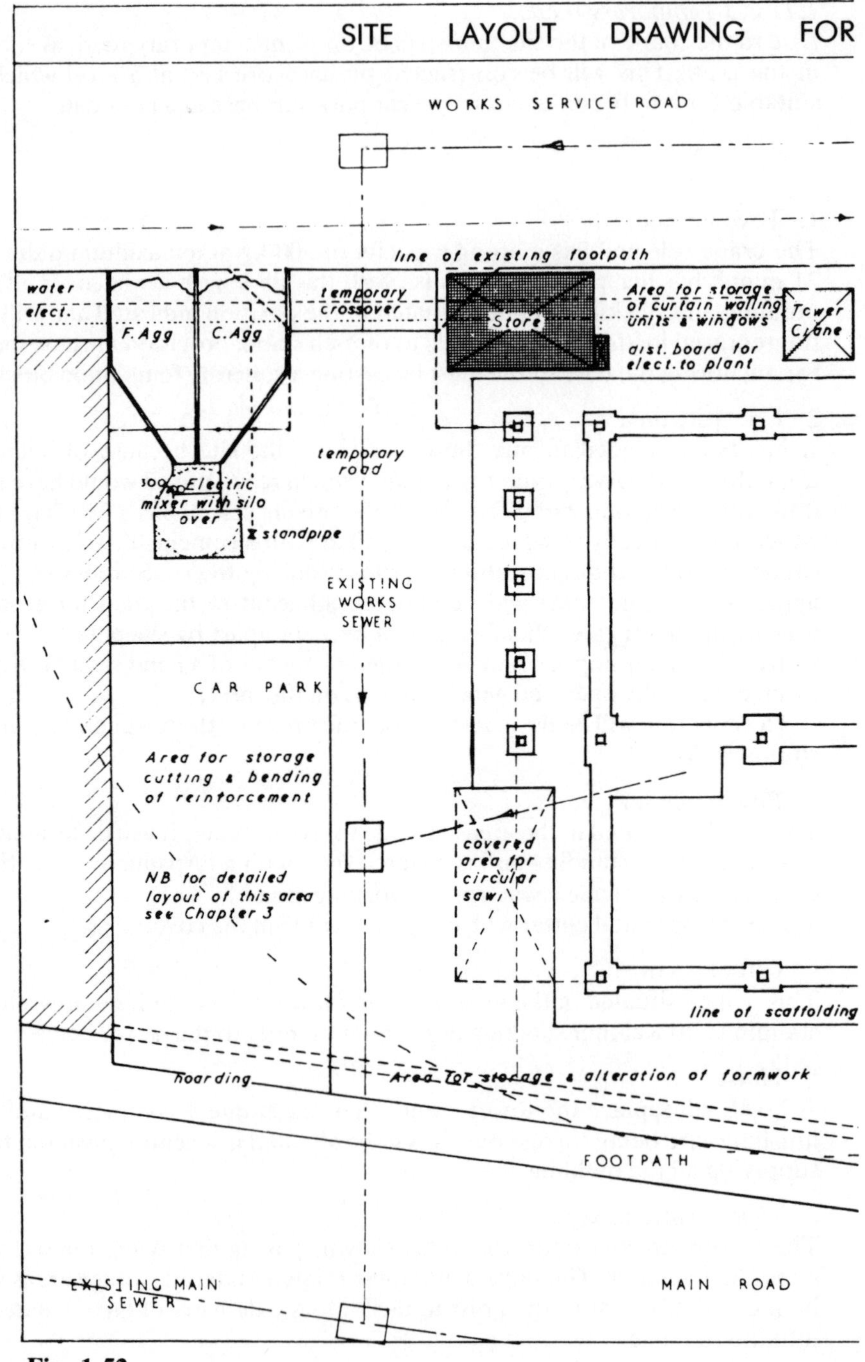
SITE LAYOUT DRAWING FOR
WORKS SERVICE ROAD
line of existing footpath
water
elect.
F. Agg
C. Agg
temporary crossover
Store
Area for storage of curtain walling units & windows
Tower Crane
dist. board for elect. to plant
temporary road
300 Electric mixer with silo over
standpipe
EXISTING WORKS SEWER
CAR PARK
Area for storage cutting & bending of reinforcement
NB for detailed layout of this area see Chapter 3
covered area for circular saw
line of scaffolding
hoarding
Area for storage & alteration of formwork
FOOTPATH
EXISTING MAIN SEWER
MAIN ROAD

Fig. 1.52

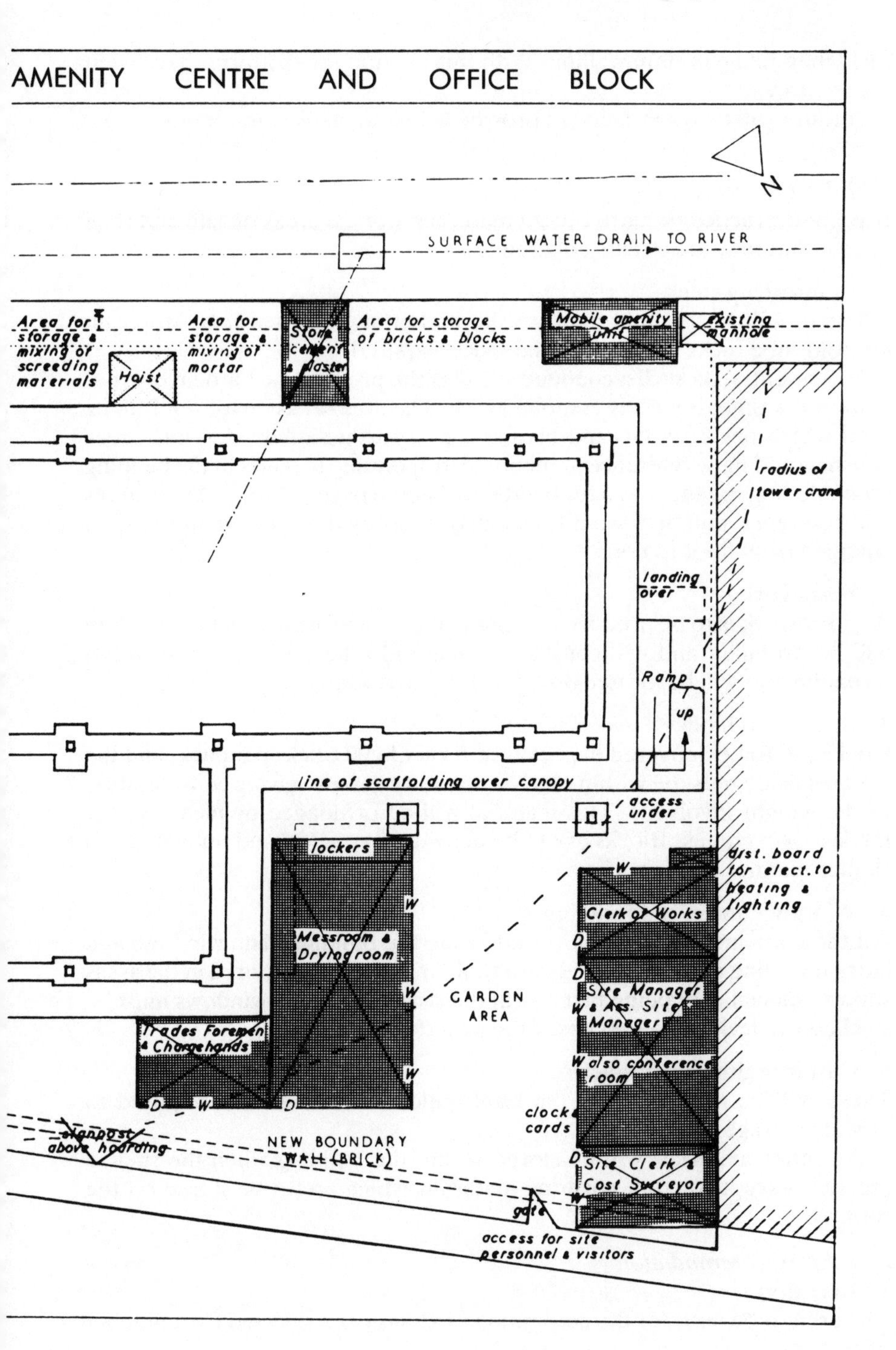
AMENITY CENTRE AND OFFICE BLOCK
SURFACE WATER DRAIN TO RIVER
Area for storage & mixing of screeding materials
Hoist
Area for storage & mixing of mortar
Store cement & plaster
Area for storage of bricks & blocks
Mobile amenity unit
existing manhole
radius of tower crane
landing over
Ramp
up
line of scaffolding over canopy
access under
lockers
Messroom & Drying room
GARDEN AREA
Trades Foremen & Chargehands
dist. board for elect. to heating & lighting
Clerk of Works
Site Manager & Ass. Site Manager
also conference room
clock cards
Site Clerk & Cost Surveyor
signpost above hoarding
NEW BOUNDARY WALL (BRICK)
gate
access for site personnel & visitors

(iii) when fixing curtain walling. With this in mind the platforms have been made 1 m wide.

A ramp will be constructed to provide access to all platform levels.

1.11.2.5 Materials

It is good practice to mark out all materials storage areas on site and then ensure that they are strictly adhered to:

1. Reinforcing steel

All steel will be cut and bent on site and straight bars will be stored in scaffold tube racks to cut down the space required.

It is essential on such a confined site that the programme for delivery, i.e. a floor at a time, is strictly complied with. The area set aside for reinforcing steel which has been cut and bent could be further subdivided into areas which would have reference numbers corresponding to pages of the bending schedule, e.g. beams 1 to 5 for bending schedule pages 22 to 23. These areas and reference numbers would obviously be altered as and when the steel changed from floor to floor.

2. Formwork

All formwork is to be produced on site in the area provided. A joiners' shop will be provided and will consist of a scaffold tube frame covered with a tarpaulin and will be located on a concrete hardstanding.

3. Bricks and blocks

Bricks are to be delivered in packaged form of 200 bricks per pack, and the area set aside must be within reach of the tower crane (giving consideration to the weight involved). This material will be off-loaded by means of the lorries' own cranes. Blocks are to be delivered on palets and unloaded in a similar manner to the bricks.

4. Windows and curtain walling

All the glass involved will be stored inside the building. The windows and curtain walling units will be delivered a floor at a time and stored in the areas shown, special care being taken with the storage, i.e. the windows must be stacked vertically on a firm base or in a scaffold tube rack.

5. Cast iron drains and stacks

These will be ordered from the local builders' merchants as required to obviate storage on site.

All other materials will be stored within the building when the shell is erected, except for roof covering materials which are to be stored on the roof.

1.11.2.6 Accommodation

1. Location

The area to be used for the accommodation has been selected for a number of reasons:

(a) It is possible to have all accommodation in one area, which helps communication between personnel.

(b) It is in an area which is likely to be less noisy than the others, i.e. away from joiners shop, mixer, etc.
(c) It efficiently utilizes an area which would be difficult to employ for any other purpose as it has poor access.
(d) The wages and time office can easily be located against the door provided for the access of workmen.
(e) People visiting the site manager can conveniently gain access to site and readily locate his office. The same point applies equally to people seeking the clerk of works.
(f) As this is a garden area which is a separate contract, it is possible to complete the site works without moving accommodation, i.e. the accommodation is in one position throughout the project.

2. Construction and fittings

The accommodation will be of sectional timber construction with weather board walls and felted roof. The inside is to be lined with insulation boarding to cut down heat loss.

All compartments will be provided with lighting in the form of fluorescent strips and heating in the form of electric fires.

3. Accommodation provided

(a) Site manager, assistant site manager, site clerk, and cost surveyor

These site staff will be kept together in one office complex for ease of communication. A first-aid box will be located in the assistant site manager's office.

(b) Clerk-of-works

The clerk-of-works' office will be situated as close to the contractor's office as possible for ease of communication.

(c) Mess room

This will be provided with a sufficient number of benches and tables for men to take their meals, water for drinking, and facilities for boiling water. It will be situated adjacent to the trades foremen's accommodation to reduce the amount of prolonged tea break.

(d) Drying and locker room

Accommodation for hanging up off-site and/or on-site clothing to dry is to be provided at the end of the mess room. Lockers will also be available for workmen's personal effects.

(e) Water closets

These should be situated away from the messroom and administrative area, but not too far away to cause inconvenience.

The facility is in the form of a mobile amenity unit having two W.C.'s, two washbasins, a urinal, and an instantaneous hot water heater in one compartment, whilst another compartment is provided with one W.C., one washbasin and an instantaneous hot water heater (this latter compartment being for the use of the staff and the clerk of works).

The amenity unit will be connected to an existing manhole and existing water and electrical service as shown.

1.11.2.7 Services

1. Water

Water will be supplied to all mixer points (terminating at a standpipe), the messroom, and the mobile amenity unit.

2. Electricity

There will be two supplies, one being a 415-volt three-phase for the plant, whilst the other is 240 volts for the heating and lighting in the accommodation. The supplies will be taken to weather-proof boxes containing distribution boards, cut-outs and meters that are located on the side of the store and the clerk-of-work's office.

The 240-volt supply would be stepped down to 110 volts or 50 volts by means of a site transformer for powering small tools used on site. The transformer can also be housed in a weather-proof box or alternatively a small weather-proof transformer can be used out on the site.

3. Telephones

These would be connected as soon as the site accommodation is completed. Telephones would be provided for the site manager, the site clerk, and the clerk-of-works.

1.12 MULTI-PROJECT PLANNING

1.12.1 Introduction

With the planning techniques used so far in this book each project has been considered in isolation. Little attention has been paid to the demand each project has placed on the total resources of the firm, providing that the maximum amount of any resource has not been exceeded. In practice many projects will be proceding at the same time and it is obvious that the total available resources and the demand for such resources by all current and future known projects should be considered when drawing up programmes for new projects. The following outlines a method which enables planning of a number of projects to take place within the total available resources of the firm or section of the firm and facilitates the coordination of a number of projects on the one chart.

1.12.2 Method of presentation

Using this planning technique all operations which follow each other within a particular project are recoded on the same line of the chart. This saves considerable space on the chart thus allowing more projects to be presented. For clarity, each block on the chart can be coloured using a particular colour for each trade and arrows of the same colour can be used to indicate the transfer of men from one project to another. If presentation is in black and white then different types of shading would be used.

1.12.3 Exceeding available resources

If the total available resources within the firm must be exceeded then adequate advanced warning is given that further recruitment of labour or hire of plant is required. Occasionally, in certain areas of the country, the demand for tradesmen such as carpenters exceeds supply and in such circumstances this planning technique is invaluable in making the best use of available tradesmen and providing them with continuity of employment, thus retaining their services.

1.12.4 Procedure

As each project is awarded to the firm it will be integrated with the previously planned projects which are all displayed on a master chart. As stated earlier, care should be taken when deciding on the timing of operations to see that the maximum available resources are not exceeded. When the new project has been added to the master chart, the programme for each projects manager and trade foreman can be taken off and presented separately. It is particularly important that the trades foremen should have a copy of this programme as they may be involved in working for more than one projects manager as well as different site managers under a particular project manager.

1.12.5 Programme for four projects (see Fig. 1.53)

As an example of the use of this technique four projects are considered. Project 1 being shops and offices, project 2 being a gymnasium, project 3 is the Amenity Centre and project 4 is the substructure to a building. The four projects are under the supervision of one projects manager.

1.12.5.1 Resource levels

The maximum level of each labour resources is as follows:

Carpenters 10
Plumbers 2 + 2
Electricians 2 + 2
Steelfixers 8
Bricklayers 8 + 4
Plasterers 4 + 2
Tilers 6 + 3
Painters 8
Scaffolders 2
General labourers no max. set.

1.12.5.2 Projects 1 and 2

The particular projects manager was given projects 1 and 2 at the same time. They start within two weeks of each other. Both these projects are planned and coordinated so as not to exceed the resources under the control of the projects manager.

1.12.5.3 Continuity of work in projects 1 and 2

To illustrate how continuity of work is achieved in projects 1 and 2 the trade of bricklayers will be used. In project 1, half way through week 2, two bricklayers are required for underpinning work, this work being almost complete by the end of week 5 when these two bricklayers are moved to

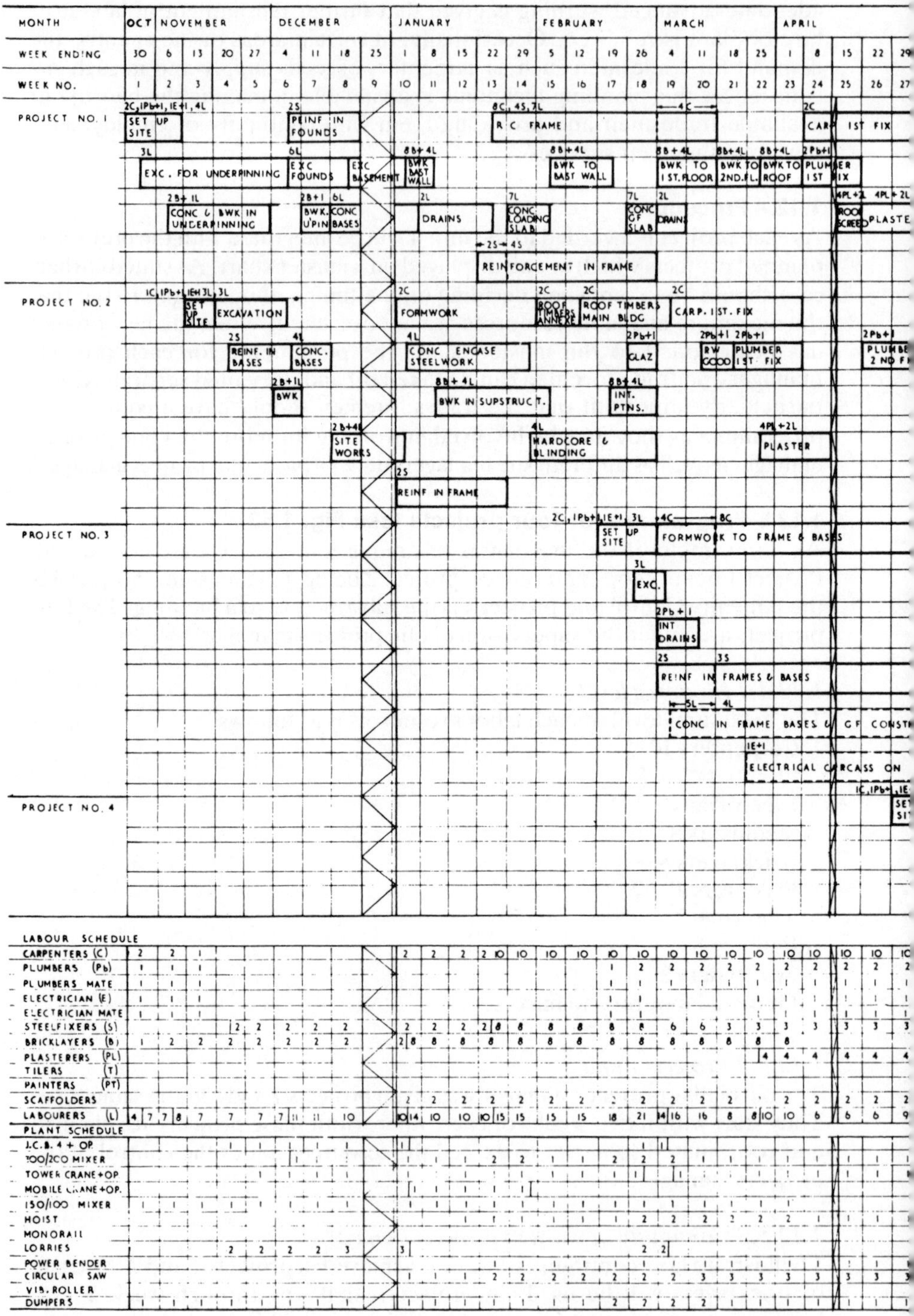

MONTH	OCT	NOVEMBER				DECEMBER				JANUARY					FEBRUARY				MARCH				APRIL				
WEEK ENDING	30	6	13	20	27	4	11	18	25	1	8	15	22	29	5	12	19	26	4	11	18	25	1	8	15	22	29
WEEK NO.	1	2	3	4	5	6	7	8	9	10	11	12	13	14	15	16	17	18	19	20	21	22	23	24	25	26	27

LABOUR SCHEDULE

WEEK NO.	1	2	3	4	5	6	7	8	9	10	11	12	13	14	15	16	17	18	19	20	21	22	23	24	25	26	27
CARPENTERS (C)	2	2	1							2	2	2	2 10	10	10	10	10	10	10	10	10	10	10	10	10	10	10
PLUMBERS (Pb)	1	1	1														1	2	2	2	2	2	2	2	2	2	2
PLUMBERS MATE	1	1	1														1	1	1	1	1	1	1	1	1	1	1
ELECTRICIAN (E)	1	1	1														1	1				1	1	1	1	1	1
ELECTRICIAN MATE	1	1	1														1	1				1	1	1	1	1	1
STEELFIXERS (S)				2	2	2	2	2		2	2	2	2 8	8	8	8	8	8	6	6	3	3	3	3	3	3	3
BRICKLAYERS (B)	1	2	2	2	2	2	2	2		2 8	8	8	8	8	8	8	8	8	8	8	8	8	8				
PLASTERERS (PL)																						4	4	4	4	4	4
TILERS (T)																											
PAINTERS (PT)																											
SCAFFOLDERS										2	2	2	2	2	2	2	2	2	2	2	2	[illegible]	2	2	2	2	2
LABOURERS (L)	4 7	7 8	7	7	7	7 11	11	10		10 14	10	10	10 15	15	15	15	18	21	14 16	16	8	8 10	10	6	6	6	9

PLANT SCHEDULE

WEEK NO.	1	2	3	4	5	6	7	8	9	10	11	12	13	14	15	16	17	18	19	20	21	22	23	24	25	26	27
J.C.B. 4 + OP				1	1	1	1	1		1								1	1								
100/200 MIXER						1	1	1		1	1	1	2	2	1	1	2	2	2	1	1	1	1	1	1	1	
TOWER CRANE + OP					1	1	1	1		1		1	1	1	1	1	1	1	1	1	1	1	1	1	1	1	
MOBILE CRANE + OP.										1	1	1	1	1													
150/100 MIXER	1	1	1	1	1	1	1	1		1	1	1	1	1	1	1	1	1	1	1	1	1	1				
HOIST												1	1	1	1	1	1	2	2	2	2	2	2	1	1	1	
MONORAIL																											
LORRIES				2	2	2	2	3		3								2	2								
POWER BENDER													1	1	1	1	1	1	1	1	1	1	1	1	1	1	
CIRCULAR SAW										1	1	1	2	2	2	2	2	2	2	3	3	3	3	3	3	3	3
VIB. ROLLER																											
DUMPERS	1	1	1	1	1	1	1	1		1	1	1	1	1	1	1	1	2	2	2	1	1	1	1	1	1	1

Fig. 1.45

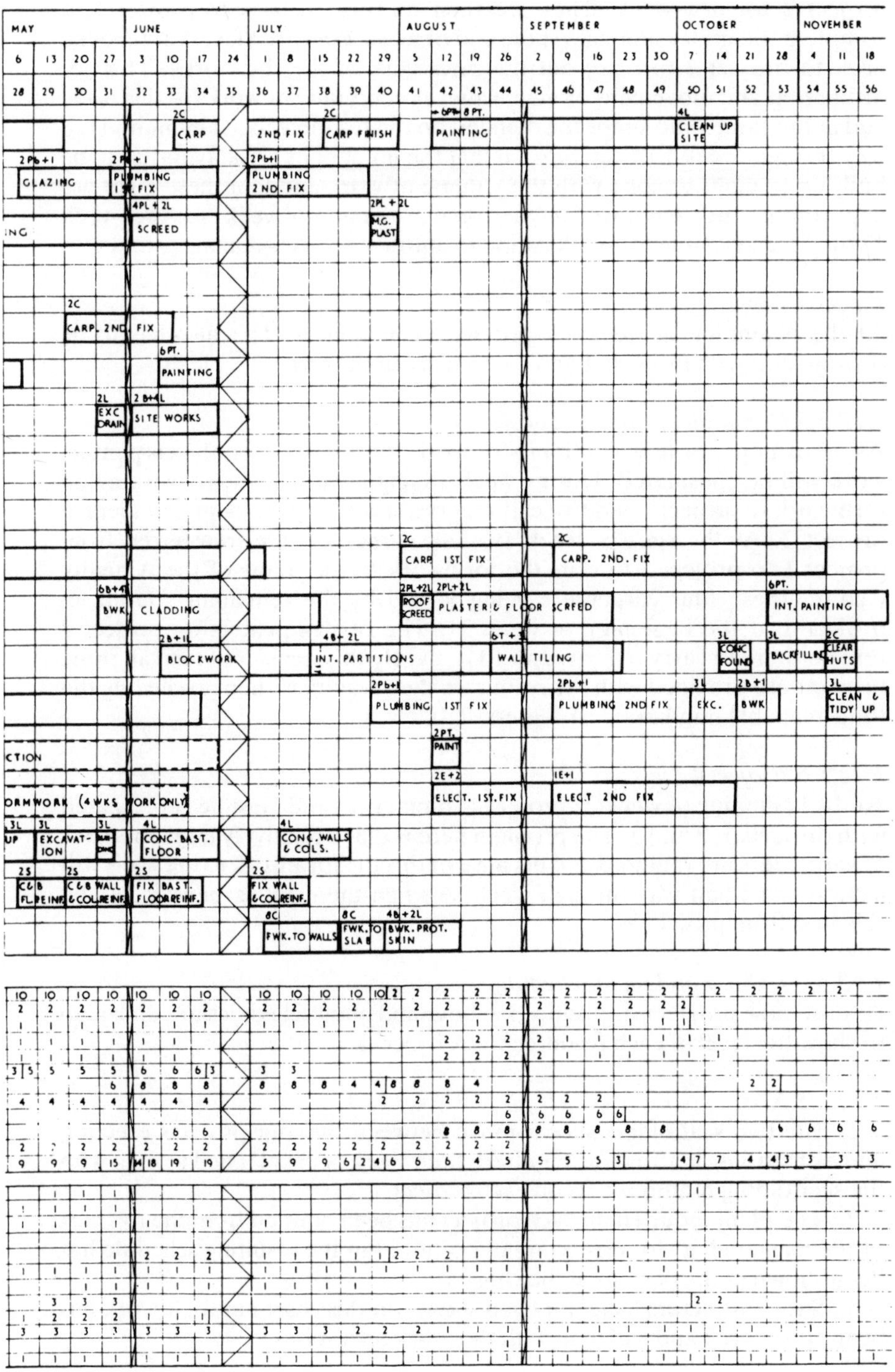

MAY				JUNE				JULY					AUGUST				SEPTEMBER					OCTOBER				NOVEMBER		
6	13	20	27	3	10	17	24	1	8	15	22	29	5	12	19	26	2	9	16	23	30	7	14	21	28	4	11	18
28	29	30	31	32	33	34	35	36	37	38	39	40	41	42	43	44	45	46	47	48	49	50	51	52	53	54	55	56

2C CARP · 2ND FIX · 2C CARP FINISH · 6PT 8PT. PAINTING · 4L CLEAN UP SITE

2Pb+1 GLAZING · 2Pb+1 PLUMBING 1ST. FIX · 2Pb+1 PLUMBING 2ND. FIX

ING · 4PL + 2L SCREED · 2PL + 2L M.G. PLAST

2C CARP. 2ND. FIX

6PT. PAINTING

2L EXC DRAIN · 2B+4L SITE WORKS

2C CARP 1ST. FIX · 2C CARP. 2ND. FIX

6B+4L BWK. CLADDING · 2PL+2L ROOF SCREED · 2PL+2L PLASTER & FLOOR SCREED · 6PT. INT. PAINTING

2B+1L BLOCKWORK · 4B+2L INT. PARTITIONS · 6T+3 WALL TILING · 3L CONC FOUND · 3L BACKFILLING · 2C CLEAR HUTS

2Pb+1 PLUMBING 1ST FIX · 2Pb+1 PLUMBING 2ND FIX · 3L EXC. · 2B+1 BWK · 3L CLEAN & TIDY UP

2PT. PAINT

CTION

2E+2 ELECT. 1ST. FIX · 1E+1 ELECT 2ND FIX

ORMWORK (4 WKS WORK ONLY)

3L UP · 3L EXCAVATION · 3L · 4L CONC. BAST. FLOOR · 4L CONC. WALLS & COLS.

2S C&B FL. REINF. · 2S C&B WALL & COL. REINF. · 2S FIX BAST. FLOOR REINF. · 2S FIX WALL & COL. REINF.

8C FWK. TO WALLS · 8C FWK. TO SLAB · 4B+2L BWK. PROT. SKIN

10 10 10 10 10 10 10 10 10 10 10 10 | 2 2 2 2 2 2 2 2 2 2 2 2 2 2 2 2 2 2

2 2 2 2 2 2 2 2 2 2 2 2 2 2 2 2 2 2 2 2 2 | 2

1 1 1 1 1 1 1 1 1 1 1 1 1 1 1 1 1 1 1 1 1 1 1 | 1

1 1 1 1 1 1 2 2 2 2 1 1 1 1 1 1 1

1 1 1 1 1 2 2 2 2 1 1 1 1 1 1 1

3 | 5 5 5 5 6 6 6 | 3 3 3

6 8 8 8 8 8 8 4 4 | 8 8 8 4 2 2

4 4 4 4 4 4 4 2 2 2 2 2 2 2 2

6 6 6 6 | 6

6 6 8 8 8 8 8 8 8 8 6 6 6 6

2 2 2 2 2 2 2 2 2 2 2 2 2 2 2

9 9 9 15 14 | 18 19 19 5 9 9 6 | 2 4 | 6 6 6 4 5 5 5 5 | 3 4 | 7 7 4 4 | 3 3 3 3

1 1 1 1

1 1 1

1 1 1 1 1 1 1 1

1 2 2 2 1 1 1 1 1 | 2 2 2 1 1 1 1 1 1 1 1 1 1 1 | 1

1 1 1 1 1 1 1 1 1 1 1 1 1 1 1 1 1 1 1 1 1 1 1 1

1 1 1 1 1 1 1 1

3 3 3

1 2 2 2 1 1 1 | 2 2

3 3 3 3 3 3 3 3 3 3 2 2 2 1 1 1 1 1 1 1 1 1 1 1 1 1

1 1

1 1 1 1 1 1 1 1 1 1 1 1 1 1 1 1 1 1 1 1 1 1 1 1 1 1 1 1

project 2 for one week. They are then transferred back to project 1 to complete the underpinning. At the end of week 7 the bricklayers are moved back to project 2 to carry out the brickwork in the site works. This co-ordination and movement of men achieves continuity of work for this trade and at the same time keeps the demand on the resource to a minimum. Half way through week 10 it can be seen that the number of bricklayers has to be built up to eight to cope with the volume of work and that these eight are kept continuously employed on these two projects until week 23. After week 23 they are available for reallocation to another projects manager.

1.12.5.4 Project 3

At the beginning of week 17, project 3 commences. This also has to be coordinated with projects 1 and 2 which are underway.

1.12.5.5 Continuity of work in project 3

As an example of how continuity of work is achieved here the carpenters work will be considered. During week 18, eight carpenters are working on formwork to project 1 and two carpenters are working on roofing timbers to project 2. At the start of week 19, four carpenters are transferred from project 1 to project 3 to start the formwork to the frame of the Amenity Centre, these four carpenters being joined by the remaining four from project 1 at the beginning of week 21. The two carpenters on project 2 remain there to carry out carpentry 1st. fix. By this transfer of men all three projects are provided with the resources they require without exceeding the total available number of carpenters.

1.12.5.6 Project 4

At the beginning of week 27 project 4 commences and must be coordinated with projects 1, 2 and 3. The problem here is that a relatively larger number of carpenters are required and because of this the formwork to the walls will have to start half way through week 36 when the eight carpenters can be released from project 3.

1.12.5.7 Labour and plant schedules

As each new project is introduced, labour and plant schedules are drawn up to show the level of each resource week by week.

1.12.5.8 Note

1. The two scaffolders shown move between the four projects erecting and dismantling scaffolds as required, their work being planned to fit in with the scaffolding requirements on each project.

2. The plant required on each project has been omitted from the top part of the master chart for clarity in publication, but the resulting schedule has been shown at the bottom of the chart.

3. Sub-contractors have not been shown on this chart but their work has been considered and will fit in with the other operations.

2 Project Network Techniques

2.1 INTRODUCTION

Project network techniques cover a number of techniques, one example being the Critical Path Method. Network techniques are particularly applicable to 'one off' projects and hence are of considerable use for many construction projects.

For small projects networks can be successfully analysed by hand, but on larger projects computers can be useful and save time in analysis, re-analysis and up-dating. This particularly applies when cost optimisation and/or resource allocation is being undertaken.

Networks can be presented as arrow diagrams (activity on arrow) or precedence diagrams (activity on node).

2.2 ADVANTAGES OF NETWORK TECHNIQUES OVER THE BAR CHART

1. When using network techniques, the inter-relationship of all operations is clearly shown. The normal bar chart does not do this and consequently requires the dependance of one operation upon another to be remembered by the planner: this is extremely difficult with large projects, and in addition the site manager (who must carry out the work) has to be informed how dependent one operation is upon another.
2. When a delay occurs and networks are being used, critical operations will stand out as requiring particular attention. When bar charts are used on a large project many operations tend to be 'crashed' unnecessarily as it is almost impossible to remember which operations are inter-dependent.
3. It is far easier for anyone taking over a partially completed project to become familiar with the progress when networks are employed.
4. When using networks it is essential to study the sequence of operations very carefully, leading to a closer understanding of the project.
5. Planning, analysing and scheduling are separated when using networks which allows a greater concentration on the planning aspect.

2.3 THE PREPARATION OF NETWORK DIAGRAMS

The first stage in the preparation of network diagrams is to make a list of the activities to be used. The amount of detail required in the breakdown depends on many factors, such as the size of the project or the stage of planning, i.e. pre-tender, pre-contact, or short term. One activity in a network drawn at the pre-tender stage may be broken down into a number of activities at a later stage. To avoid misunderstanding, diagrams are arranged so that time flows left to right. When drawing network diagrams it is important to remember that 'off site' activities such as delivery of plasterboard can be critical and must therefore be included on the diagram.

As each activity is considered the following questions should be asked.

1. Which activity must be completed before this activity can start?
2. Which other activities cannot start until this activity is completed?

A common mistake made here is for the planner to show an activity where he thinks it will be done, rather than the earliest time at which it can be done.

2.4 THE PREPARATION OF ARROW DIAGRAMS

2.4.1 Activities and events

An arrow diagram consists of activities and events. An activity is an operation or process. All activities start and finish at an event, which is a point in time and may be the junction of two or more activities (Fig. 2.1). Activities take time whereas events do not, and an activity cannot be started until all the activities leading into its preceding event are completed.

Activities are represented by arrows and the sequence of the arrows represents the sequence of activities; events are normally represented by circles.

'Dummy' activities are sometimes necessary in an arrow diagram. These do not take time to perform and are used either to make the sequence clear (Fig. 2.2) or to give a unique numbering system (Figs 2.3.1, 2.3.2 and 2.3.3). If a 'dummy' is not used here, two activities would have the same reference, e.g. joinery 1st fix and elect work 1st fix (Fig. 2.3.1) would have the same preceding and succeeding event. This is not acceptable when using a computer for analysis and it is therefore essential that a unique numbering system be used (Figs 2.3.2 and 2.3.3).

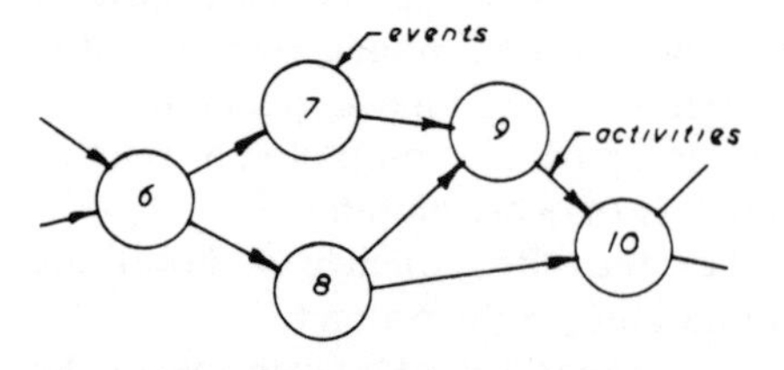

Fig. 2.1 Activity 9–10 cannot start until activities 7–9 and 8–9 are completed.

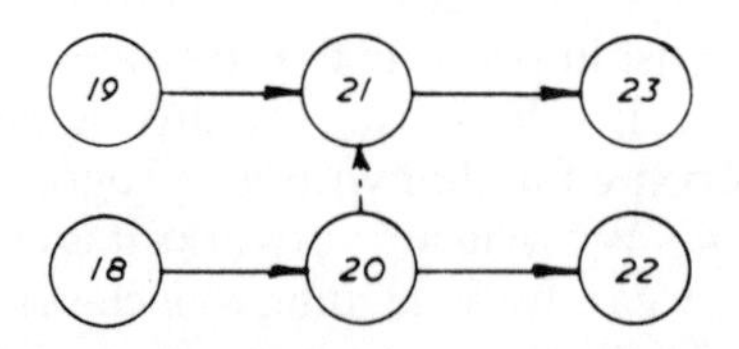

Fig. 2.2 Activity 21–23 cannot start until activities 19–21 and 18–20 are completed, but activity 20–22 can start when activity 18–20 is completed and is not dependent on activity 19–21.

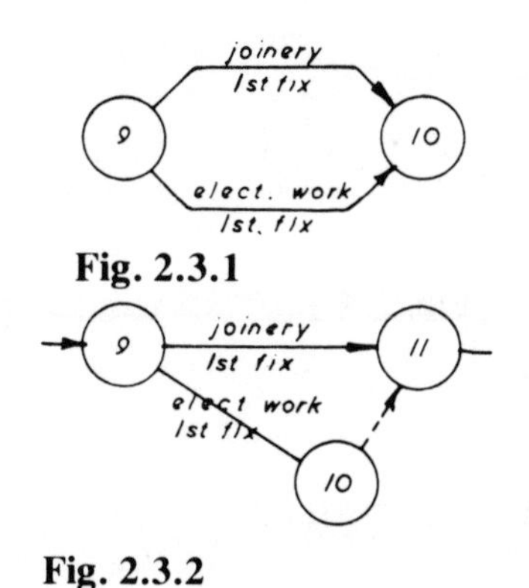

Fig. 2.3.1

Fig. 2.3.2

Figs 2.3.1, 2.3.2 and **2.3.3** represent the same situation.

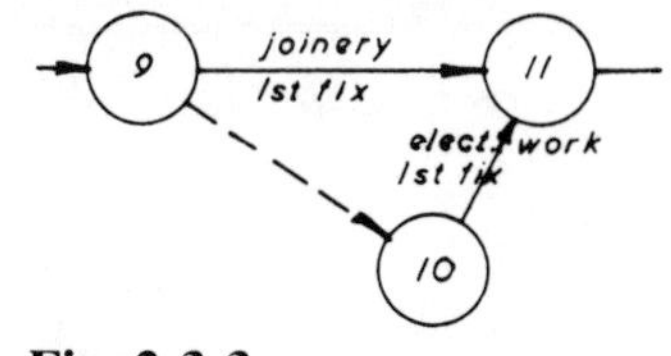

Fig. 2.3.3

2.4.2 Overlap of activities

Many activities in a building project can overlap, one activity starting before the previous activity is completed. For example, in a large concreting operation involving numerous columns, steelfixing would begin first and then formwork can be started when perhaps two columns had been completed; concreting then follows when the formwork is sufficiently advanced. To show this in a network it is necessary to split each activity.

1. Divide each activity into smaller equal portions (Fig. 2.4). This method of dividing each activity into the same number of equal parts is the only method which truly portrays the situation, but if each activity is divided into many parts the diagram becomes unduly complicated and in practice one of the following methods is often used as an alternative.

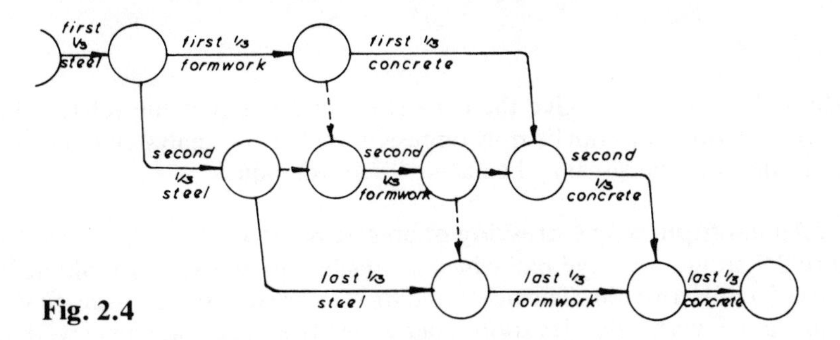

Fig. 2.4

2. Divide each activity into three sections; begin, continue and complete and again show the sequence in the normal way (Figs 2.5.1 and 2.5.2).

3. Use 'ladders' and 'time restraints' (Fig. 2.6). Activities 9–11 and 11–13 are known as 'lead time' and 10–12 and 12–14 as 'lag time', activity 9–11 represents the amount of reinforcement that must be fixed before formwork can start, activity 11–13 represents the amount of formwork to be erected before concrete can be poured, activity 10–12 represents the last of the formwork to be erected after steelfixing is completed, and activity 12–14 represents the last of the concrete to be poured after the formwork is completed.

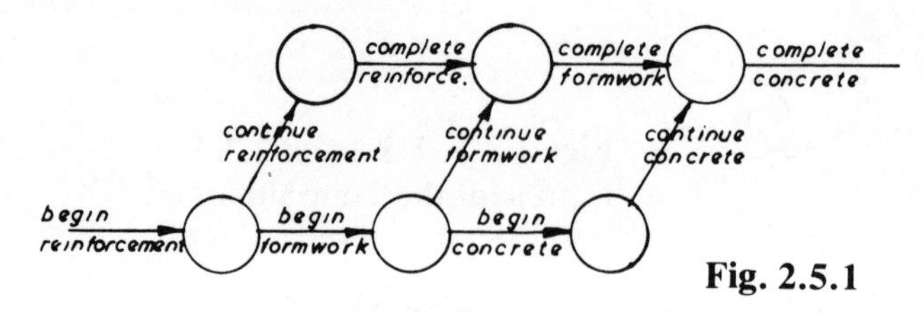

Fig. 2.5.1

THIS CAN BE SIMPLIFIED THUS

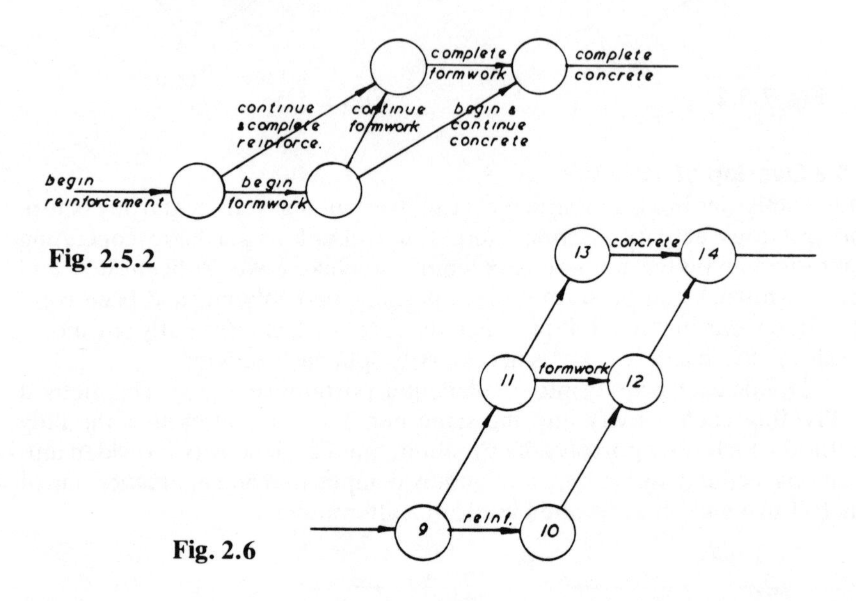

Fig. 2.5.2

Fig. 2.6

Methods 2 and 3 will give the correct event times, but the relationship between activities may not be truly represented when the analysis is tabulated. Care is therefore necessary when using these techniques.

2.4.3 An example of the drawing of an arrow diagram

Figure 2.7 shows a village hall which is attached to an existing building by means of a link corridor. The new building is to be constructed using timber portal frames with side elevations clad with brickwork and end elevations clad with timber frame units incorporating windows and doors. In this example the activities are limited to those listed below, and for simplification no overlap has been allowed.

The activities involved are: foundations, deliver frames, erect frames, brickwork up to damp-proof course, brickwork up to eaves, roof construction, roof finish, floor construction, end panels, glazing, internal partitions, electrician first fixing, joiner first fixing, deliver plasterboard, plasterer, electrician second fixing, joiner second fixing, make good plaster, floor screeds, floor finish, painting and decorating, electrician finishings, joiner finishings, rainwater goods, allow floor screed to dry, paving, drainage, clean-up and hand over.

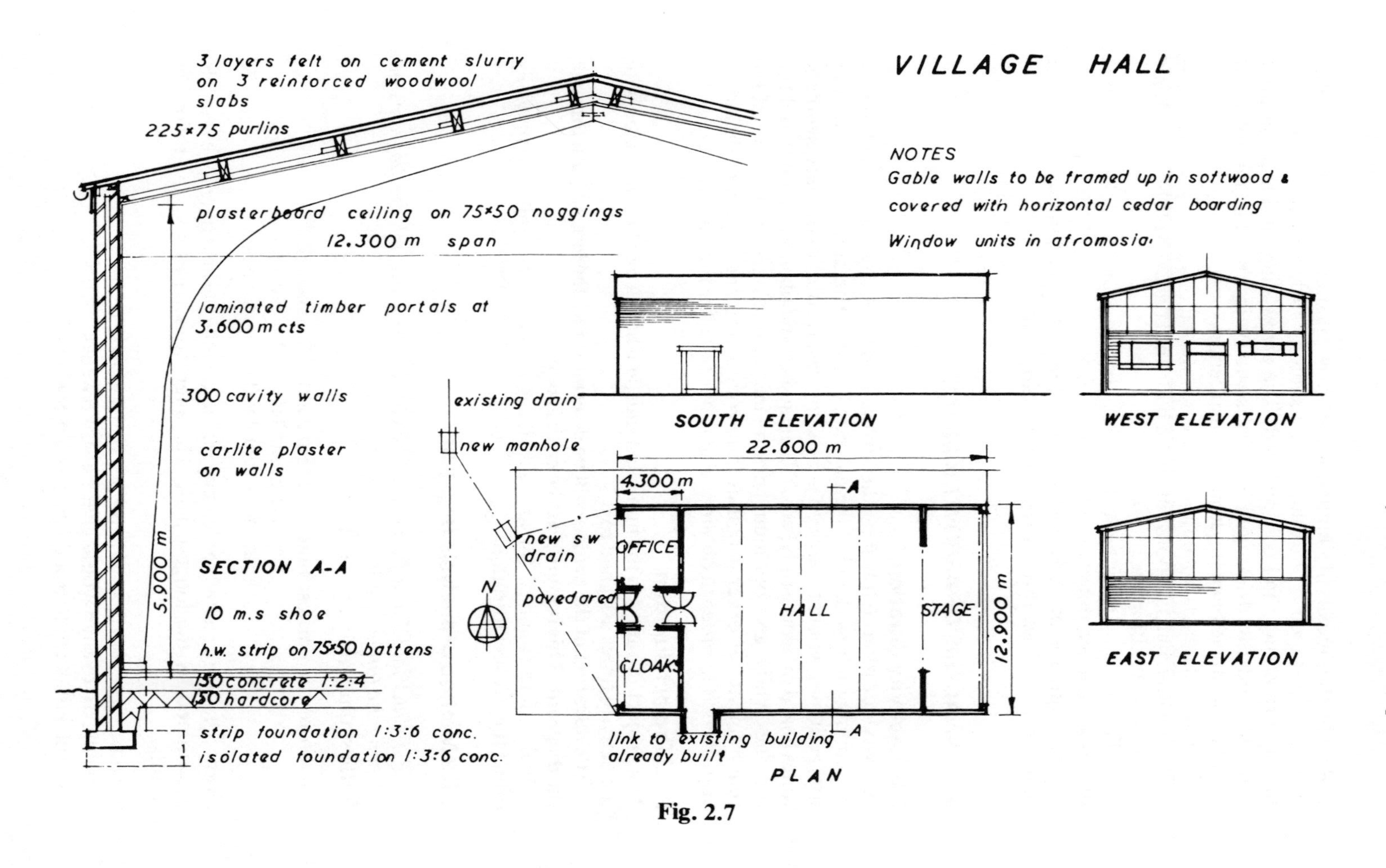

VILLAGE HALL
NOTES
Gable walls to be framed up in softwood &
covered with horizontal cedar boarding
Window units in afromosia
3 layers felt on cement slurry on 3 reinforced woodwool slabs
225×75 purlins
plasterboard ceiling on 75×50 noggings
12.300 m span
laminated timber portals at 3.600 m cts
300 cavity walls
carlite plaster on walls
SECTION A-A
10 m.s shoe
h.w. strip on 75×50 battens
150 concrete 1:2:4
150 hardcore
strip foundation 1:3:6 conc.
isolated foundation 1:3:6 conc.
5.900 m
existing drain
new manhole
new sw drain
paved area
N
SOUTH ELEVATION
WEST ELEVATION
EAST ELEVATION
22.600 m
4.300 m
12.900 m
A
A
OFFICE
CLOAKS
HALL
STAGE
link to existing building already built
PLAN

Fig. 2.7

As a further means of simplifying the arrow diagram some of the activities have been grouped. For example, 'foundations' includes excavation, levelling and ramming, and concrete bed. The arrow diagram based on these activities and the drawing (Fig. 2.7) is shown in Fig. 2.8, the following assumptions have been made in its construction:

1. The end panels are built off the concrete slab.
2. The internal partitions cannot be built until the external brickwork and end panels are complete.
3. The building must be weatherproof before joiner and electrician first fixings are carried out.

The sequence of activities shown in Fig. 2.8 is not unique and other planners may draw this network slightly differently.

2.5 ANALYSIS OF ARROW DIAGRAMS

2.5.1 Activity duration

Activity durations (shown on the arrows) can be in days, weeks or months, depending on how detailed the diagram is to be. The durations for contractors activities would be arrived at by using past records of outputs, synthetic times based on work study, etc. The gang sizes and plant to be used would be the ones which give optimum performance at this stage irrespective of the requirements of other activities. Sub-contractors and suppliers should be approached in order to obtain realistic durations for their activities.

2.5.2 Project duration

The project duration is the minimum time in which it can be completed with the activity times assigned to it.

The duration of the project will be determined by the longest path through the diagram. This is known as the critical path.

If the time required for any activity is affected, this will automatically effect the project duration.

2.5.3 Method of analysis (Fig. 2.9)

2.5.3.1 Calculate the earliest event times by working through the diagram and selecting the longest path.

Earliest time for event 1 is day 0:

Therefore the earliest time for event 2 is $0 + 3 = 3$ weeks
for event 3 is $0 + 2 = 2$ weeks
and for event 4 is $0 + 5 = 5$ weeks

Event 5 has three activities leading into it, and the earliest time for event 5 is determined by the longest path. The alternative paths give the following results:

event 2 to event 5 would give $3 + 0 = 3$ weeks
event 3 to event 5 would give $2 + 4 = 6$ weeks
event 4 to event 5 would give $5 + 3 = 8$ weeks

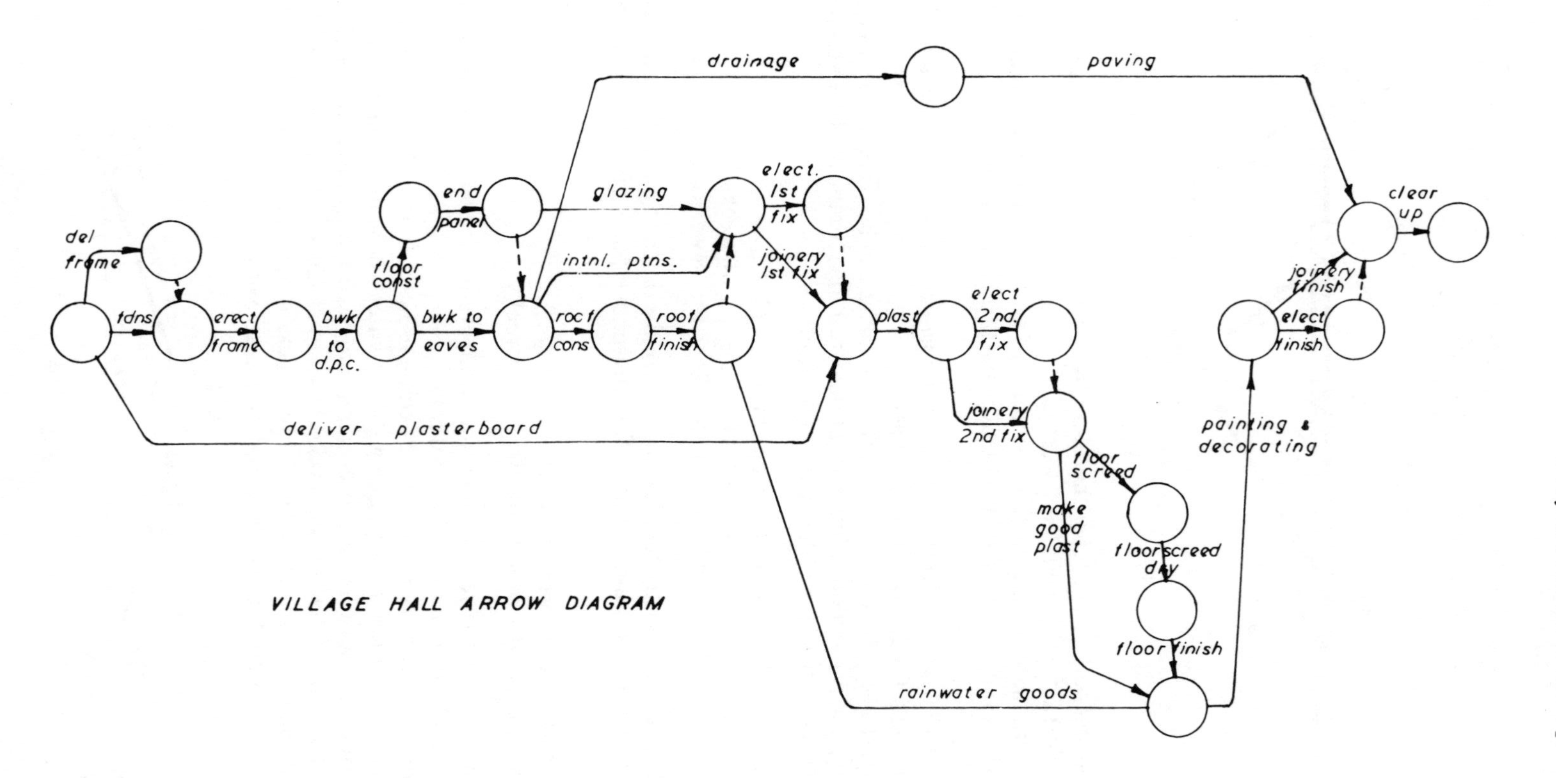
del frame
fdns
erect frame
bwk to d.p.c.
floor const
end panel
bwk to eaves
glazing
intnl. ptns.
roof cons
roof finish
drainage
paving
elect. 1st fix
joinery 1st fix
plast
elect 2nd. fix
joinery 2nd fix
floor screed
make good plast
floorscreed dry
floor finish
rainwater goods
painting & decorating
elect finish
joinery finish
clear up
deliver plasterboard
VILLAGE HALL ARROW DIAGRAM

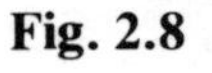
Fig. 2.8

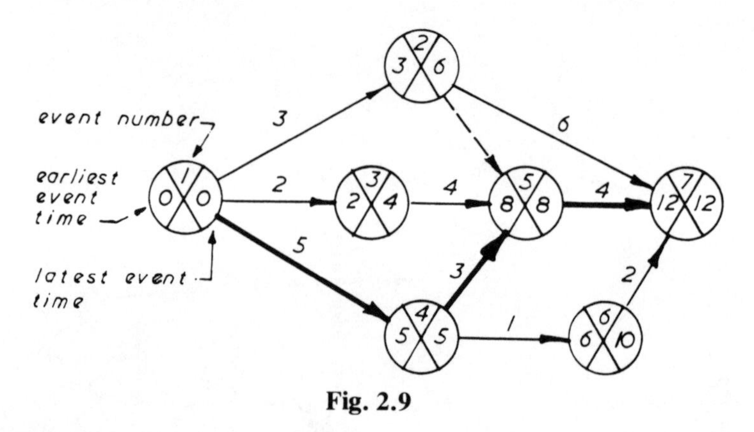

Fig. 2.9

The longest path to event 5 is therefore from event 4 to event 5 which means the earliest time for event 5 is 8 weeks.

The remainder of the earliest event times are calculated in this way, and the earliest time for the final event gives the project duration.

2.5.3.2 Calculate the latest event times by working backwards through the diagram again, selecting the longest path.

The latest event time is the latest time an event can be completed without affecting the project duration, so that the latest time for the final event is the same as its earliest time (the project duration).

Therefore the latest time for event 7 is week 12
for event 6 is 12 − 2 = 10 weeks
and for event 5 is 12 − 4 = 8 weeks

Event 4 has two activities leaving it, so the latest time for event 4 is determined by the longest path. The alternative paths give the following results:

event 5 to event 4 would give 8 − 3 = 5 weeks
event 6 to event 4 would give 10 − 1 = 9 weeks

The longest path to event 4 is therefore from event 5 to event 4, which means that the latest time for event 4 is 5 weeks.

Critical events have earliest times equal to latest times. The *critical path* passes through these events, along the activities whose duration is equal to the difference between the preceding and succeeding event times (Fig. 2.10). The *critical path* passes along activities 25–26 and 26–27 but 25–27 is not critical as activity duration of three weeks is not equal to the difference between the times for events 25 and 27, i.e. 10 weeks.

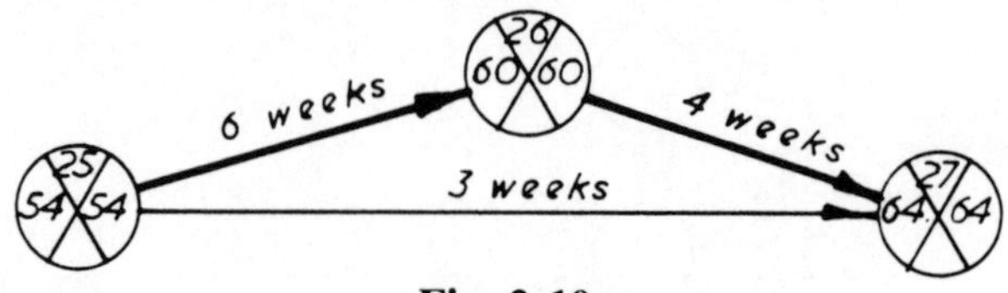

Fig. 2.10

2.5.4 Float

Float is the name given to the spare time available on non-critical activities. There are various types of float (Fig. 2.11) as follows.

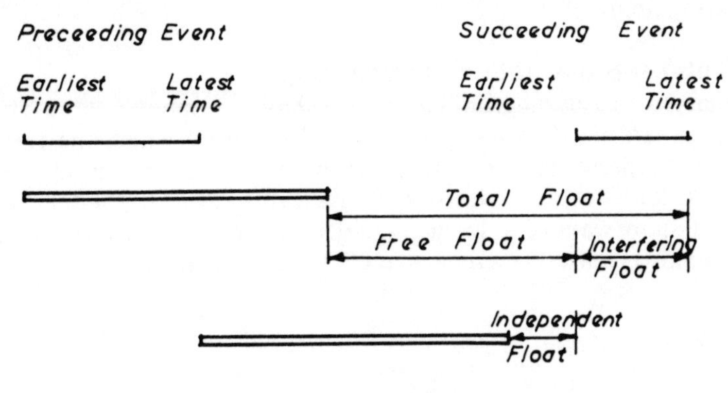

Fig. 2.11

2.5.4.1 Total float

Total float is the total amount of spare time available in an activity. It is calculated by: latest succeeding event time − earliest preceding event time − activity duration. It should be noted that total float can be the amount of float in a chain of activities.

2.5.4.2 Free float

Free float is a part of total float and represents the amount of spare time which can be used without affecting subsequent activities, providing the activity starts at its earliest time. It is calculated by: earliest succeeding event time − earliest preceding event time − activity duration.

2.5.4.3 Interfering float

Interfering float is the amount of spare time available which, if used, will affect subsequent activities, and may be calculated by: total float − free float.

2.5.4.4 Independent float

Independent float is the amount of spare time available which can be used without affecting any succeeding activity and which cannot be affected by any preceding activity. The calculation is: earliest succeeding event time − latest preceding event time − activity duration.

2.5.5 Significance of the various types of float

If additional resources were required on a critical activity they could be obtained from:

1. Activities with independent float without affecting the rest of the network at all.
2. Activities with free float without affecting the float of subsequent activities.

3. Activities with interfering float only, which will affect the float of previous and subsequent activities.

If only the total float is calculated, this may include independent, free and interfering float.

2.5.6 Tabular presentation of arrow diagrams

The analysis of the diagram can be presented in tabular form, and when computers are used in analysis this is the way in which the results will normally be presented. The table shown in Fig. 2.12 presents the information in terms of activities in numerical order, but the sequence can be varied in many ways (e.g. by arranging in the order of least total float). The table can also be used as a means of controlling the project.

Fig. 2.12

Activity number	*Duration*	*Earliest*		*Latest*		*Total float*
		Start	*Finish*	*Start*	*Finish*	
1–2	3	0	3	3	6	3
1–3	2	0	2	2	4	2
1–4	5	0	5	0	5	0
2–5	0	3	3	8	8	5
2–7	6	3	9	6	12	3
3–5	4	2	6	4	8	2
4–5	3	5	8	5	8	0
4–6	1	5	6	9	10	4
5–7	4	8	12	8	12	0
6–7	2	6	8	10	12	4

2.5.7 Method of analysis

1. List the activities in numerical order.
2. Fill in the earliest start time, which is the earliest time for the preceding event (obtainable from the network; Fig. 2.9).
3. Calculate the earliest finish time, which is the earliest start time plus activity duration.
4. Fill in the latest finish time, which is the latest time for the succeeding event.
5. Calculate the latest start time, which is the latest finish time minus the activity duration.
6. Calculate the total float, which is either the latest start time minus the earliest start time or alternatively the latest finish time minus the earliest finish time. Both methods give the same result.

It should be noted that critical activities have no float.

2.5.8 Analysis of the village hall project

Figure 2.13 shows the activity durations and resources for the village hall in Fig. 2.7. The analysed arrow diagram for this project is shown in Fig. 2.14 and the completed table of results with total float in Fig. 2.15.

Fig. 2.13

Activity	*Man-hours*	*Plant-hours*	*Resources*	*Time req'd. (days)*
Foundations				
Excavate oversite strip / Excavate for foundations	40	20	J.C.B. 4 + operator + 2 labourers	5
Concrete foundations	160	20	8 labourers 400/300 litre mixer	
Frame				
Delivery				10
Erect	16	8	Crane + operator + 2 labourers	1
Brickwork to D.P.C.	96	16	4 bricklayers 2 labourers 150/100 litre mixer	2
Floor construction				
Hardcore / Concrete	208	24	8 labourers 400/300 litre mixer	3
Brickwork to eaves	528	88	4 bricklayers 2 labourers 150/100 litre mixer	11
End panels	208		2 joiners	13
Glazing	112		Plumber and mate	7
Roof construction				
Carcass / Woodwool	240		2 joiners	15
Internal partitions	96	16	4 bricklayers 2 labourers 150/100 litre mixer	2
Drainage				
Excavation and backfill / Concrete bed	136	6	8 labourers 400/300 litre mixer	4
Lay drains / Manholes	32	16	Bricklayer and mate 150/100 litre mixer	
Roof finish	144		2 roofers	9

Fig. 2.13 *continued*

Activity	*Man-hours*	*Plant-hours*	*Resources*	*Time req'd. (days)*
Rainwater goods	32		Plumber and mate	2
Electrician first fixing	112		Electrician and mate	7
Joiner first fixing	96		2 joiners	6
Deliver plasterboard				70
Plastering	410	80	4 plasterers 2 labourers 150/100 litre mixer	10
Electrician second fixing	48		Electrician and mate	3
Joinery second fixing	80		2 joiners	5
Make good plaster	8		Plasterer	1
Floor screed	160	32	4 plasterers 1 labourer 150/100 litre mixer	4
Floor screed dry				10
Floor finish	192		2 joiners	12
Painting and decorating	256		4 painters	8
Paving	144		Bricklayer and mate	9
Electrical finishing	16		Electrician and mate	1
Joinery finishing	48		2 joiners	3
Clean and hand over	40		1 labourer	5

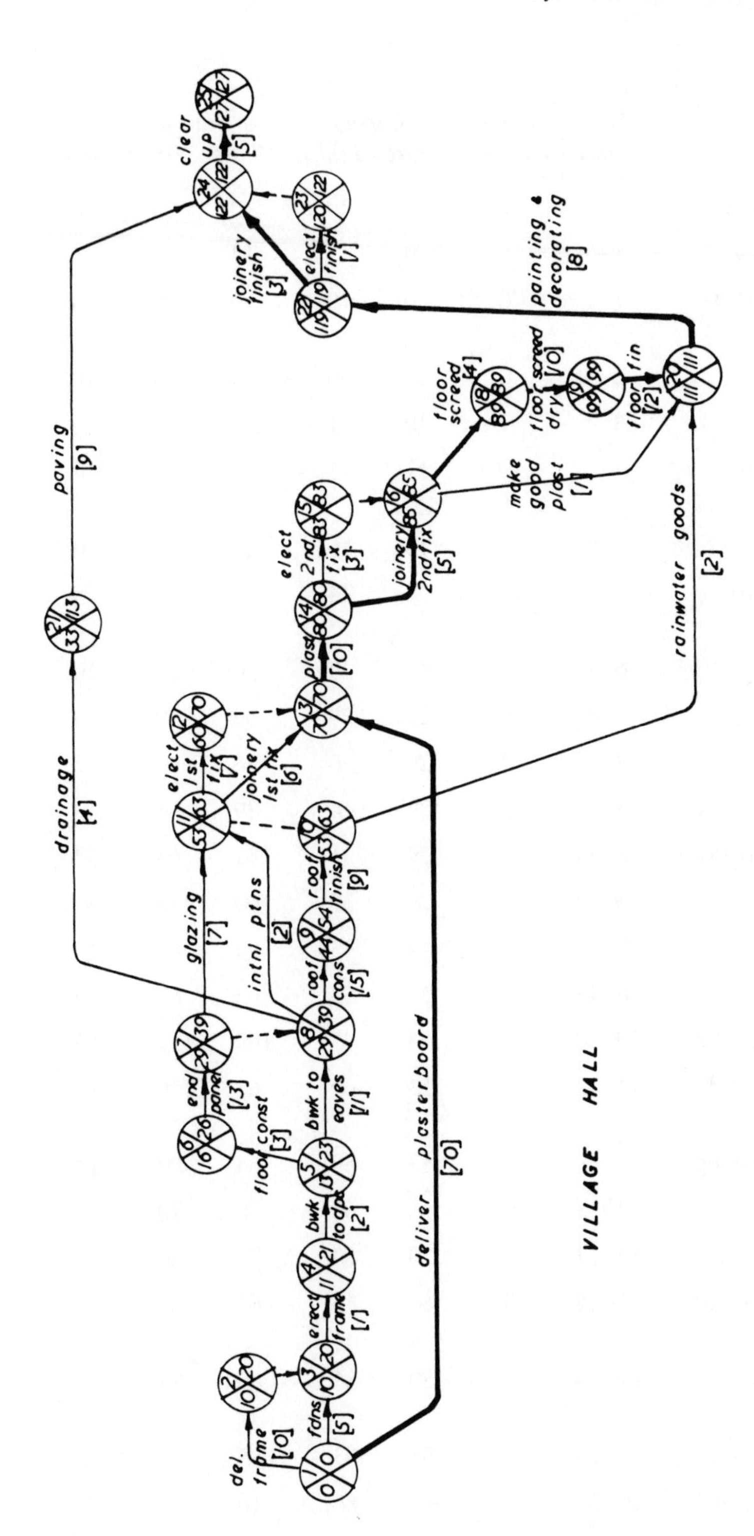
VILLAGE HALL
del. frame [10]
fdns [5]
erect frame [11]
bwk to dpc [2]
floor const [3]
end panel [13]
bwk to eaves [11]
roof cons [15]
roof finish [9]
glazing [7]
intnl ptns [2]
drainage [4]
deliver plasterboard [70]
elect 1st fix [7]
joinery 1st fix [6]
plast [10]
paving [9]
elect 2nd fix [3]
joinery 2nd fix [5]
rainwater goods [2]
floor screed [4]
floor screed dry [10]
make good plast [1]
floor fin [12]
painting & decorating [8]
joinery finish [3]
elect finish [1]
clear up [5]

Fig. 2.14

Fig. 2.15

Activity	*Activity number*	*Duration*	*Earliest Start*	*Earliest Finish*	*Latest Start*	*Latest Finish*	*Total float*
Deliver frame	1–2	10	0	10	10	20	10
Foundations	1–3	5	0	5	15	20	15
Deliver plaster-board	1–13	70	0	70	0	70	0
Dummy	2–3	0	10	10	20	20	10
Erect frame	3–4	1	10	11	20	21	10
Brickwork to D.P.C.	4–5	2	11	13	21	23	10
Floor construction	5–6	3	13	16	23	26	10
Brickwork to eaves	5–8	11	13	24	28	39	15
End panels	6–7	13	16	29	26	39	10
Dummy	7–8	0	29	29	39	39	10
Glazing	7–11	7	29	36	56	63	27
Roof construction	8–9	15	29	44	39	54	10
Internal partitions	8–11	2	29	31	61	63	32
Drainage	8–21	4	29	33	109	113	80
Roof finish	9–10	9	44	53	54	63	10
Dummy	10–11	0	53	53	63	63	10
Rainwater goods	10–20	2	53	55	109	111	56
Electrician first fixing	11–12	7	53	60	63	70	10
Joiner first fixing	11–13	6	53	59	64	70	11
Dummy	12–13	0	60	60	70	70	10
Plastering	13–14	10	70	80	70	80	0
Electrician second fixing	14–15	3	80	83	82	85	2
Joiner second fixing	14–16	5	80	85	80	85	0
Dummy	15–16	0	83	83	85	85	2
Make good plaster	16–20	1	85	86	110	111	25

Fig. 2.15 *continued*

Activity	*Activity number*	*Duration*	*Earliest Start*	*Earliest Finish*	*Latest Start*	*Latest Finish*	*Total float*
Floor screed	16–18	4	85	89	85	89	0
Floor screed dry	18–19	10	89	99	89	99	0
Floor finish	19–20	12	99	111	99	111	0
Painting and decorating	20–22	8	111	119	111	119	0
Paving	21–24	9	33	42	113	122	80
Electrician finish	22–23	1	119	120	121	122	2
Joiner finish	22–24	3	119	122	119	122	0
Dummy	23–24	0	120	120	122	122	2
Clean and hand over	24–25	5	122	127	122	127	0

2.6 SCHEDULING

Scheduling is the process of determining the actual time periods during which the activities are planned to take place i.e. start and finish dates for each activity.

2.6.1 Resource allocation

Resource allocation is concerned with scheduling activities and their resources within predetermined constraints. Initially when a network is drawn, no account is taken of the limit of availability of resources for any particular activity and it is assumed that resources are always available when required. In some cases, however, the resources may be required on different activities which are on parallel paths in the network.

2.6.2 Resource levelling

If availability of resources is a critical factor, then the project duration may well be influenced by insufficient resources. Resource levelling is used to ensure availability is not exceeded.

2.6.3 Resource smoothing

If the project duration is of prime importance, the first analysis will often show an excessive duration. The critical path method is extremely useful in these circumstances as it highlights those activities which must be examined in order to reduce the project duration. The aim will be to reduce the time required for those activities which will result in least cost overall.

Resource smoothing is used when some smoothing of resources is carried out within the activity floats to limit fluctuations in demand.

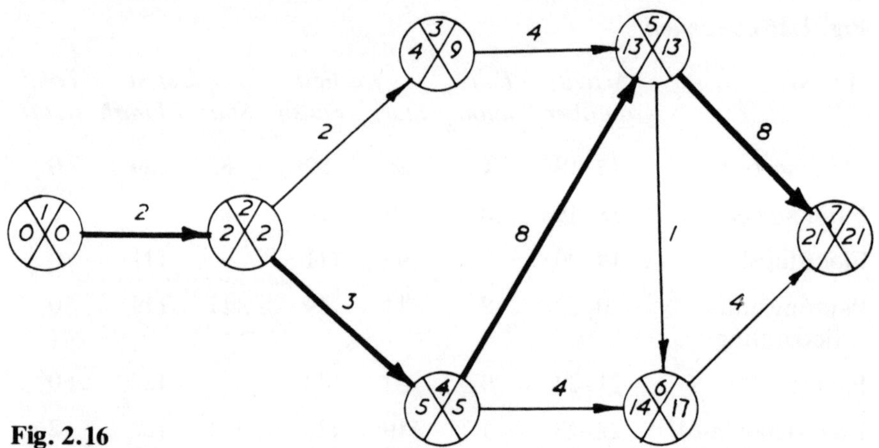

Fig. 2.16

2.6.4 Example of resource levelling

Fig. 2.16 represents a small project, the duration of each activity, and labour required being given in the first three columns in Fig. 2.17. From this information the critical path and the project duration can be found. For simplicity it is assumed that the operatives are totally interchangeable, although the number of operatives on the activities must be as stated. The maximum number of operatives available is 10.

Fig. 2.17

Activity	*Duration*	*Gang size*	*Earliest Start*	*Earliest Finish*	*Latest Start*	*Latest Finish*	*Total float*
1–2	2	10	0	2	0	2	0
2–3	2	2	2	4	7	9	5
2–4	3	5	2	5	2	5	0
3–5	4	6	4	8	9	13	5
4–5	8	3	5	13	5	13	0
4–6	4	4	5	9	13	17	8
5–6	1	2	13	14	16	17	3
5–7	8	8	13	21	13	21	0
6–7	4	4	14	18	17	21	3

2.6.4.1 Solution

It can be seen that the diagram as analysed gives a project duration of 21 weeks.

A bar chart showing all activities at their earliest times drawn first is as shown in Fig. 2.18. The activities are then moved within their float to level out requirements and to avoid exceeding the number available as shown in Fig. 2.19.

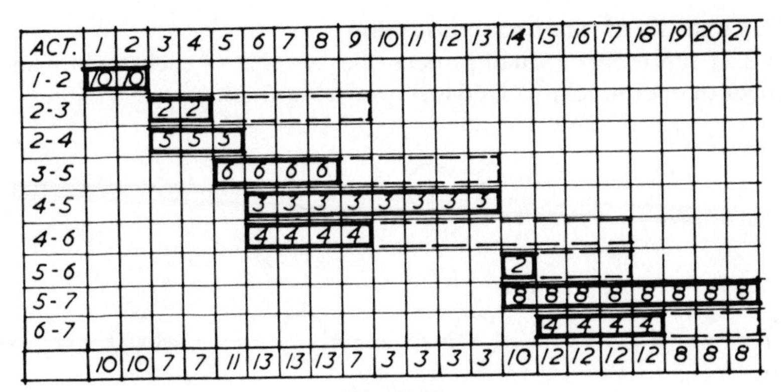

ACT.	1	2	3	4	5	6	7	8	9	10	11	12	13	14	15	16	17	18	19	20	21
1-2	10	10																			
2-3			2	2																	
2-4			5	5	5																
3-5					6	6	6	6													
4-5						3	3	3	3	3	3	3	3								
4-6						4	4	4	4												
5-6														2							
5-7														8	8	8	8	8	8	8	8
6-7															4	4	4	4			
	10	10	7	7	11	13	13	13	7	3	3	3	3	10	12	12	12	12	8	8	8

Fig. 2.18

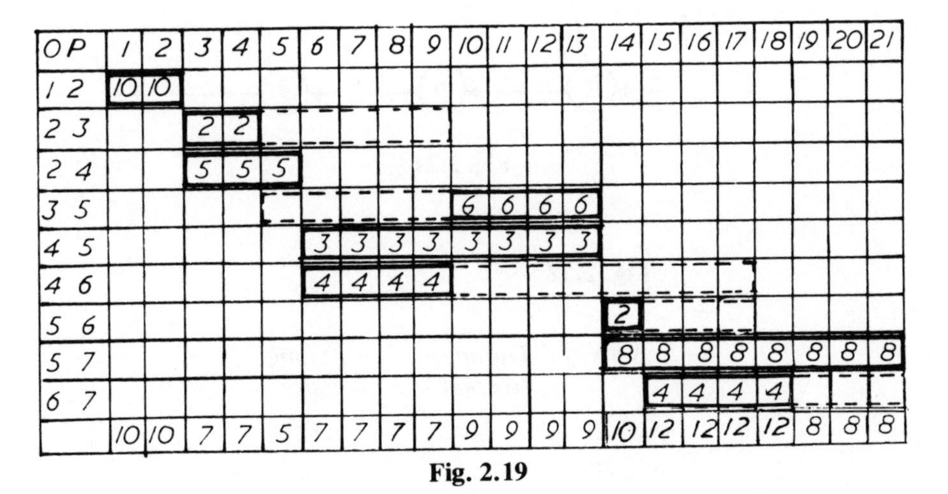

OP	1	2	3	4	5	6	7	8	9	10	11	12	13	14	15	16	17	18	19	20	21
1 2	10	10																			
2 3			2	2																	
2 4			5	5	5																
3 5										6	6	6	6								
4 5						3	3	3	3	3	3	3	3								
4 6						4	4	4	4												
5 6														2							
5 7														8	8	8	8	8	8	8	8
6 7															4	4	4	4			
	10	10	7	7	5	7	7	7	7	9	9	9	9	10	12	12	12	12	8	8	8

Fig. 2.19

As critical activities have no float these are drawn first. As the maximum number of men available is 10, some adjustment to Fig. 2.19 must be made. In this case the project duration has been increased to 25 weeks by moving on activity 6–7 as shown in Fig. 2.20.

OP	1	2	3	4	5	6	7	8	9	10	11	12	13	14	15	16	17	18	19	20	21	22	23	24	25
1–2	10	10																							
2–3			2	2																					
2–4			5	5	5																				
3–5										6	6	6	6												
4–5						3	3	3	3	3	3	3	3												
4–6						4	4	4	4																
5–6														2											
5–7														8	8	8	8	8	8	8	8				
6–7																						4	4	4	4
	10	10	7	7	5	7	7	7	7	9	9	9	9	10	8	8	8	8	8	8	8	4	4	4	4

Fig. 2.20

2.6.5 Example of resource smoothing

Fig. 2.21 represents a small project, the duration of each activity and the number of operatives involved is shown in Fig. 2.22.

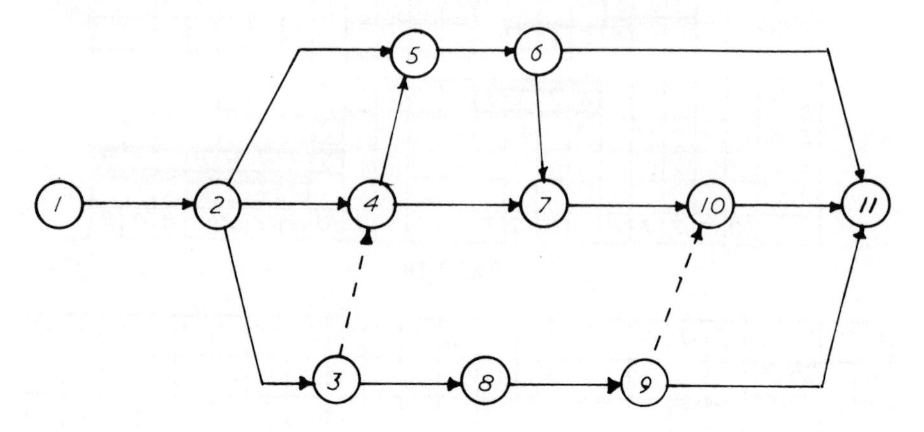

Fig. 2.21

Fig. 2.22

Activity	*Duration (weeks)*	*Gang size*
1–2	2	12
2–3	5	4
2–4	8	2
2–5	4	3
3–8	2	4
3–4	0	0
4–7	1	4
4–5	3	3
5–6	2	6
6–7	1	5
6–11	3	2
7–10	2	8
8–9	4	3
9–10	0	0
9–11	7	6
10–11	4	3

2.6.5.1 General information

The contract has been planned to employ local casual labour working a 5 day week. The labour force can be recruited as required and employed on a week-to-week basis at a weekly rate of £240, with an 'Engagement' Establishment Charge of £64 per person for each new appointment. Although no working is permitted on Saturday or Sunday, overtime can be worked in order to meet the requirement of the employer, which is to complete the project within 20 weeks. The employer also wishes to know the cost of carrying out the work as set out in the arrow diagram and without overtime.

The rules for working overtime are:
The overtime can be worked in periods of 1 week and will account for $33\frac{1}{3}\%$ increase in output on the relevant activities. The increased cost of overtime working is calculated at the rate of £120 per week per operative. Site overheads are calculated at £700 per week.

NOTE: Each activity must be continuous, i.e. once started it must be completed without a break. The gang sizes cannot be altered.

2.6.5.2 Solution

It can be seen that the diagram as analysed gives a project duration of 22 weeks (see Fig. 2.23). Fig. 2.24 shows the network converted to a bar chart with all the activities at their earliest times and Fig. 2.25 shows the bar chart after smoothing the resources. The cost of the project as shown would be:

Engagement 21 × £64	£ 1 344.00
Labour costs 198 × £240	£47 520.00
Overheads 22 × £700	£15 400.00
Cost	£64 264.00

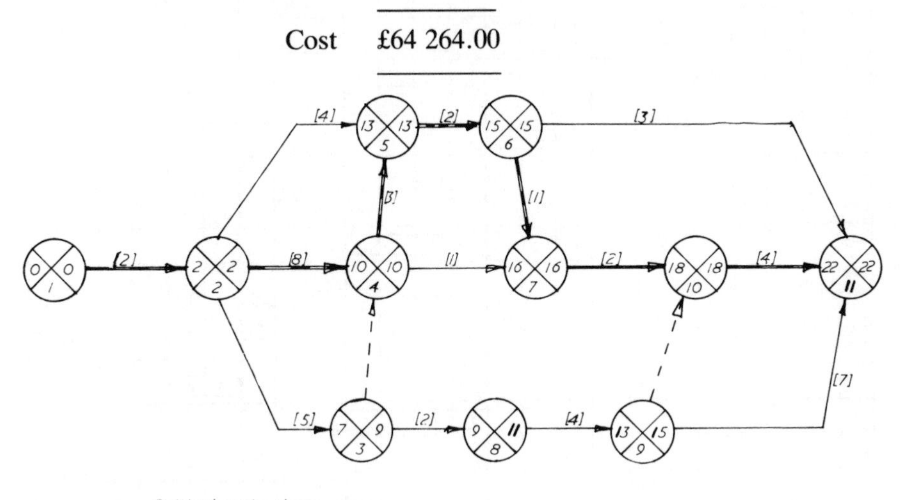

Fig. 2.23

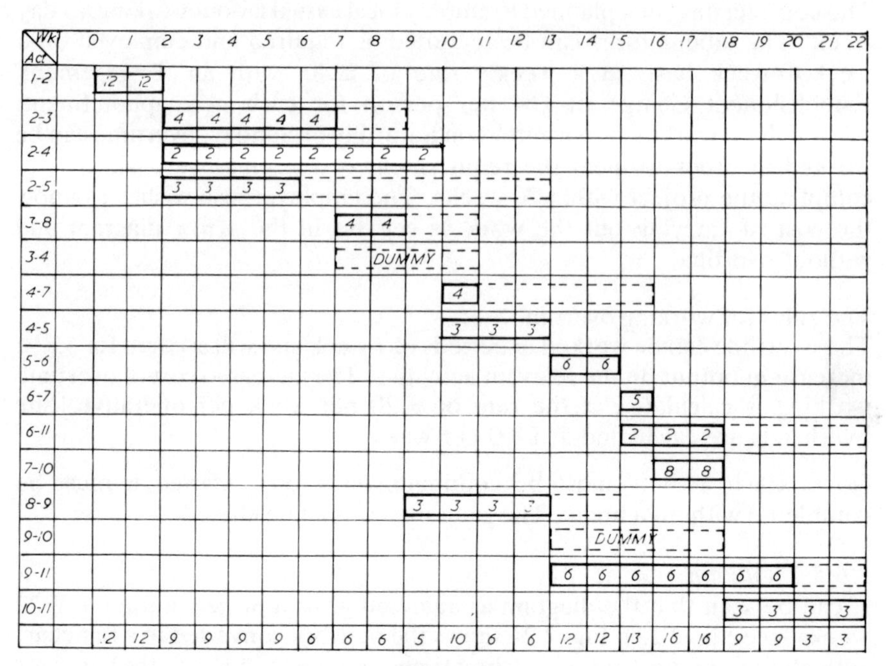

Fig. 2.24

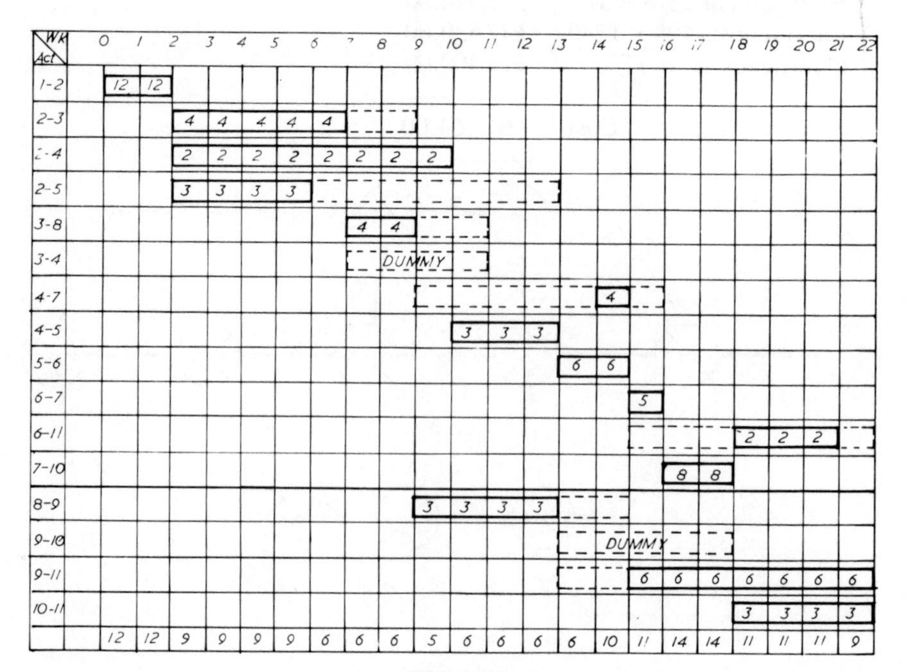

Fig. 2.25

As the employer requires the project to be completed in 20 weeks, the duration must be reduced by 2 weeks. To do this, overtime will have to be worked, and in order to keep the cost to a minimum it will be worked on the critical activity which will cost the least, i.e., the activity with the least labour (in this case 2–4). If overtime is worked for 6 weeks, 8 weeks work will be done (due to $33\frac{1}{3}\%$ increase in output).

Fig. 2.26 shows the amended bar chart with all activities at their earliest

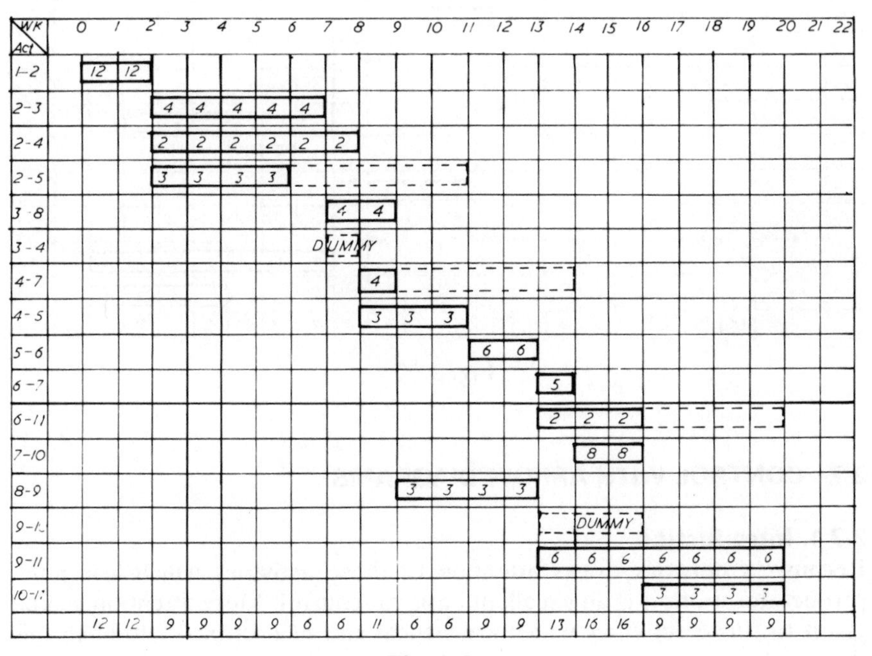

Fig. 2.26

times and Fig. 2.27 shows the amended bar chart after smoothing the resources. The cost of the project as shown would be:

Engagements 22 × £64	£ 1 408.00
Labour costs 194 × £240	£46 560.00
Overtime 2 × 6 × £120.00	£ 1 440.00
Overheads 20 × £700.00	£14 000.00
Cost	£63 408.00

It can be seen that both project duration and cost are reduced in this example.

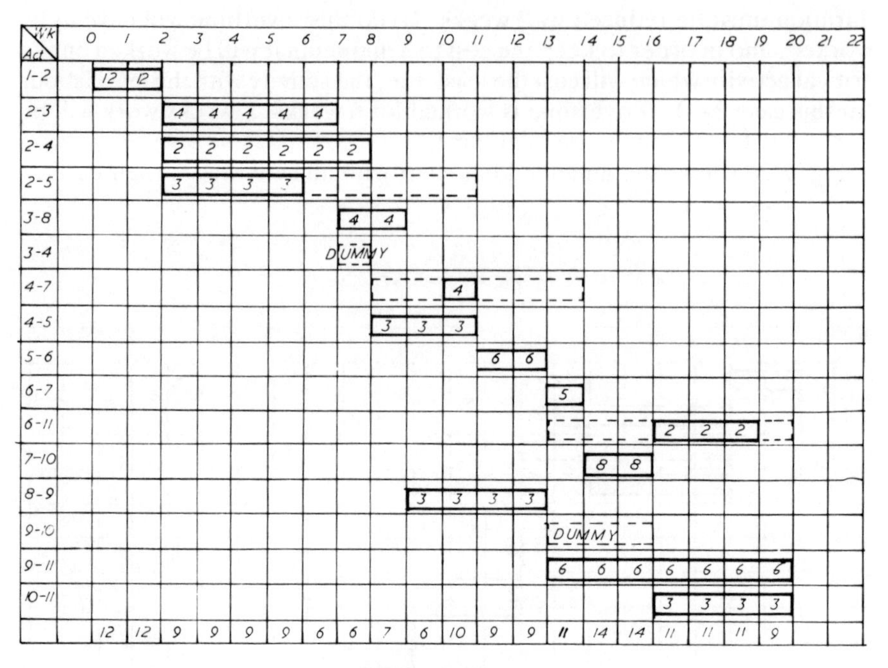

Fig. 2.27

2.7 CONTROL WITH ARROW DIAGRAMS

2.7.1. Introduction

Because a network draws attention to those activities which affect the project duration, it is an excellent basis for control. More attention can be given to those activities which are critical and near critical, although this does not mean that activities with a reasonable amount of float can be ignored altogether or they themselves may then become critical.

For site use the arrow diagram is usually converted into a bar chart, which is used for short term control and shows operations at scheduled times; progress is then marked onto this and the information from the bar chart can be transferred back onto the arrow diagram for re-analysis or up-dating of the network for the whole project.

An alternative method of presentation is to draw the arrow diagram on a time scale showing activities at scheduled times and to mark the progress directly onto this (see section 2.11). It is possible to use the diagram in its original form and to enter the actual durations on the diagram for re-analysis: if the results are tabulated and the actual durations are entered on the schedule this will form a method of control.

Re-analysis of the diagram should be carried out at regular intervals (say fortnightly) so that any change in critical activities is brought to the attention of the management and corrections can be made in reasonable time.

If the arrow diagram is used to show progress, and it is not drawn on a time scale, elasticated string can be used to give an impression of progress. By inserting pins into activities in progress (the position being proportionate to the amount of work done on each activity) and then stretching the string from pin to pin, it would be obvious that all activities to the right of the string are still outstanding.

2.7.2 Re-analysis of a project partially completed

One method of re-analysing a project partially completed, is as follows:

1. Insert 'lead time' to represent the amount of time passed.
2. Insert duration zero on all completed activities.
3. Insert estimated time to complete partially completed activities.
4. Carry out analysis in the normal way.

For example, assume that in the diagram in Fig. 2.9 a period of 5 weeks has passed and the position is:

Activities 1–2 and 1–4 are completed.
Activity 1–3 requires an estimated 1 week to complete.
Activity 2–7 requires an estimated 5 weeks to complete.
All other activities not yet started.

The re-analysed diagram would now appear as shown in Fig. 2.28. It can be seen that the critical path has changed and the the project duration is now 14 weeks. If it is necessary to complete in 12 weeks, the activities requiring attention (i.e. critical activities) are clearly shown. All activities running behind time can be demonstrated by inserting the original project duration as the latest finish time for the end event before calculating the latest finish times for the other events. From this the negative float can be calculated, which will indicate the activities that are in arrears.

A second method of showing progress on the diagram is to remove all completed activities, leaving only partially completed activities and those not yet started. Real time dummies are then inserted and given a value equal to the time which as passed. This replaces 1 and 2 in the preceding method, and constitues a particularly useful alternative when the project is well

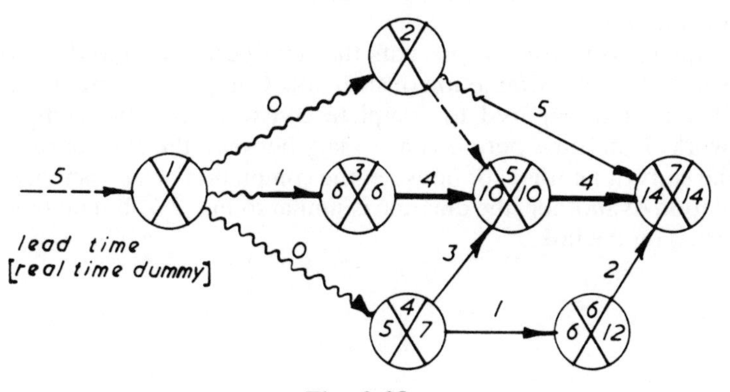

Fig. 2.28

advanced. For example, assume in the village hall project that after 100 days the following activities are partially completed and the estimated time necessary to complete is as shown:

Floor finish, 6 days
Rainwater goods, 1 day
All subsequent activities have not yet started.

Using the alternative method, the arrow diagram will be as in Fig. 2.29.

2.8 THE PREPARATION OF PRECEDENCE DIAGRAMS

2.8.1 Activities

Activities are represented by boxes and are linked together by lines of dependency. The general direction of time flow is from left to right. Activities are assumed to start at the left-hand end of the box and finish at the right hand end of the box. When the start of an activity has more than one line of dependency it is dependent on *all* the activities to which it is connected and therefore all the preceeding activities must be completed before it can start.

2.8.2 Dummy activities

Activities used for convenience, e.g. at the beginning and end of the project.

2.8.3 Delays

When there is to be a delay between activities, e.g. hardening of concrete, this is shown on the link between them (see Fig. 2.30).

2.8.4 Overlap

1. Divide each activity into the same number of equal parts (Fig. 2.31), this is the only method which truly portrays the situation but it can increase the size of the network unnecessarily in practice.

2. Use lead and lag restraints (Fig. 2.32). 'Begin steel' represents the time period required between start of steel and start of formwork. 'Begin formwork' represents the time period required between start of formwork and start of concrete.

'Complete formwork' represents the time period required to complete formwork after the completion of steel and 'Complete concrete' represents the time period required to complete concrete after the completion of formwork. If no time period is necessary between the start of one activity and the start of another, or between the completion of one activity and the completion of another, the diagram is similar to Fig. 2.32 but no time period is inserted on the link.

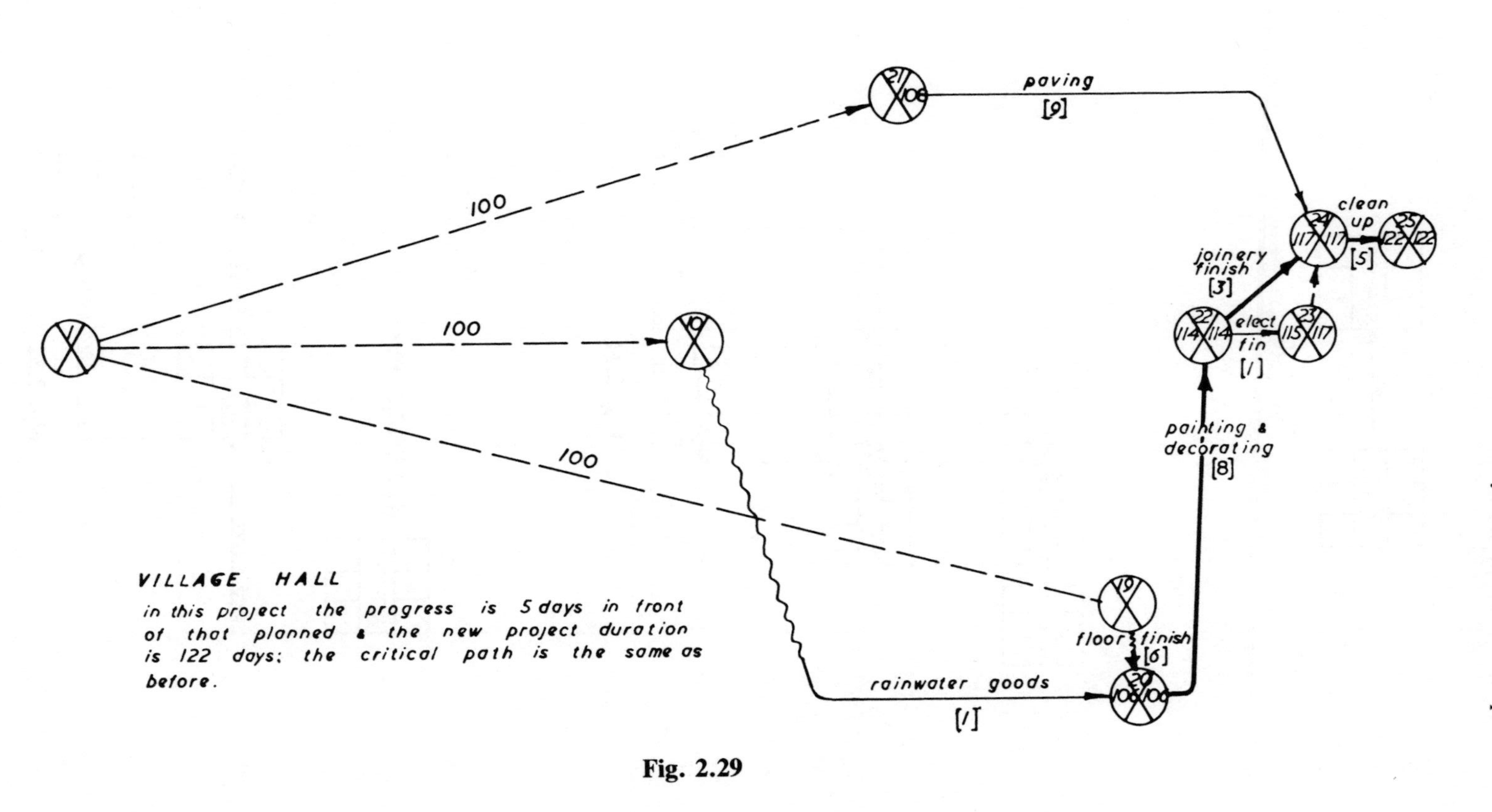
VILLAGE HALL
in this project the progress is 5 days in front of that planned & the new project duration is 122 days; the critical path is the same as before.
100
100
100
paving
[9]
rainwater goods
[1]
floor finish
[6]
painting & decorating
[8]
joinery finish
[3]
elect fin
[1]
clean up
[5]

Fig. 2.29

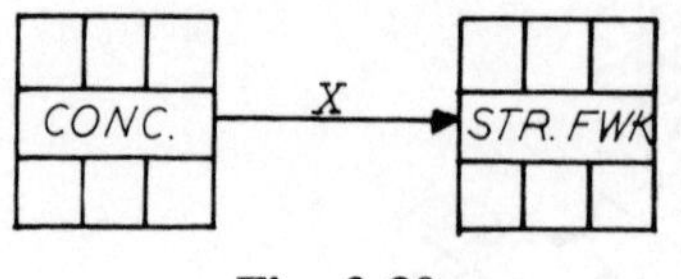

Fig. 2.30

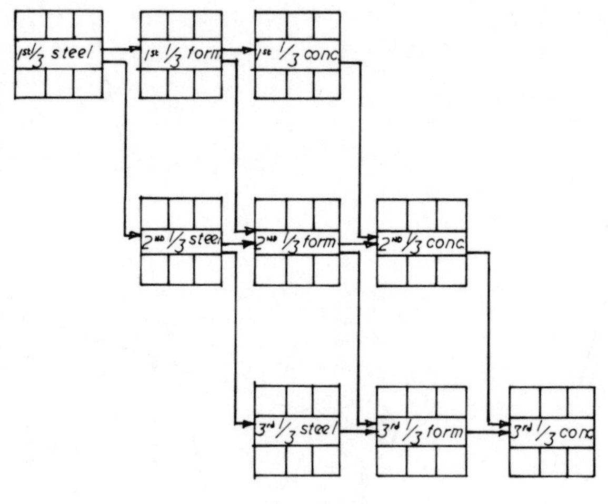

Fig. 2.31

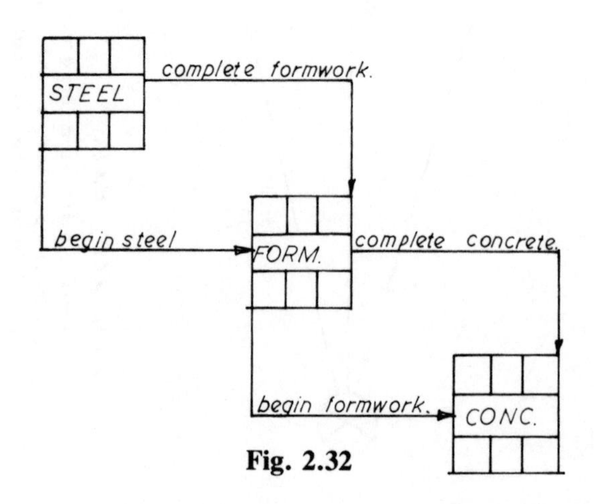

Fig. 2.32

2.9 ANALYSIS OF PRECEDENCE DIAGRAMS

2.9.1 Method of analysis where no overlap or delay is present (Fig. 2.33)

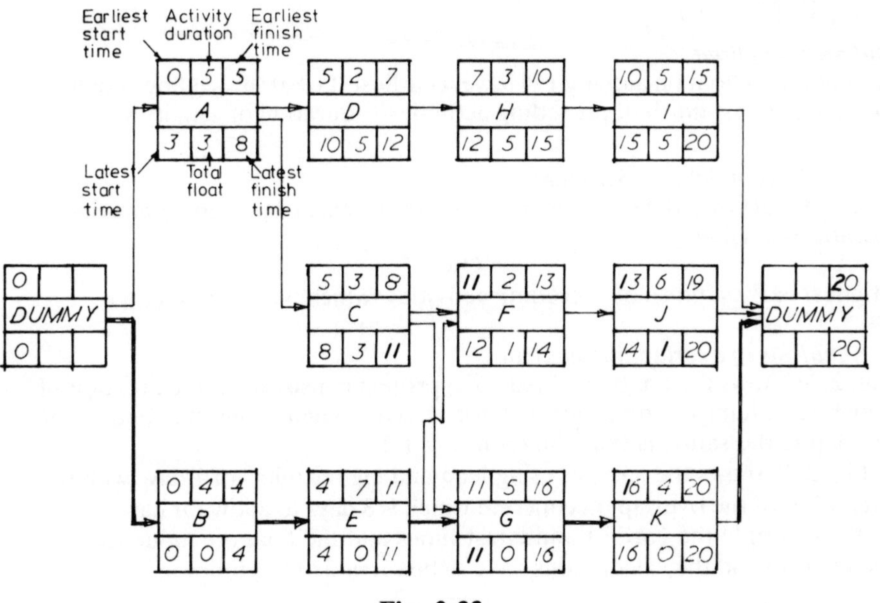

Fig. 2.33

1. *Calculate the earliest activity times by working through the diagram and selecting the longest path.*
In order that the diagram can start and end with a single activity, a dummy has been introduced at the beginning and at the end.

The earliest start time of an activity is the highest earliest finish of preceeding activities. The earliest finish time is earliest start + duration.

2. *Calculate the latest activity times by working backwards through the diagram again selecting the longest path.*
The latest finish time is the latest time by which the activity can be completed without affecting the project duration. The latest finish time for the end activity is therefore the same as the earliest finish time.

The latest finish time for other activities is the lowest latest start time of their succeeding activities. The latest start time = latest finish − duration.

3. *Critical activities*
For critical activities, the earliest and latest times are the same.

4. *Float*
Total float
Latest start time − earliest start time.

or

Latest finish time − earliest finish time.
Free float
Free float activity x = lowest earliest start of succeeding activities − earliest finish of activity x.
Interfering float
Total float − free float.
Independent float
Independent float on activity x = lowest earliest start of succeeding activities − highest latest finish of preceding activities − duration of activity x.

5. *Analysis of delayed activities*
Earliest start of a delayed activity = earliest finish of preceding activity + duration of delay.

OR

Highest earliest finish of preceding activities, whichever is the greater.

6. *Analysis of overlapping activities*
Fig. 2.34 shows the analysis of part of a project consisting of the erection of concrete columns. The procedure for analysis when using this method of overlap is the same as that shown in 2.9.1.1.

Fig. 2.35 shows the analysis using lead and lag restraints of 1 day, with the exception of the overlap on concrete which is 3 days to allow for hardening. In the example the earliest and latest times and the float are influenced by the overlap and the calculations have to be carried out as follows:

The earliest start of an activity = earliest start of preceding activity + duration of lead

OR

highest earliest finish of other preceding activities, whichever is the greater.

The earliest finish of an activity = earliest finish of preceding activity + duration of lag

OR

earliest start of activity + duration of activity, whichever is the greater.

The latest finish of an activity = latest finish of succeeding activity − duration of lag

OR

lowest latest start time for other succeeding activities, whichever is the lesser.

The latest start of an activity = latest start of succeeding activity − duration of lead.

OR

latest finish of activity − duration, whichever is the lesser.

7. *Total float on overlapping activities*
Starting total float = latest start − earliest start. Finishing total float = latest finish − earliest finish. If the activity cannot be split, the float has to be the lesser of the two values.

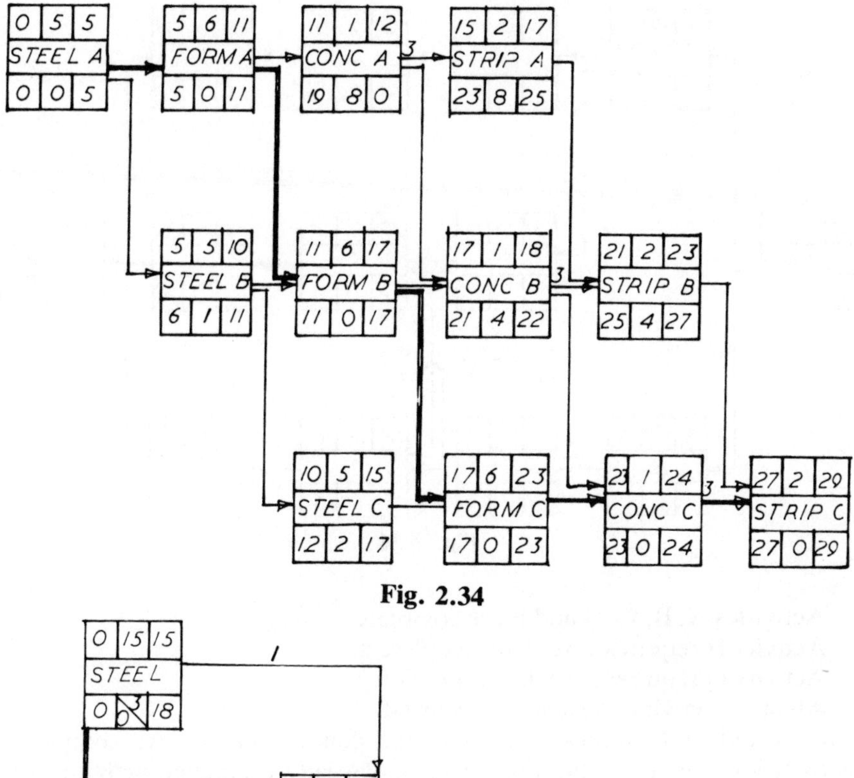

Fig. 2.34

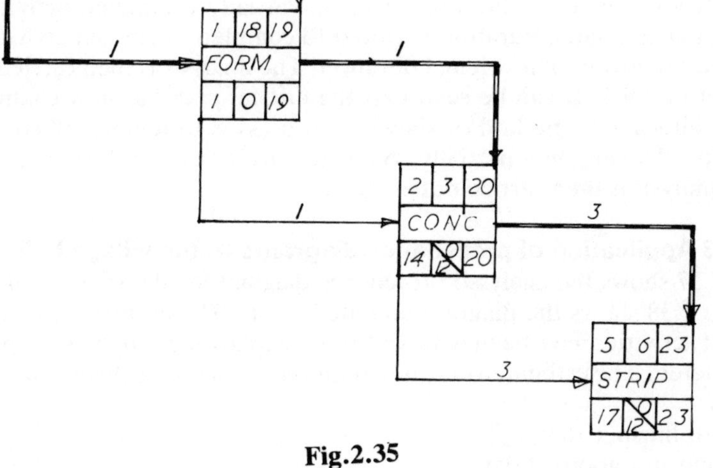

Fig.2.35

2.10 CONTROL WITH PRECEDENCE DIAGRAMS

2.10.1 Re-analysis of a project partially completed

Assume that after a period of 12 weeks the progress on the project shown in Fig. 2.36 is as follows:

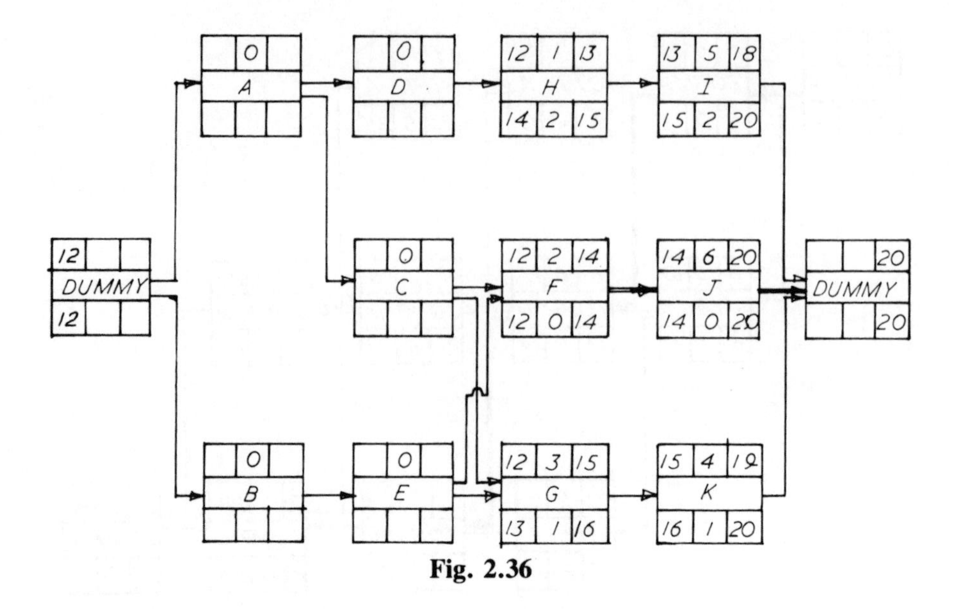

Fig. 2.36

Activities A, B, C, D and E are complete
Activity H requires 1 week to complete it
Activity G requires 3 weeks to complete it
All other activities have not yet started.

The period of 12 weeks is put into the dummy activity. All completed activities are given a duration of zero. Partially completed activities are given the estimated duration required to complete them and activities not started are given their original duration. The analysis is then carried out as set out in 2.9.1. It can be seen that the critical path has now changed.

An alternative method of showing progress is to remove all completed activities leaving only partially completed activities and those not started. The analysis is then carried out as above.

2.10.2 Application of precedence diagrams to the village hall project

Fig. 2.37 shows the analysed precedence diagram for the village hall project

Fig. 2.38 shows the diagram up dated after 100 days progress. All completed activities have been removed and the following activities are partially completed. The estimated duration required to complete them is also shown below:

Floor finish, 6 days
Rainwater goods, 1 day
All subsequent activities have not yet started.

2.11 TIME SCALE PRESENTATION OF ARROW DIAGRAMS

2.11.1 Introduction

Networks can be drawn in the form of a bar chart or time bar diagram in

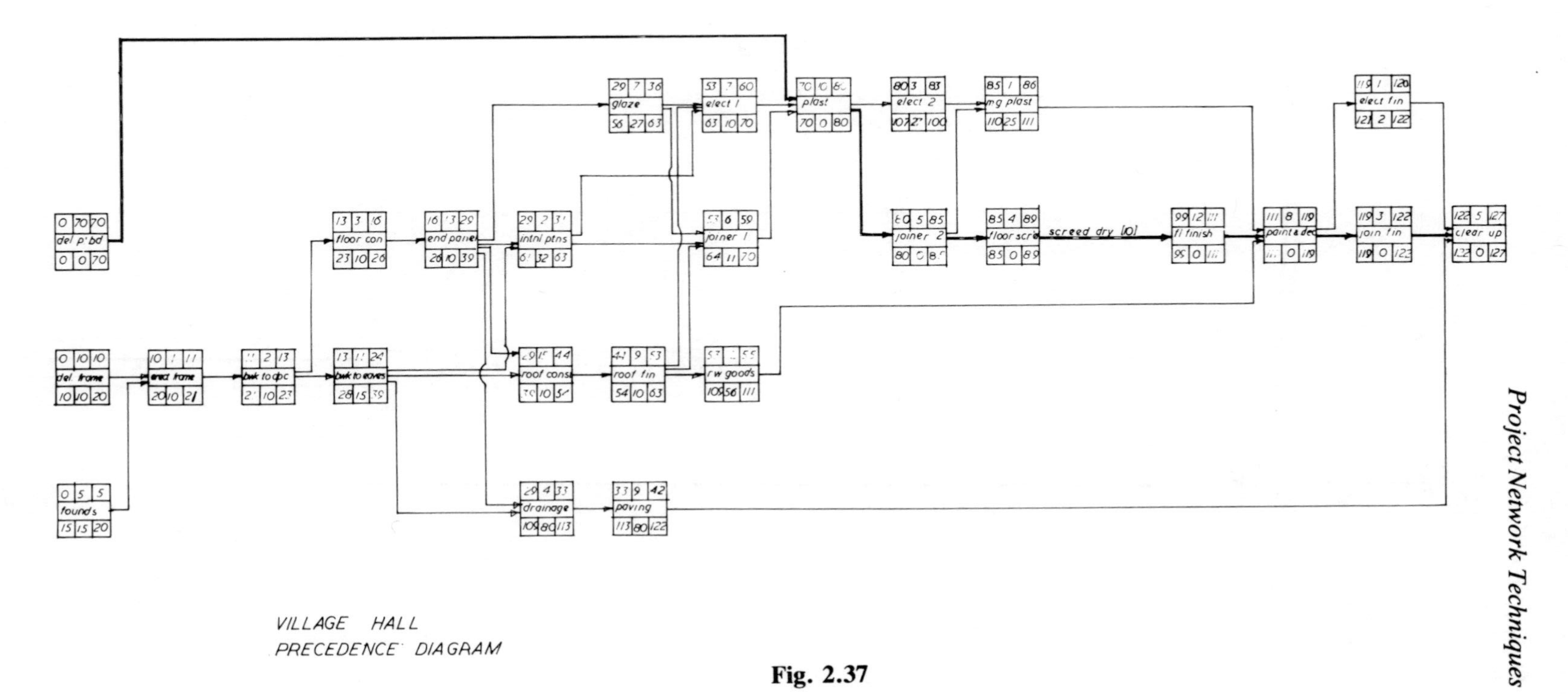
del p'bd
del frame
found's
erect frame
bwk to dpc
floor con
bwk to eaves
end panel
intnl ptns
roof cons
drainage
glaze
roof fin
paving
elect 1
joiner 1
rw goods
plast
elect 2
joiner 2
mg plast
floor scr'd
screed dry [10]
fl finish
paint & dec
elect fin
join fin
clear up
VILLAGE HALL
PRECEDENCE DIAGRAM

Fig. 2.37

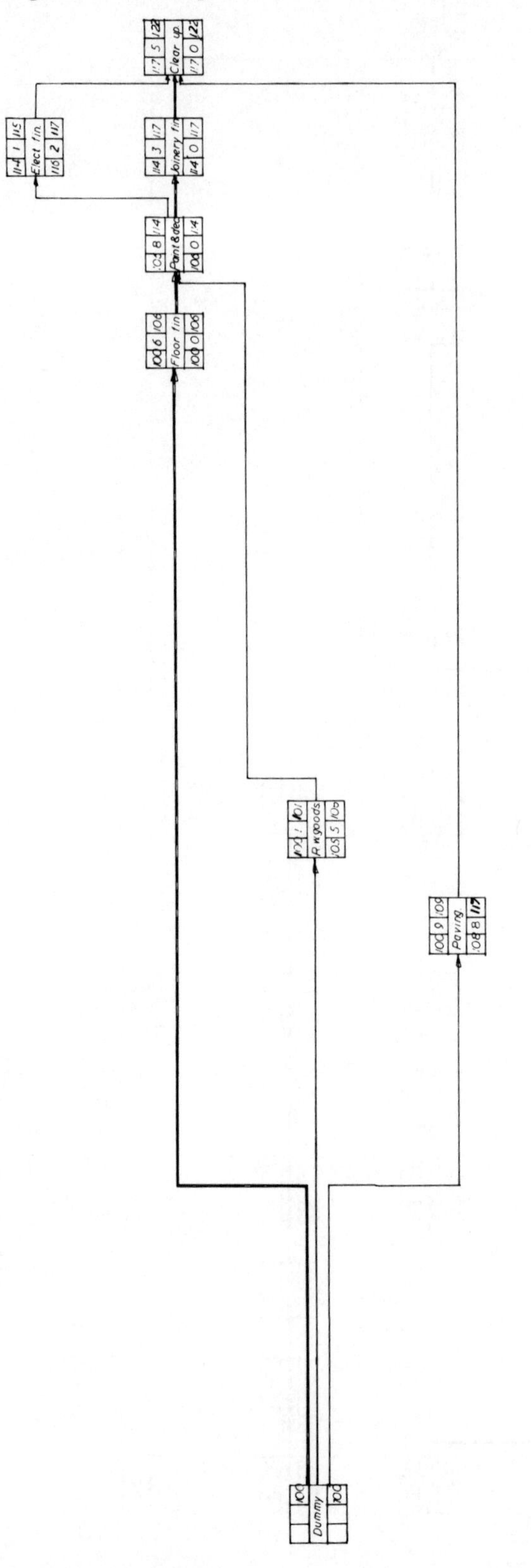
Clear up
Elect fin.
Joinery fin
Paint & dec
Floor fin
R.w goods
Paving
Dummy
VILLAGE HALL

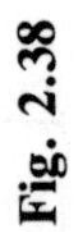
Fig. 2.38

which case the activities are normally drawn as parallel lines to a suitable time scale. This form of presentation defines the time limitations of each activity and the relationship between activities.

The first stage when constructing an activity progress chart is to compile an activity sheet which records the sequence in which the work is to be carried out and the duration of the individual activities. The information on this sheet is then transferred to the time bar diagram for diagrammatic presentation.

2.11.2 Activity sheet (see fig. 2.39) (taken from village hall project Fig. 2.15)

This is a list of all the activities, their durations and any restraints involved in the project.

The first step is to list all the activities involved in the project, preferably in the order in which they are to be carried out, but this is not essential; these activities are then numbered. The next step is to provide the information regarding the relationship between activities by means of a sequence of numbers alongside the activity number. These numbers indicate which activities must be completed before the activity under consideration can be started. For example in Fig. 2.39 in the sequence column for activity 1, are the figures 0;1 this indicates that there are no other activities to be completed before activity 1 can start. Another example is activity 4, in the sequence column are the figures 1, 2;4 which indicates that activity 1 (deliver frame) and activity 2 (foundations) must be completed before activity 4 (erect frame) can start. The other column records the duration of each activity which would be arrived at via the work content and resources applied to it. This information can now be recorded diagrammatically in time bar diagram form and the critical path ascertained.

2.11.3 Time bar diagram (see Fig. 2.40)

In this chart, each activity is represented by a straight line drawn to a suitable scale. The length of each line is proportional to the duration of the activity being represented. The chart can either be drawn vertically or horizontally. If vertically then the chart is started at the top of the sheet, if horizontally then the chart is drawn from left to right as is common with most forms of bar chart.

Assuming that the chart is to be drawn horizontally then the first step is to set up a suitable time scale on the horizontal axis and then to erect a perpendicular at time zero, this line being called the 'start' line. The job lines are now drawn on the chart in the sequence indicated in the activity sheet (see Fig. 2.39), each activity being given its respective number for identification purposes. A short vertical line is introduced at the end of each activity to indicate the end of that particular activity and to separate it from the next in sequence. The relationship with activities in other parts of the chart is indicated by a longer vertical line. If there is a place in the project where a number of activities must be completed so that another activity can be started then the vertical line is drawn at the end of the activity line which is latest in time and produced to connect all activities involved. These vertical

Fig. 2.39 Activity sheet – village hall

Activity number	*Activity*	*Activity sequence*	*Activity duration (days)*
1	Deliver frame	0; 1	10
2	Foundations	0; 2	5
3	Deliver plasterboard	0; 3	70
4	Erect frame	1, 2; 4	1
5	Brickwork to D.P.C	4; 5	2
6	Floor construction	5; 6	3
7	Brickwork to eaves	5; 7	11
8	End panels	6; 8	13
9	Glazing	8; 9	7
10	Roof construction	7, 8;10	15
11	Internal partitions	7, 8;11	2
12	Drainage	7, 8;12	4
13	Roof finish	10;13	9
14	Rainwater goods	13;14	2
15	Electrician first fixing	9, 11, 13;15	7
16	Joiner first fixing	9, 11, 13;16	6
17	Plastering	3, 16, 15;17	10
18	Electrician second fixing	17;18	3
19	Joiner second fixing	17;19	5
20	Make good plaster	18, 19;20	1
21	Floor screed	18, 19;21	4
22	Floor screed dry	21;22	10
23	Floor finish	22;23	12
24	Painting and decorating	14, 20, 23;24	8
25	Paving	12;25	9
26	Electrician finish	24;26	1
27	Joiner finish	24;27	3
28	Clean and handover	25, 27, 26;28	5

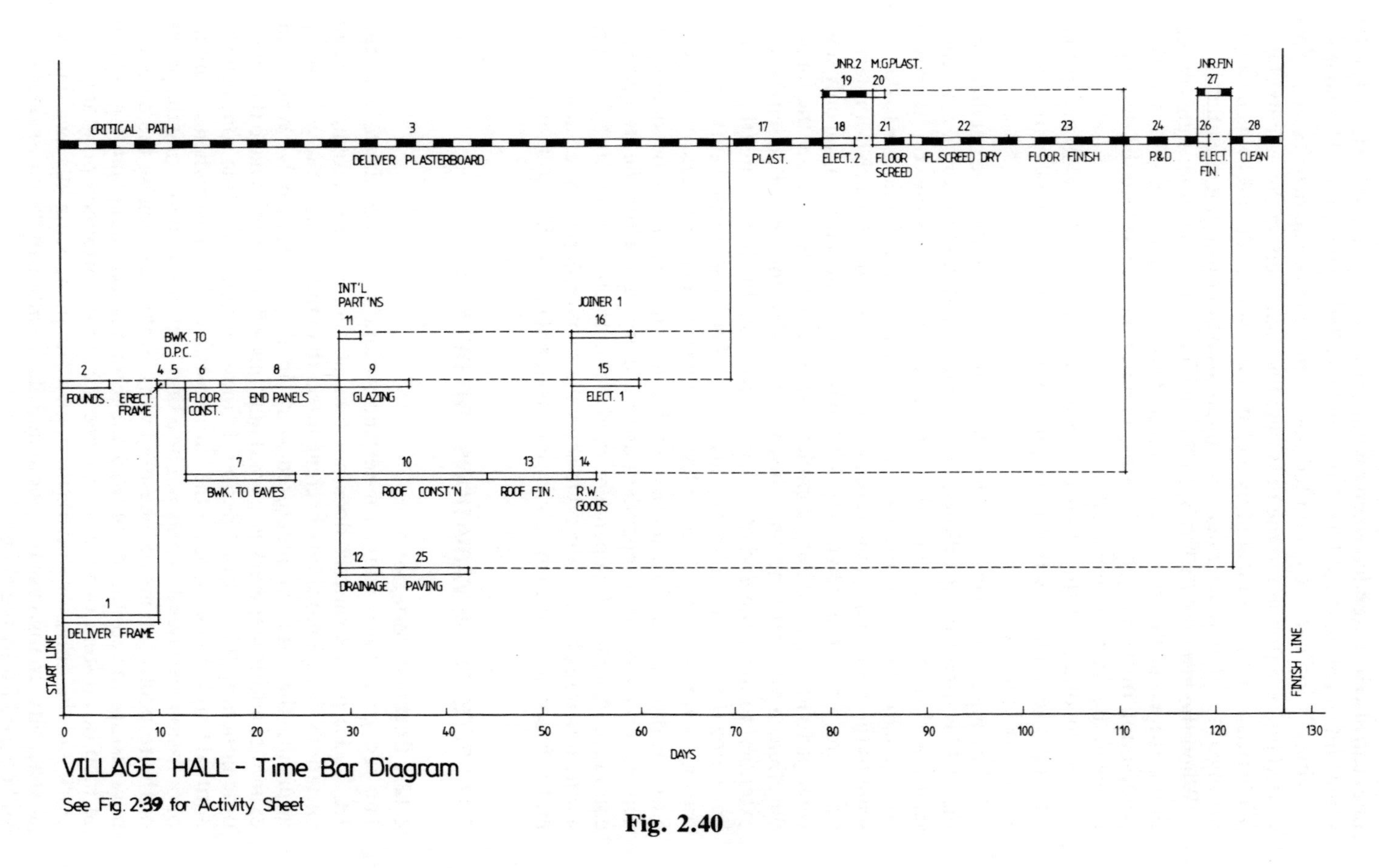

VILLAGE HALL - Time Bar Diagram

See Fig. 2·39 for Activity Sheet

Fig. 2.40

lines can also be extended when a number of activities are to be started at the same time. The vertical lines are usually called 'time lines'. The float on activities is indicated by a dotted line at the end of the activity line. This dotted line is extended to the next time line indicated on the activity sheet. The critical path is indicated by a continuous set of activities which can be arranged to fall in one straight line if required but this is not essential.

When the last activity has been drawn on the chart, another long vertical line is drawn at this point. This is called the finish line and represents the completion of the project. The project duration is therefore the dimension between the start and finish lines.

The advantage of this chart is that it shows clearly the relationship that exists between all activities in the project, the sequence of all activities and which are concurrent.

Following these basic principles, the construction of the time bar diagram, Fig. 2.40 is therefore as follows: having calibrated the horizontal axis (time scale) and drawn the start line, the next step is to draw on activity 1 commencing the line at the start line (time zero) and extending it 10 days to scale to the right. The same is done for activities 2 and 3 which both start at time zero. From the activity sheet it can be seen that activity 4 can only start when activities 1 and 2 are complete. Due to activity 1 having the longer duration, a vertical time line must be erected at the end of activity 1 and extended to meet the end of activity 2 produced by dotted line (the dotted line being the amount of float on activity 2). The time line prevents activity 4 starting before activity 1 is completed. Activity 4 can now be drawn. Activity 5 can be drawn straight after activity 4 as activity 5 is only dependent upon activity 4. Activity 6 is dependent upon activity 5 therefore the straight line continues. Activity 7 is dependent upon the completion of activity 5 therefore a time line must be introduced to show the relationship between activities 5 and 7. From here on each activity is considered in a similar fashion.

2.12 EFFECTS OF VARIATIONS AND DELAYS

2.12.1 Settling variations

The recognised procedure for settling variations in accordance with the J.C.T. Form of Contract does not take all factors into account. Certain variations have a more far-reaching affect than those taken into account by adjusting the rate for measured work in the contract bills. When, for example, additional work is carried out, this will sometimes have the effect of lengthening the contract period. In these circumstances the contractor is entitled to reimbursement for the extra costs involved. Even when variations of omission are issued, it may not be a simple matter of deducting the prices of the respective items of measured work in the bill because the omission may disrupt the continuity of work for certain labour and plant, e.g. a gang of men or an item of plant may have a carefully arranged programme of work and the omission may result in men and plant being idle awaiting the next activity. In this case the labour and plant content of the omission should still be paid for by the client.

Provision is made for the architect to allow payment to the contractor in these circumstances. It will of course be necessary for the contractor to prove his case and to do this a programme is extremely useful.

2.12.2 Other factors causing delays

Many other factors can cause delays to the contractor besides variations, for which he does not, in many cases, receive full payment due to his inability to substantiate his claim, even though provision is made for payment under the J.C.T. form of contract. Examples of these factors are:

Late arrival of drawings, details and levels even when prior notice of these requirements has been given.
Opening up work for inspection or testing.
Discrepancies between drawings and contract bills.
Delays caused by persons employed direct by the client etc.

The contractor can claim payments for losses caused by these factors where his progress has been disturbed.

2.12.3 Delays where no payment is warranted

Some delays do not of course warrant extra payment to the contractor, but it is nevertheless useful to have a means available by which the amount of extension of time can be justified. Typical delays in this category are exceptionally inclement weather, fire, strikes, shortage of materials or labour which could not have been foreseen. Extension of time can be granted for these and other reasons under the J.C.T. form of contract.

2.12.4 Assessing the effects of delays

It is often difficult to assess realistically the effect of delays on costs to the contractor, particularly when extension of time is involved. Without a programme it is virtually impossible. When a programme has been prepared and updated the problem of price adjustment, where appropriate, is made much easier. This particularly applies when network techniques have been used. If the network is a contract document, it will almost certainly be used as a basis for calculating these adjustments. Even when the programme is not a contract document, providing the contractor updates it, he can present it to the quantity surveyor or architect as appropriate, whenever a variation order has the affect of interrupting the programme and it will prove to be formidable evidence in support of a claim.

On a network programme the effect can be clearly seen and variation orders or other causes which affect activities on the critical path will often warrant an extension of time and, possibly, subsequent additions to the contract sum.

2.12.5 Settling a variation

Figure 2.14, which is the analysed arrow diagram for the village hall project will be used to illustrate the effects of a variation order on costs and contract period.

Assume a variation order is issued for additional screed (activity 16–18) and floor finish (activity 19–20) under the stage area. (This extra flooring is considered necessary because the client will require the stage to be moved on some occasions and wishes therefore to have the whole floor covered.)

The additional screed and floor finish will increase the total amount in the bill by one fifth and direct costs to the contractor will also increase by one fifth. These costs will be covered by adjusting the price included in the contract bills which will also increase by one fifth. However, both these activities are on the critical path and the time required for each will also increase by one fifth. This will result in an extension of time of $(\frac{1}{5} \times 4) + (\frac{1}{5} \times 12) = 3\frac{1}{5}$ days. The contractor may therefore require this extension of time plus the extra cost of overheads included in the preliminaries and related to the contract period.

It may of course be possible to avoid extending the contract period by overtime working or bringing in extra men. This will however increase costs and if the contractor agrees to take this action he is entitled to payment for the extra costs. This would probably be a matter for negotiation between the contractor and the quantity surveyor.

The example given is of course a very simple one, and in practice the problem may be more difficult. However the above procedure can be followed and differences of opinion should be much easier to resolve.

2.12.6 Assessing costs using bar charts

Bar charts can be used in a similar manner to that shown above but it may be more difficult to 'prove' that a particular activity is on the critical path and will therefore affect the project duration. A good updated bar chart is however far superior to no programme at all.

2.13 COST OPTIMISATION

It may be possible to vary the duration of some activities in a network and this will normally affect the cost of the activities and consequently the cost of the project.

2.13.1 The relationship between project duration and cost

Generally speaking certain assumptions can be made about the relationship between project duration and cost as follows:

(i) Direct costs rise as project duration decreases.
(ii) Indirect costs rise as project duration increases.
(iii) There is a minimum duration for any project beyond which further reduction is not feasible.

2.13.2 Determining optimum project cost

At each stage in network optimisation calculations the following procedure can be adopted.

(a) Prepare a table of activities on the critical path showing normal

activity duration and cost, and minimum (crash) activity duration and cost.
(b) Calculate the cost slope of each activity in the table.
(c) List the activities in order of minimum cost slope.
(d) Omit activities which cannot be compressed (and those fully crashed from previous compressions).
(e) Compress the activity (or activities if two are being compressed) with the least cost slope, the maximum amount possible, or until some other activity becomes critical.
(f) Calculate the new project duration and direct cost.
(g) When minimum duration is reached, calculate indirect costs for each project duration.
(h) Add indirect costs to direct costs and calculate total costs for each project duration.
(i) Determine optimum project cost.

2.13.3 Example of simple cost optimisation

Fig. 2.41 shows a network for the construction of a bridge. The normal cost and duration and the crash cost and duration for each activity is shown in Fig. 2.42. The indirect costs are £400.00 per week.
The all normal solution is given in Fig. 2.41. The critical activities are 0–2, 2–8, 8–9, 9–10, 10–12 and 12–13. The project duration is 24 weeks and the cost is £20920.

The list of critical activities in order of minimum cost slope after omitting activities which cannot be compressed is as follows:

Activity	*Maximum compression*	*Cost per week*
10–12	1	£125
0–2	5	£150
9–10	1	£200
8–9	2	£500

2.13.3.1 First compression

Compress activity 10–12 by 1 week at a cost of £125. The project duration is now 23 weeks and the direct cost is £20920 + £125 = £21045. Fig. 2.43 shows the network after the first compression. The shared float on activities 3–5, 5–7, 7–11 and 11–12 is reduced to 8 weeks.

2.13.3.2 Second compression

Compress activity 0–2 by 4 weeks at a cost of 4 × £150 = £600. The project duration is now 19 weeks and the direct cost is £21045 + £600 = £21645. Fig. 2.44 shows the network after the second compression. Activities 0–3, 3–4, 4–6 and 6–8 are now also critical.

The shared float on activities 3–5, 5–7, 7–11 and 11–12 is further reduced

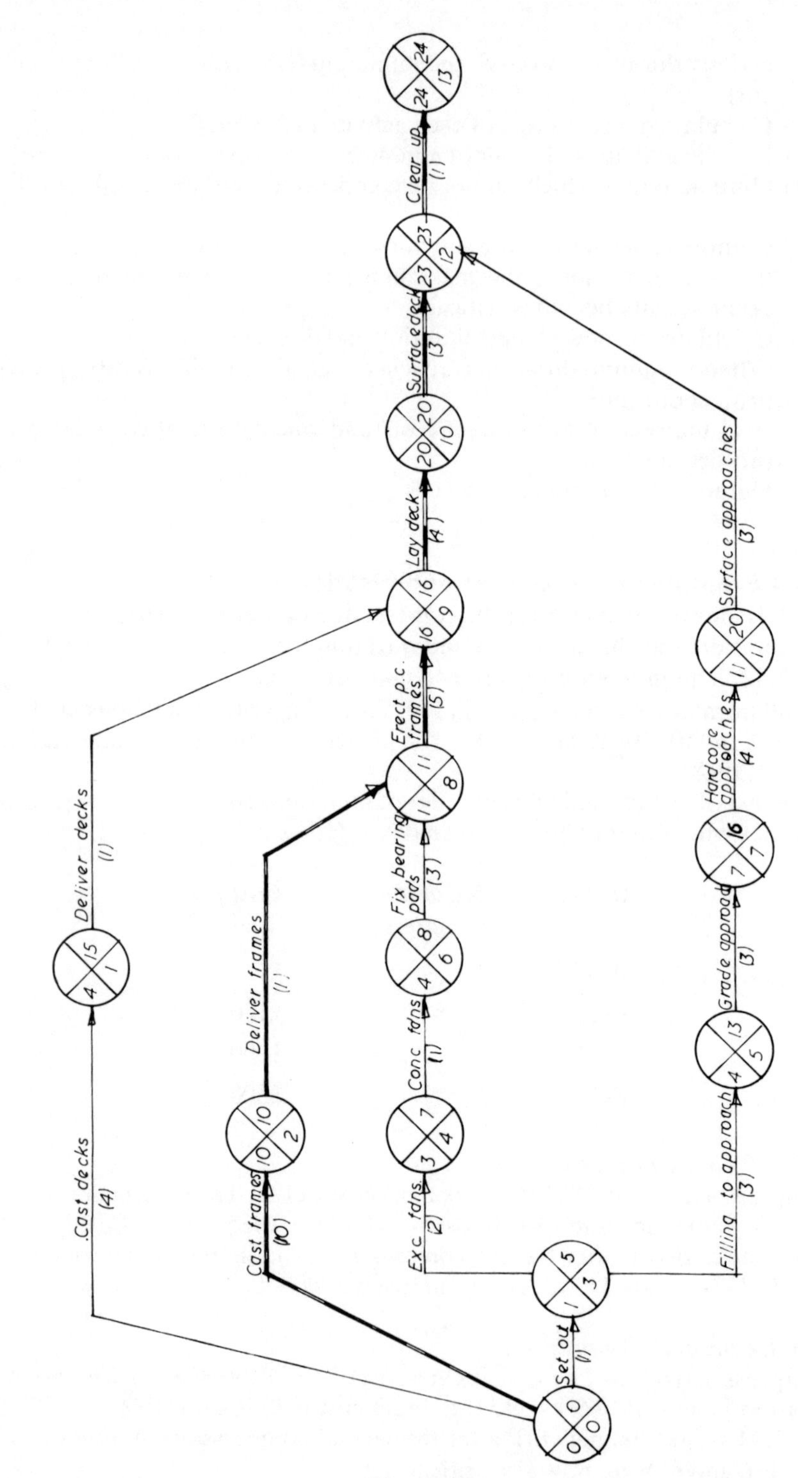

Cast decks
(4)
Deliver decks
(1)
Cast frames
(10)
Deliver frames
(1)
Exc. fdns.
(2)
Conc fdns
(1)
Fix bearing pads
(3)
Erect p.c. frames
(5)
Lay deck
(4)
Surface deck
(3)
Clear up
(1)
Set out
(1)
Filling to approach
(3)
Grade approach
(3)
Hardcore approaches
(4)
Surface approaches
(3)

Fig. 2.41

Fig. 2.42

Activity reference	*Normal* *Duration*	*Cost £*	*Crash* *Duration*	*Cost £*
0–1	4	3000	3	3300
0–2	10	9000	5	9750
0–3	1	100	1	100
1–9	1	120	1	120
2–8	1	100	1	100
3–4	2	1500	2	1500
3–5	3	500	2	600
4–6	1	500	1	500
5–7	3	350	3	350
6–8	3	450	2	525
7–11	4	550	3	600
8–9	5	1500	3	2500
9–10	4	1200	3	1400
10–12	3	750	2	875
11–12	3	800	2	1100
12–13	1	500	1	500

to 4 weeks and the shared float in activities 0–1 and 1–9 is reduced to 7 weeks.

As two critical paths have now emerged a revised list of critical activities is required in order of minimum cost slope omitting activities which have been fully compressed and omitting activities which cannot be compressed as follows:

Activity	*Maximum compression*	*Cost per week*
6–8	1	£75
0–2	1	£150
9–10	1	£200
8–9	2	£500

2.13.3.3 Third compression

Although 6–8 has the least cost slope of the activities under consideration, compressing this activity will have no effect on the project duration unless activity 0–2 is also compressed. The combined effect of compressing both of these activities is £75 + £150 = £225 per week. It is therefore cheaper to compress activity 9–10.

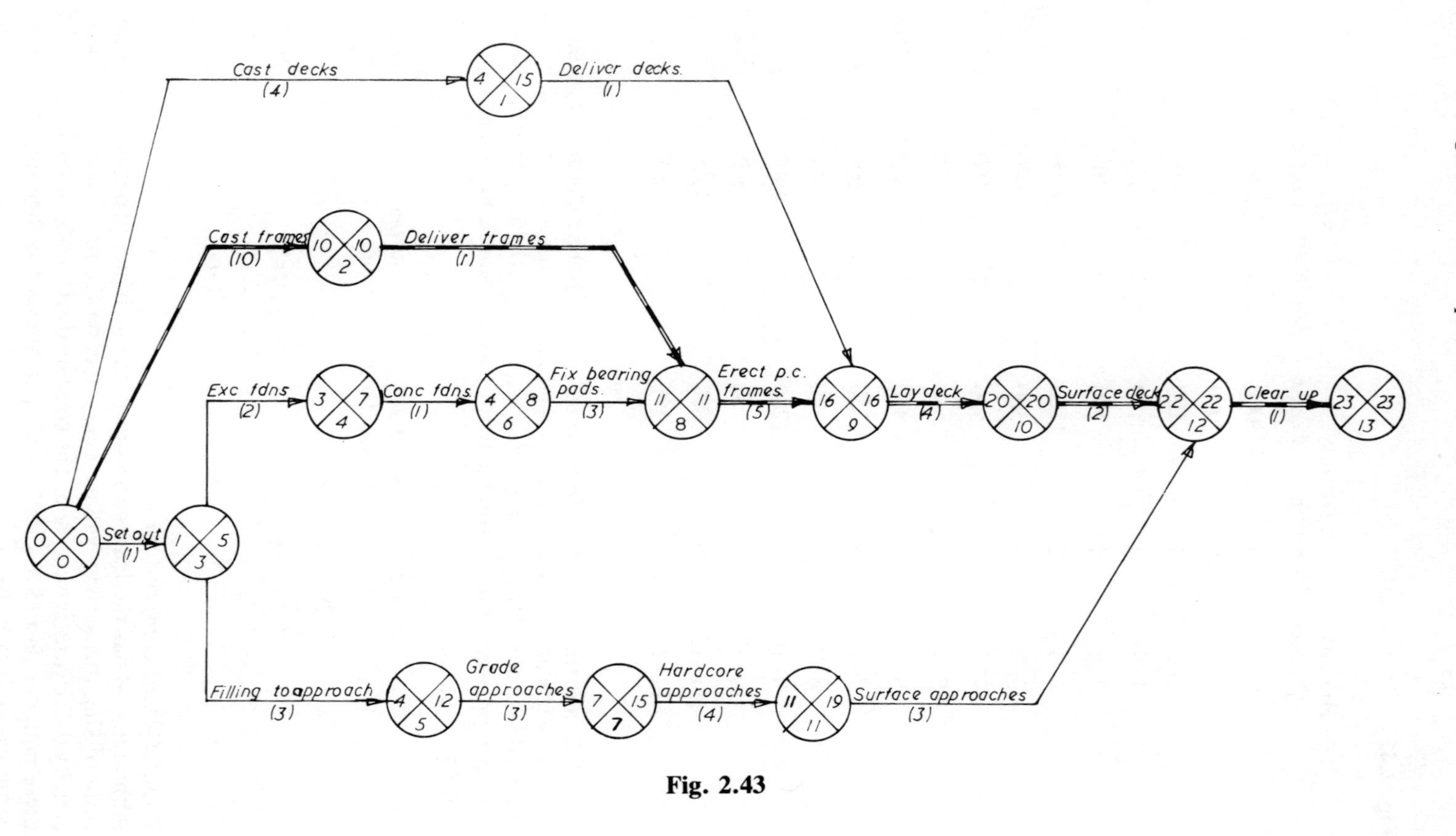

Cast decks
(4)
Deliver decks.
(1)
Cast frames
(10)
Deliver frames
(1)
Exc fdns
(2)
Conc fdns.
(1)
Fix bearing pads.
(3)
Erect p.c. frames.
(5)
Lay deck
(4)
Surface deck
(2)
Clear up
(1)
Set out
(1)
Filling to approach
(3)
Grade approaches
(3)
Hardcore approaches
(4)
Surface approaches
(3)

Fig. 2.43

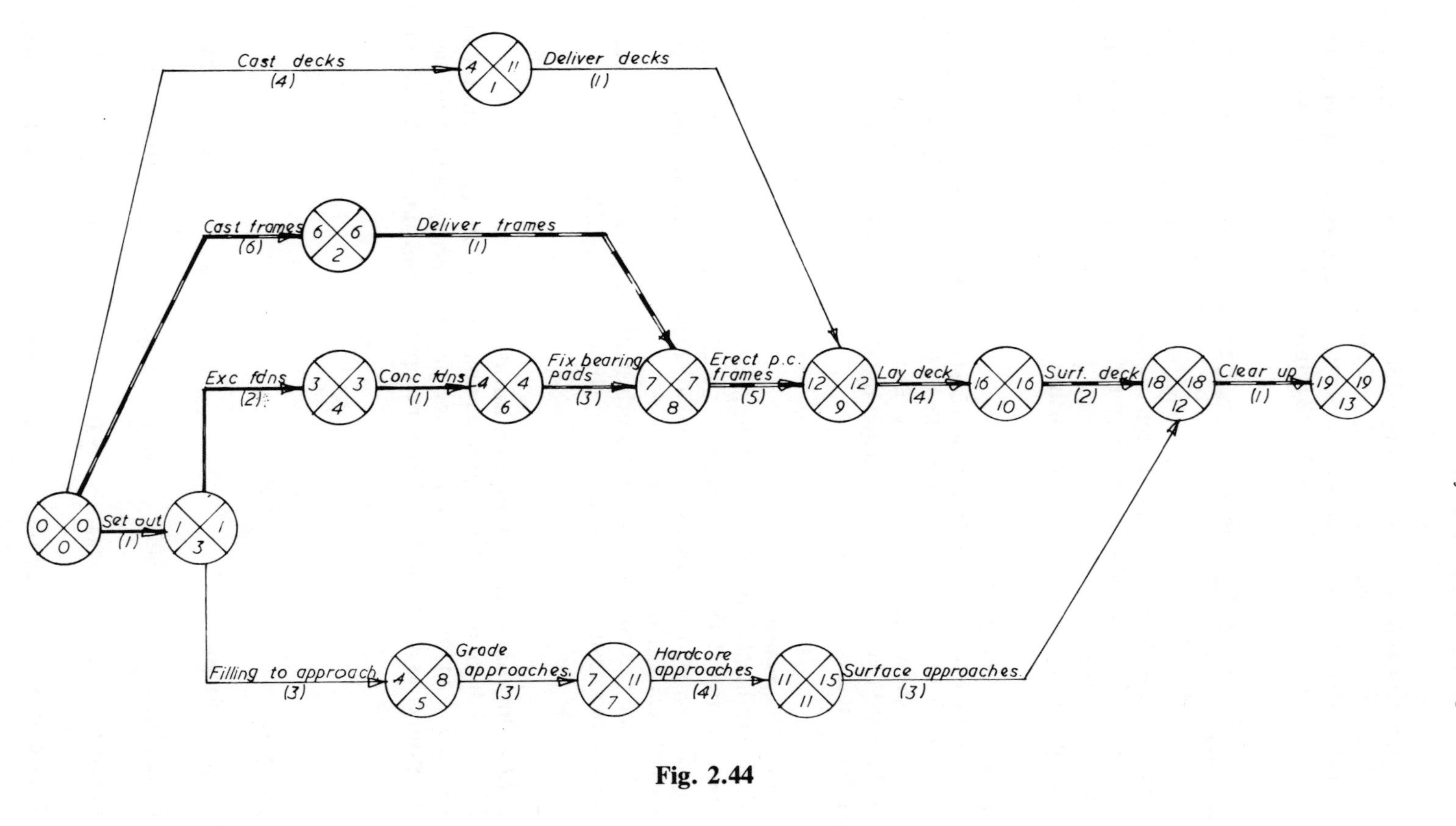

Cast decks
(4)
Deliver decks
(1)
Cast frames
(6)
Deliver frames
(1)
Exc fdns
(2)
Conc fdns
(1)
Fix bearing pads
(3)
Erect p.c. frames
(5)
Lay deck
(4)
Surf. deck
(2)
Clear up
(1)
Set out
(1)
Filling to approach
(3)
Grade approaches.
(3)
Hardcore approaches
(4)
Surface approaches.
(3)

Fig. 2.44

Compress activity 9–10 by one week at a cost of £200. The project duration is now 18 weeks and the direct cost is £21645 + £200 = £21845.

Fig. 2.46 shows the network after the third compression. The shared float in activities 3–5, 5–7, 7–11 and 11–12 is further reduced to 3 weeks.

2.13.3.4 Fourth compression

Compress activities 0–2 and 6–8 by 1 week at a combined cost of £150 + £75 = £225 (as this is cheaper than compressing 8–9). The project duration is now 17 weeks and the direct cost is £21845 + £225 = £22070.

Fig. 2.47 shows the network after the fourth compression. The shared float on activities 3–5, 5–7, 7–11 and 11–12 is further reduced to 2 weeks and the shared float on 0–1 and 1–9 is further reduced to 6 weeks.

2.13.3.5 Fifth compression

Compress activity 8–9 by 2 weeks at a cost of £500 × 2 = £1000. The project duration is now 15 weeks and the direct cost is £22070 + £1000 = £23070. Fig. 2.48 shows the network after the fifth comparison. Activities 3–5, 5–7, 7–11 and 11–12 are now criticial and the shared float on activities 0–1 and 1–9 is reduced to 4 weeks.

2.13.3.6 Optimising cost

To determine the optimum cost of the project, indirect cost must be added as shown in Fig. 2.45. The optimum cost is £28870 at a duration of 17 weeks. Fig. 2.49 shows the network analysed for 17 weeks duration and the weekly cost of each activity is also shown.

A graph can now be produced which shows the project direct cost curve, the indirect costs and the total costs as shown in Fig. 2.50.

2.13.4 Optimal cost at minimum duration

If project duration must be kept to a minimum then the approach could be varied and an 'all-crash' solution found first by crashing all the activities in the project as shown in Fig. 2.51. The optimal least time solution is then

Fig. 2.45

Project durations	15	17	18	19	23	24
Direct costs £	23070	22070	21845	21645	21045	20920
Indirect costs £	6000	6800	7200	7600	9200	9600
Total costs £	29078	28870	29045	29245	30245	30520

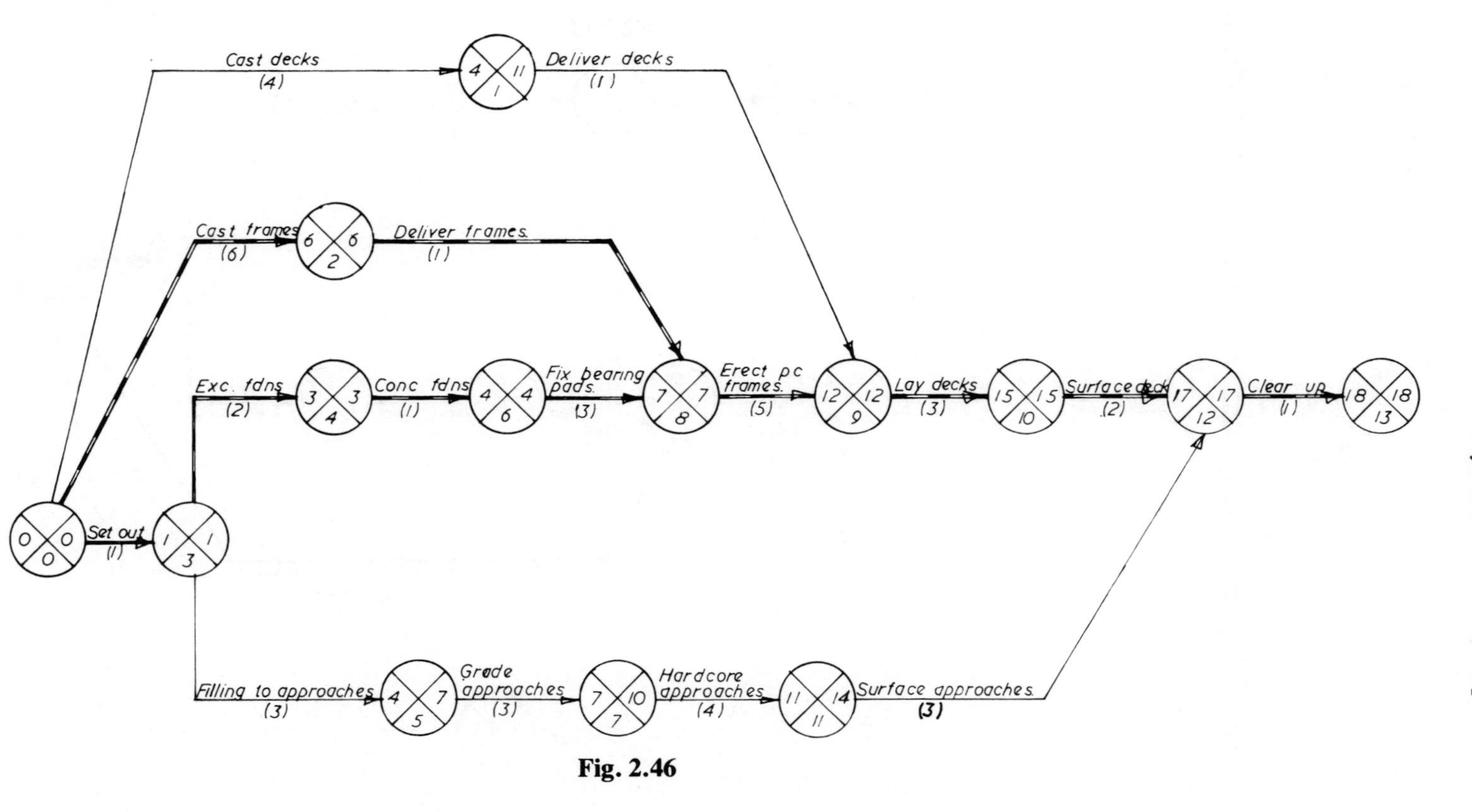
Cast decks
(4)
Deliver decks
(1)
Cast frames
(6)
Deliver frames.
(1)
Set out
(1)
Exc. fdns
(2)
Conc fdns
(1)
Fix bearing pads.
(3)
Erect pc frames.
(5)
Lay decks
(3)
Surface deck
(2)
Clear up
(1)
Filling to approaches
(3)
Grade approaches
(3)
Hardcore approaches
(4)
Surface approaches.
(3)

Fig. 2.46

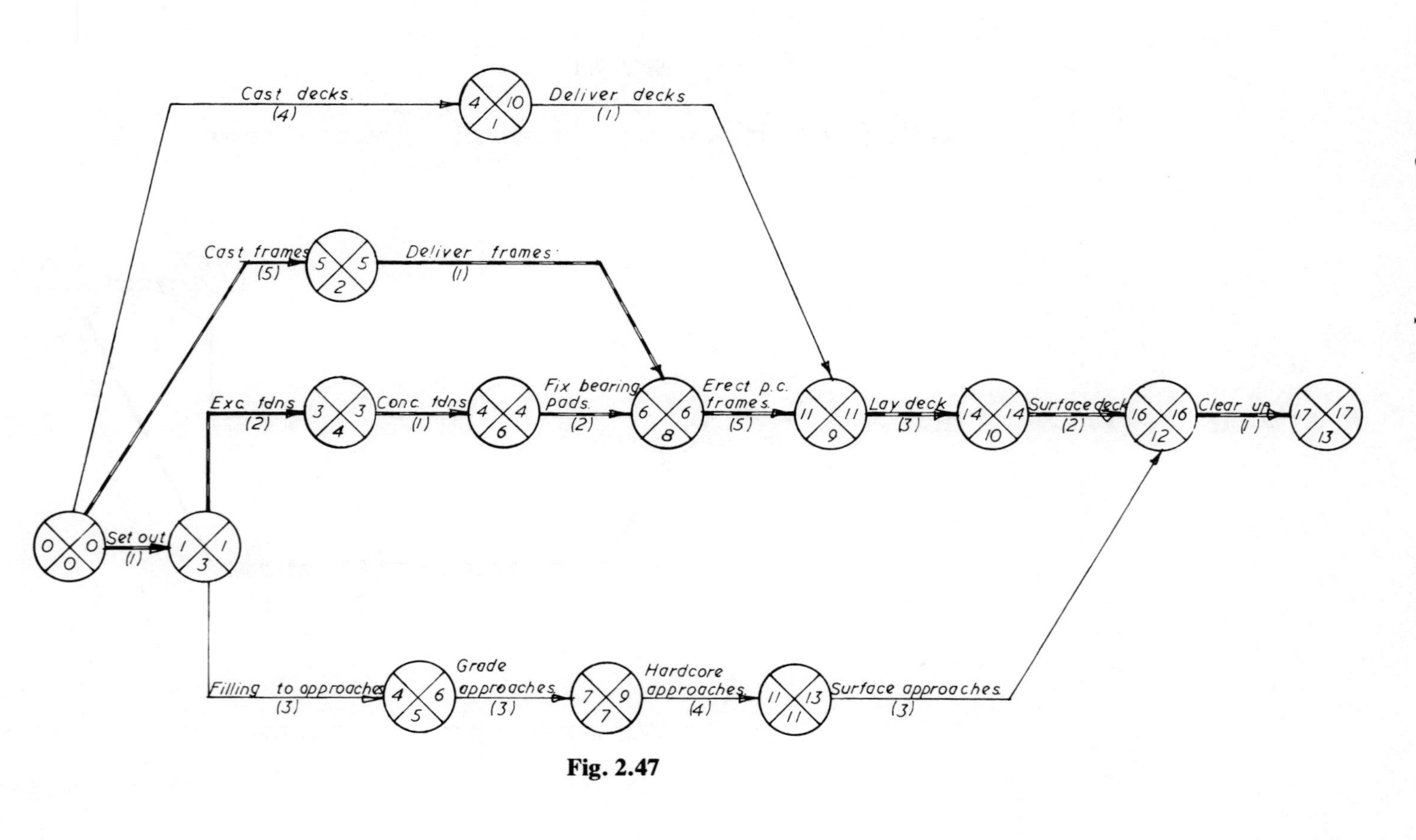

Cast decks. (4)
Deliver decks (1)
Cast frames (5)
Deliver frames (1)
Exc. fdns (2)
Conc. fdns (1)
Fix bearing pads. (2)
Erect p.c. frames. (5)
Lay deck (3)
Surface deck (2)
Clear up (1)
Set out (1)
Filling to approaches (3)
Grade approaches (3)
Hardcore approaches (4)
Surface approaches. (3)

Fig. 2.47

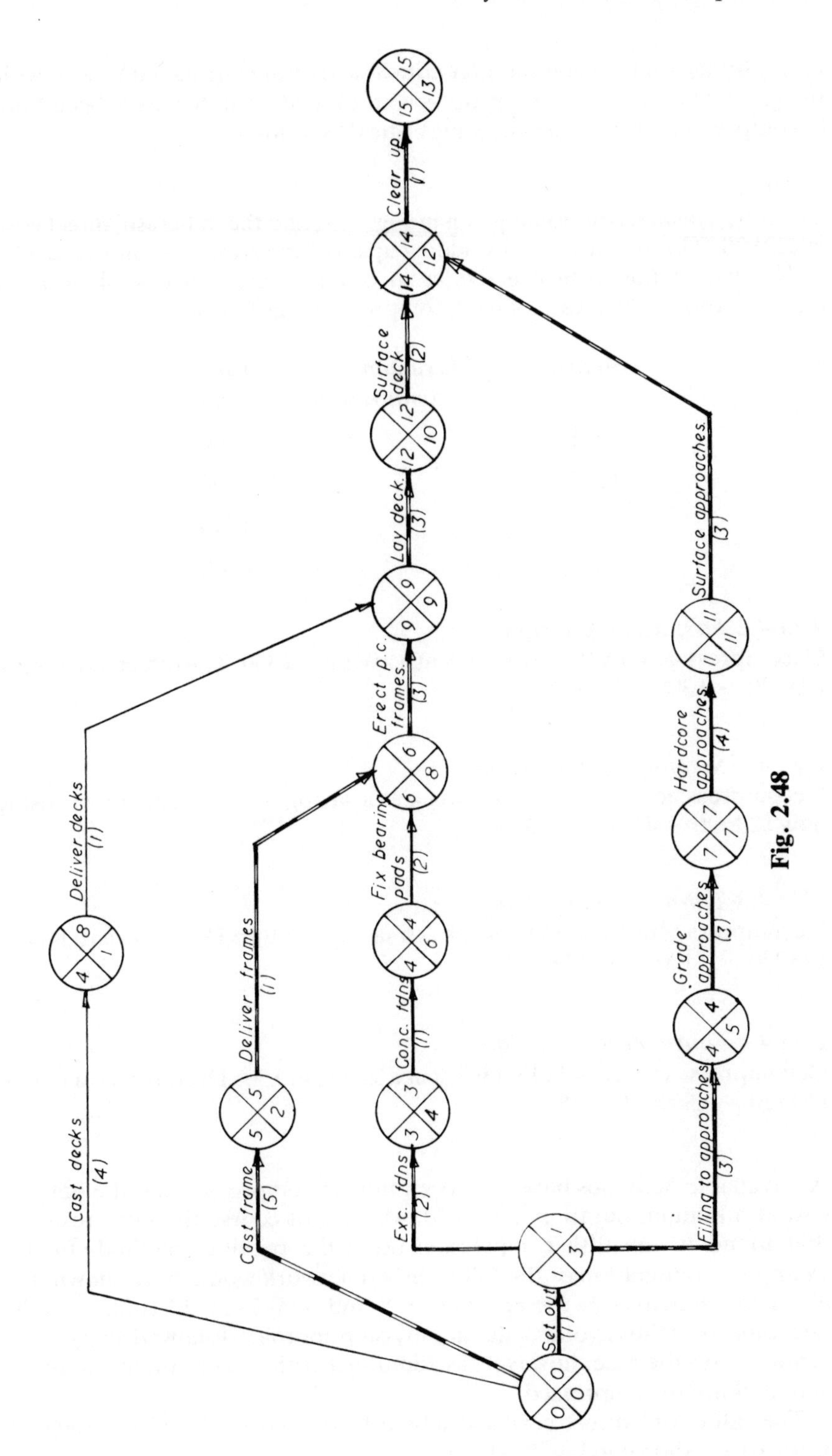
Cast decks
(4)
Deliver decks
(1)
Cast frame
(5)
Deliver frames
(1)
Set out
(1)
Exc. fdns
(2)
Conc. fdns
(1)
Fix bearing pads
(2)
Erect p.c. frames.
(3)
Lay deck.
(3)
Surface deck
(2)
Clear up.
(1)
Filling to approaches
(3)
Grade approaches
(3)
Hardcore approaches
(4)
Surface approaches.
(3)

Fig. 2.48

found by decompressing the non-critical activities starting with those with the greatest cost slope. This process is continued until they have been fully decompressed or until they become critical as follows:

Example
The fully crashed solution is given in Fig. 2.51 and the 'all-crash' direct cost is £23820. This is calculated by adding up all of the crash costs in Fig. 2.42.

The list of non-critical activities in order of greatest cost slope after omitting those which cannot be decompressed is as follows:

Activity	*Maximum decompression*	*Cost slope*
0–1	1	£300
11–12	1	£300
3–5	1	£100
7–11	1	£ 50

2.13.4.1 First decompression
Decompress activity 0–1 by 1 week at a saving of £300. The direct cost is now £23820 − £300 = £23520.

2.13.4.2 Second decompression
Decompress activity 11–12 by 1 week at a saving of £300. The direct cost is now £23520 − £300 = £23220.

2.13.4.3 Third decompression
Decompress activity 3–5 by 1 week at a saving of £100. The direct cost is now £23220 − £100 = £23120.

2.13.4.4 Fourth decompression
Decompress activity 7–11 by 1 week at a saving of £50. The direct cost is now £23120 − £50 = £23070.

All available activities have now been fully decompressed and the optimal cost at minimum duration is £23070, which is of course the same result as that found in the fifth compression using the previous method. In this example, examination of the fully crashed network would have shown that all of the activities in the chains 0–1–9 and 3–5–7–11–12 could be fully decompressed thus avoiding the step-by-step approach followed above. This is not always the case, however, as additional critical paths often emerge as activities are decompressed.

The 'all-crash' direct cost and all-crash total costs (£23820 + £6000 = £29820) are shown in Fig. 2.50.

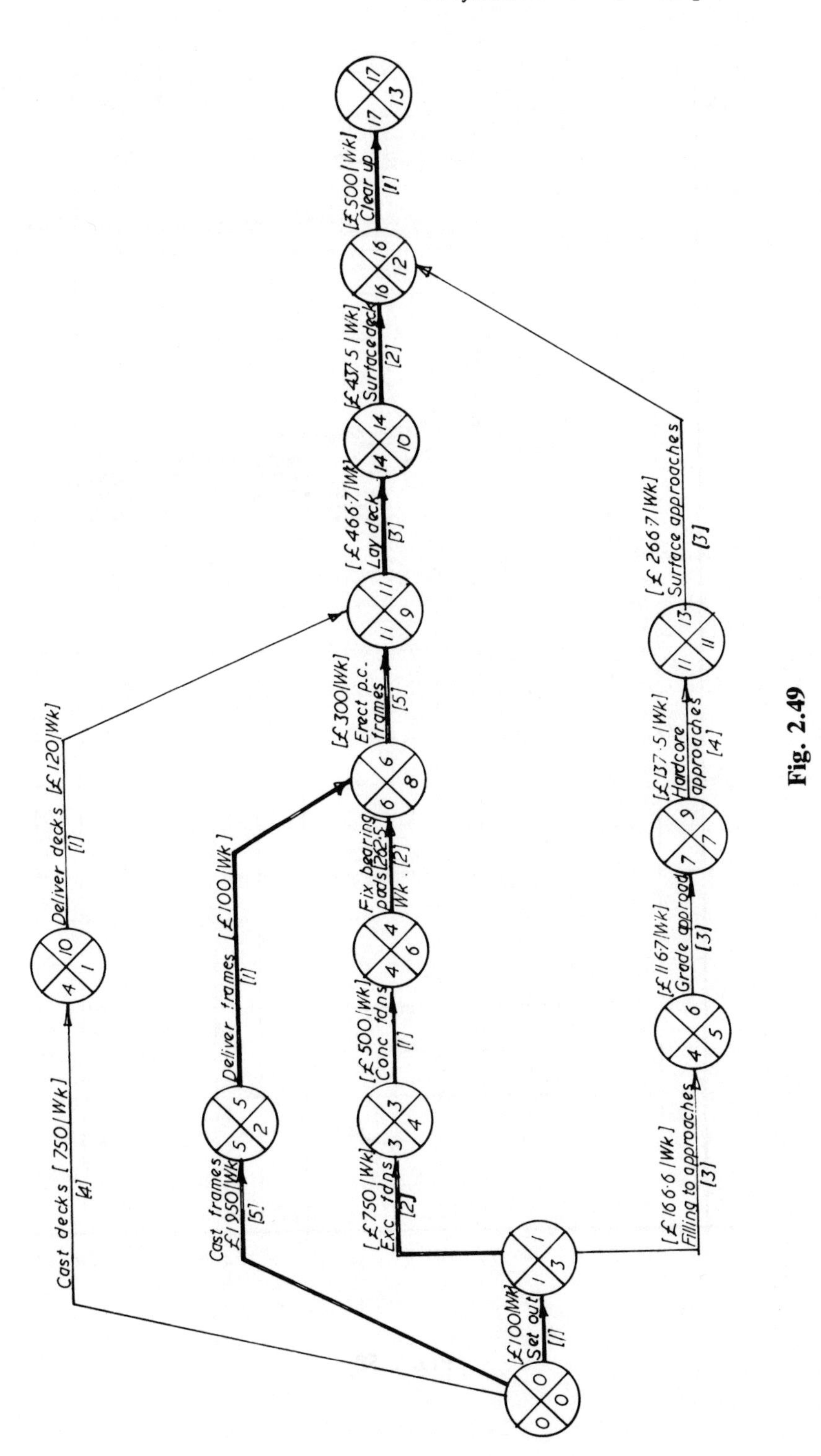

Cast decks [750/Wk]
[4]
Deliver decks [£120/Wk]
[1]
Cast frames
£1950/Wk
[5]
Deliver frames [£100/Wk]
[1]
[£750/Wk]
Exc fdns
[2]
[£500/Wk]
Conc fdns
[1]
Fix bearing
pads[202.5]
Wk [2]
[£300/Wk]
Erect p.c.
frames
[5]
[£466.7/Wk]
Lay deck
[3]
[£437.5/Wk]
Surface deck
[2]
[£500/Wk]
Clear up
[1]
[£100/Wk]
Set out
[1]
[£166.6/Wk]
Filling to approaches
[3]
[£11.67/Wk]
Grade approach
[3]
[£137.5/Wk]
Hardcore
approaches
[4]
[£266.7/Wk]
Surface approaches
[3]

Fig. 2.49

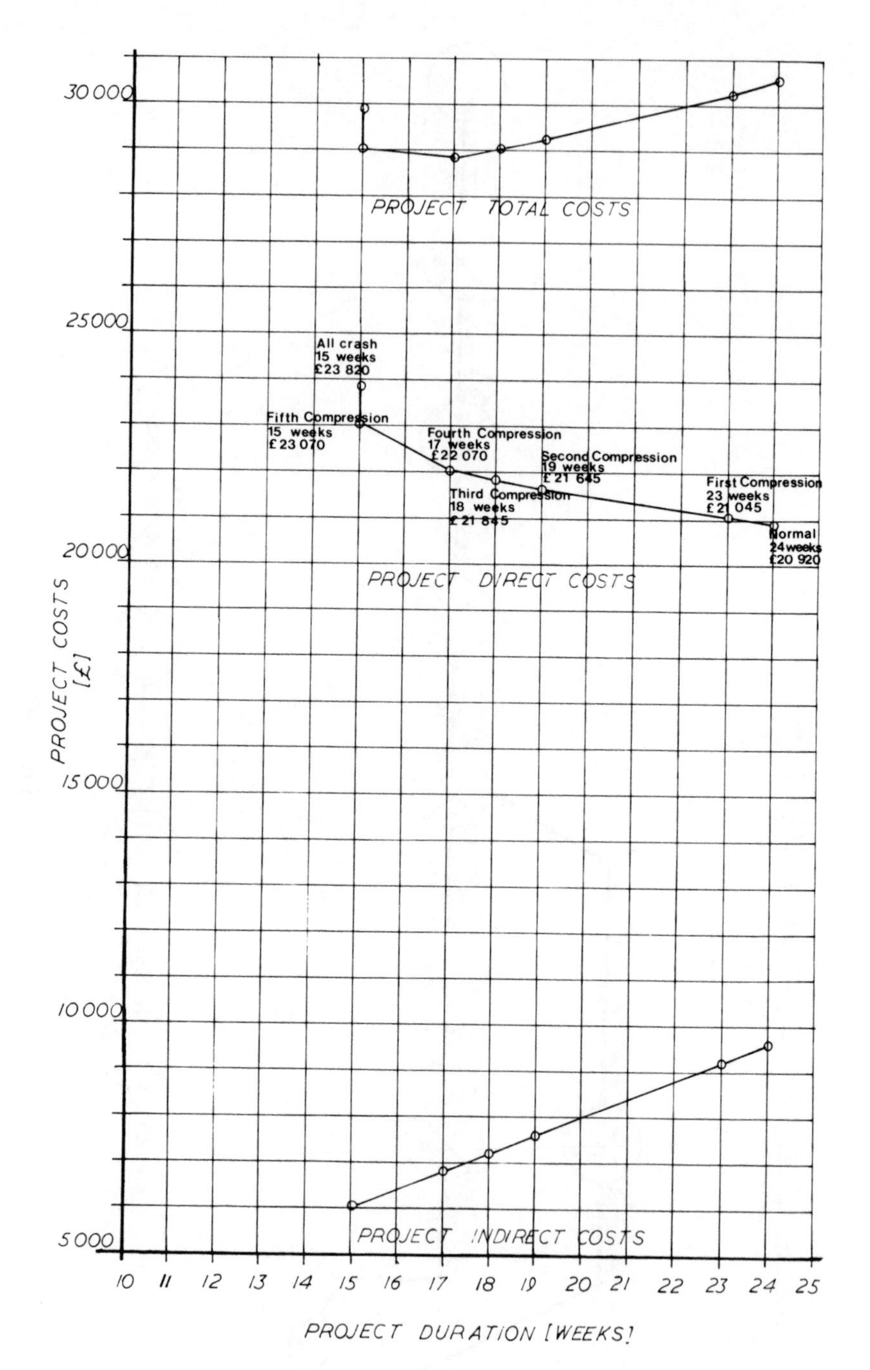

30 000
25000
20000
15 000
10 000
5 000
PROJECT COSTS [£]
PROJECT TOTAL COSTS
All crash
15 weeks
£23 820
Fifth Compression
15 weeks
£ 23 070
Fourth Compression
17 weeks
£22 070
Third Compression
18 weeks
£ 21 845
Second Compression
19 weeks
£ 21 645
First Compression
23 weeks
£ 21 045
Normal
24 weeks
£20 920
PROJECT DIRECT COSTS
PROJECT INDIRECT COSTS
10 11 12 13 14 15 16 17 18 19 20 21 22 23 24 25
PROJECT DURATION [WEEKS]

Fig. 2.50

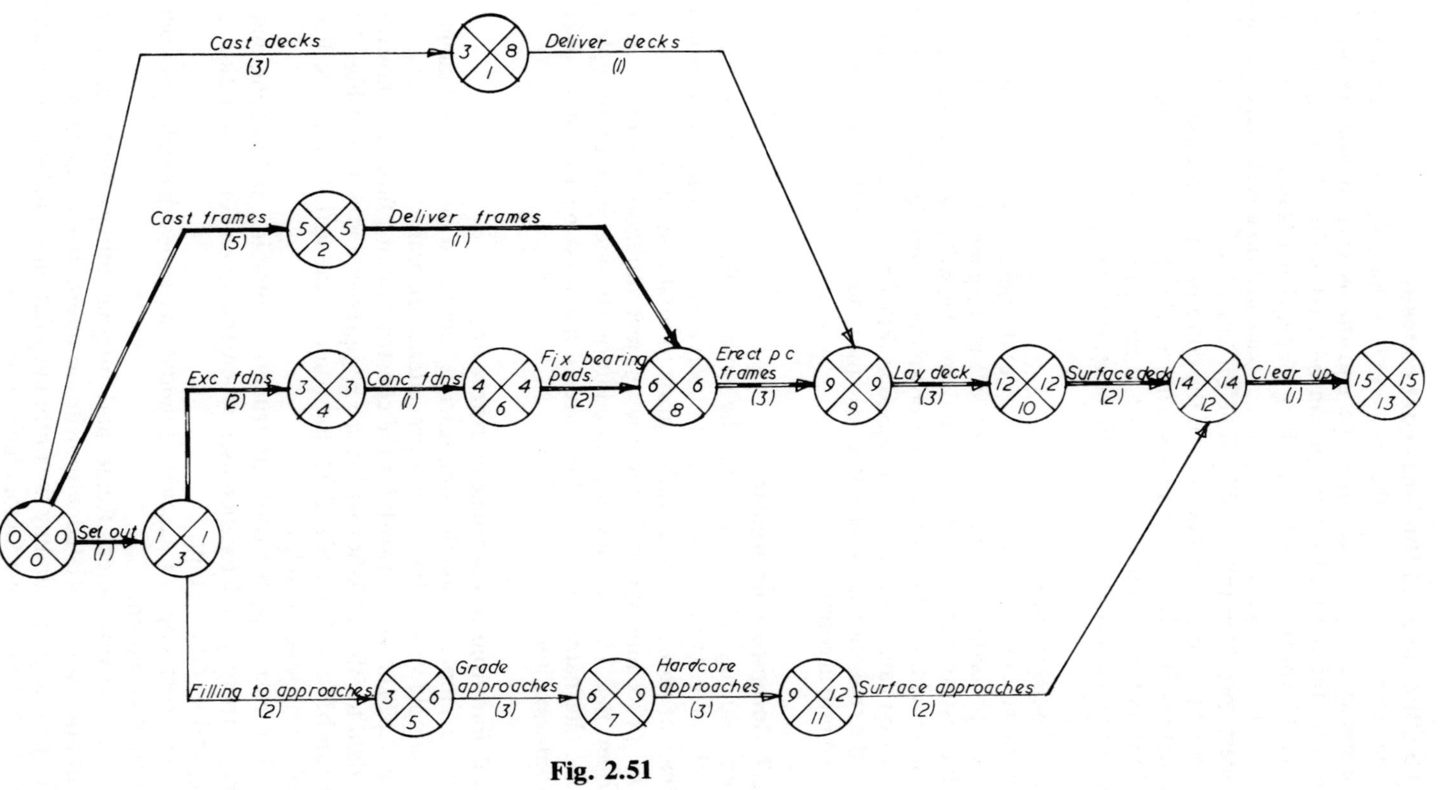
Cast decks
(3)
Deliver decks
(1)
Cast frames
(5)
Deliver frames
(1)
Exc fdns
(2)
Conc fdns
(1)
Fix bearing pads.
(2)
Erect p c frames
(3)
Lay deck
(3)
Surface deck
(2)
Clear up
(1)
Set out
(1)
Filling to approaches
(2)
Grade approaches
(3)
Hardcore approaches
(3)
Surface approaches
(2)

Fig. 2.51

2.13.5 Other uses of the least cost concept

In order to make full use of the least cost concept almost *all* the information is necessary at the tendering stage and this is very unusual in practice. However, the least cost concept can be used during the progress of the project. If critical activities are taking longer than planned, then *one* of the considerations in deciding how to gain back lost time is to make savings at the least cost. Naturally, saving time on activities with the least cost slope would achieve this. Another use of the concept is to meet a stated project completion date. If, on first analysis, the project duration is longer than that set down in the conditions of contract, it can be reduced more economically by saving time on the activities with the least cost slope.

2.13.6 Decision rules

Decision rules can be used to help select the best course of action and one effective procedure is to increase the size of the gang for the trade which has the highest total time on the critical path. This will have an effect on many critical activities and sometimes cuts down the project duration considerably. Very often gang sizes can be increased at little or no extra cost, particularly when the reduction in overheads due to the reduction in the project duration is taken into account.

2.13.7 Complex compression

When reducing project duration, the network diagram can be re-analysed at each stage. However, by careful observation, it is possible to perform a number of reductions in duration prior to total re-analysis.

The procedure for reducing the project duration is similar to that decribed in 2.13.2, but includes in addition consideration of multiple cost slopes, alternative methods of construction and taking account of discrete cost information.

2.13.8 Example of complex compression

As the example on simple compression was based on an arrow diagram, this example will be based on a precedence diagram.

Fig. 2.52 shows a network for the construction of a furniture showroom. The data for the activities which are compressible is shown in Fig. 2.53.

It can be seen from Fig. 2.53 that some of the activities do not have a simple cost slope as follows:

1. The cost slope for roof construction, plumbing and heating 2nd fix and carpentry 2nd fix increases when the reduction in duration goes beyond a certain point.
2. The cost slope of excavate foundations varies depending on which method is chosen.
3. The activities erect frame and internal fittings have two specific durations and associated costs with no interpolation between.

The procedure for finding the optimum solution is similar to the example in 2.13.3 but the above factors have to be taken into account.

The all-normal solution is shown in Fig. 2.52. The project duration is 115

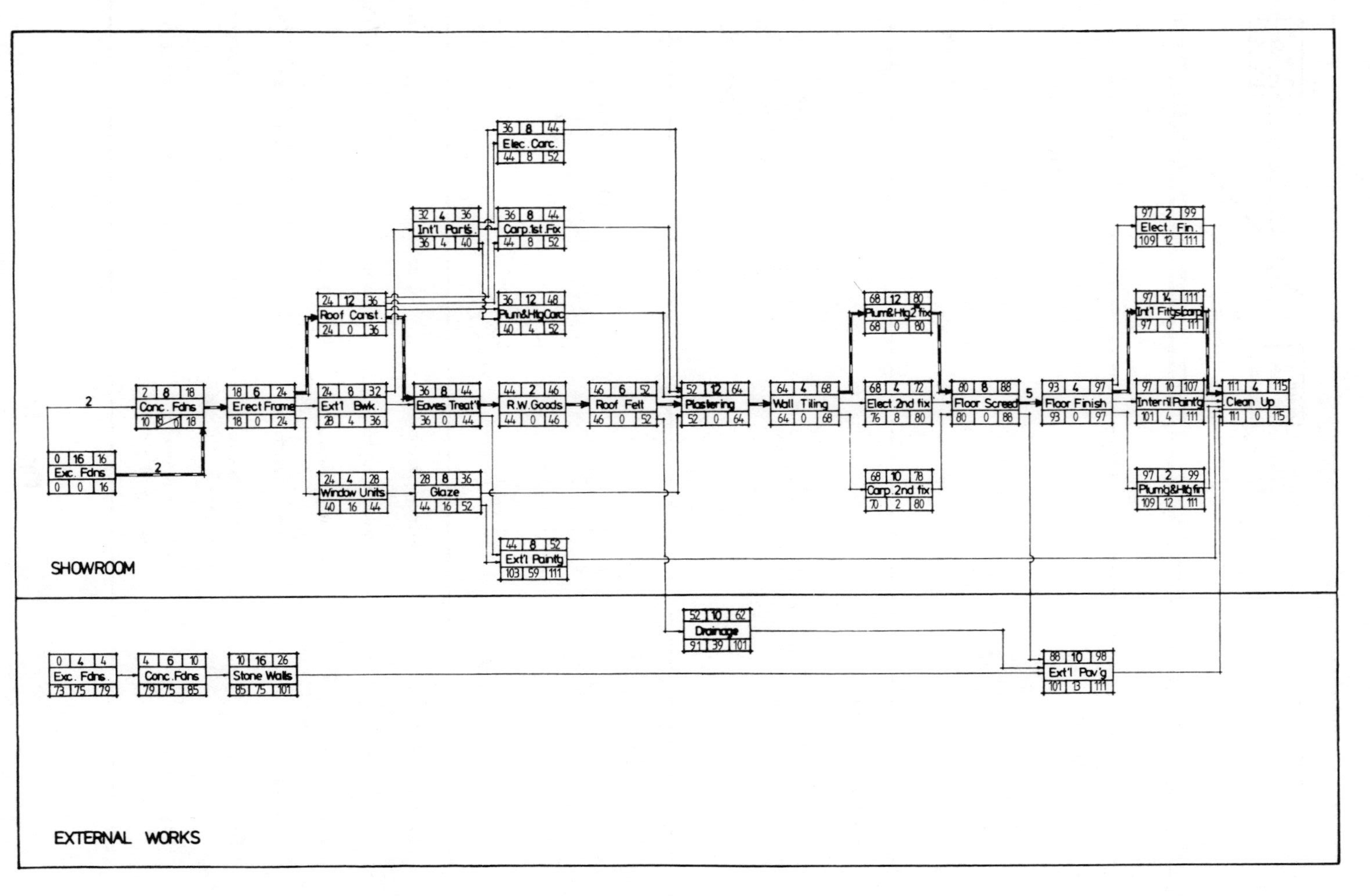
Exc. Fdns
Conc. Fdns
Erect Frame
Roof Const.
Ext1 Bwk.
Window Units
Int'l Parts.
Eaves Treat't
Glaze
Elec. Conc.
Carp. 1st Fix
Plum&Htg Conc
R.W. Goods
Ext'l Paintg
Roof Felt
Plastering
Wall Tiling
Plum&Htg 2nd fix
Elect 2nd fix
Carp 2nd fix
Floor Screed
Floor Finish
Elect. Fin.
Int'l Fittgs&carp
Intern'l Paint'g
Plumb&Htg fin
Clean Up
Drainage
Ext'l Pav'g
Exc. Fdns.
Conc. Fdns
Stone Walls
SHOWROOM
EXTERNAL WORKS

Fig. 2.52

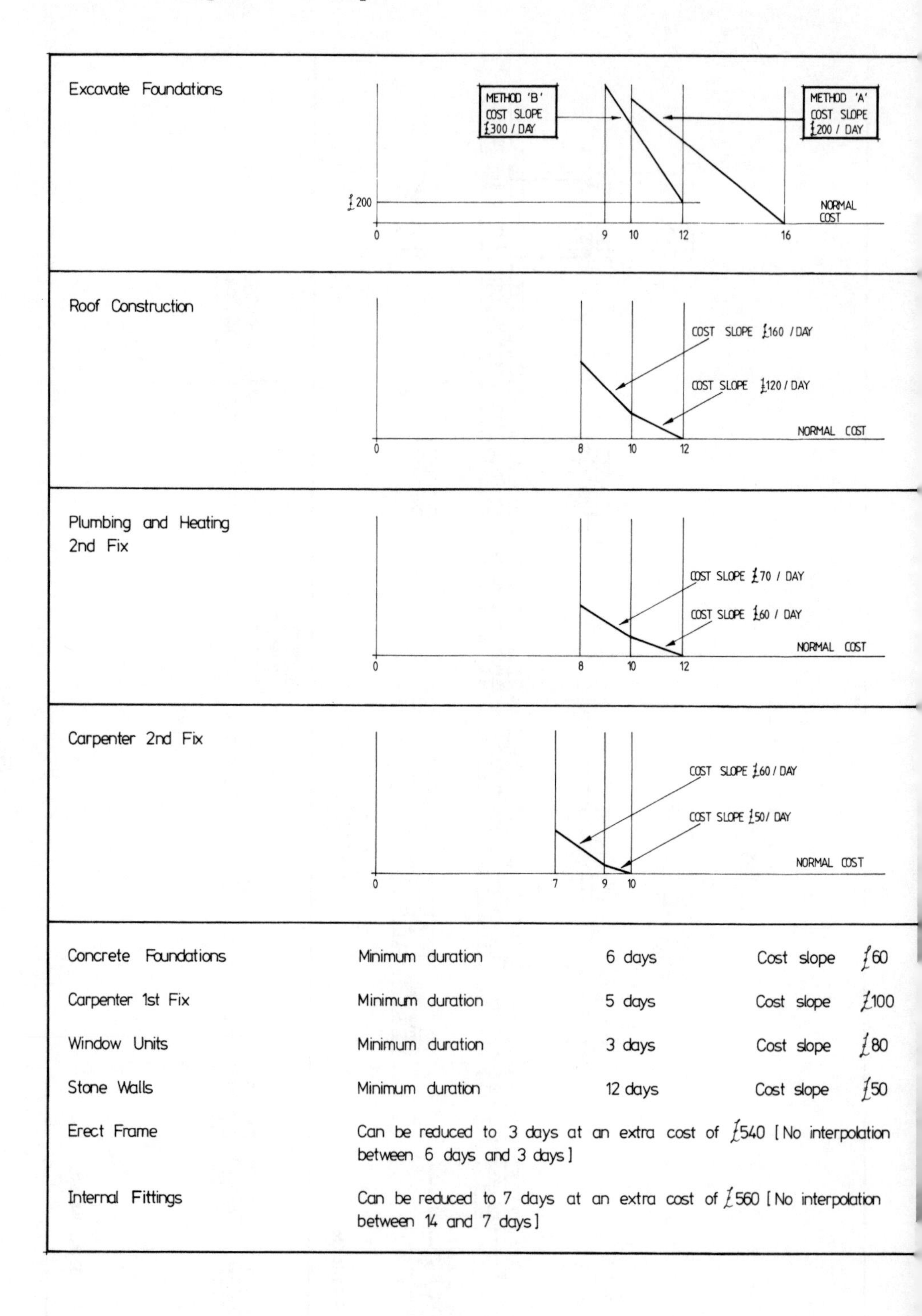
Excavate Foundations
METHOD 'B'
COST SLOPE
£300 / DAY
METHOD 'A'
COST SLOPE
£200 / DAY
£200
NORMAL COST
0
9
10
12
16
Roof Construction
COST SLOPE £160 / DAY
COST SLOPE £120 / DAY
NORMAL COST
0
8
10
12
Plumbing and Heating 2nd Fix
COST SLOPE £70 / DAY
COST SLOPE £60 / DAY
NORMAL COST
0
8
10
12
Carpenter 2nd Fix
COST SLOPE £60 / DAY
COST SLOPE £50/ DAY
NORMAL COST
0
7
9
10
Concrete Foundations
Minimum duration
6 days
Cost slope
£60
Carpenter 1st Fix
Minimum duration
5 days
Cost slope
£100
Window Units
Minimum duration
3 days
Cost slope
£80
Stone Walls
Minimum duration
12 days
Cost slope
£50
Erect Frame
Can be reduced to 3 days at an extra cost of £540 [No interpolation between 6 days and 3 days]
Internal Fittings
Can be reduced to 7 days at an extra cost of £560 [No interpolation between 14 and 7 days]

Fig. 2.53

days and the direct cost of labour and plant is £38 000. Indirect costs are to be charged at £140 per day.

2.13.8.1 Priority of critical activities

Activity	*Max. Comp.*	*Cost*	*Notes*
Alternate method for excavation	4	£200	£50 per day equiv. cost slope
Plumbing and Heating 2nd fix	2	£60/day	
Roof construction	2	£120/day	
Internal fiittings	4	£560 tot	£140 per day equiv. cost slope
Erect frame	3	£540 tot	£180 per day equiv. cost slope
Excavate foundations – Method B	3	£300/day	

The equivalent cost slope has been given for some activities and it should be noted that although internal fittings can be reduced by 7 days for £560, another path becomes critical after a reduction of 4 days and therefore the equivalent cost slope is based on 4 days.

2.13.8.2 First compression
Compress excavate foundations by 4 days using method B at a cost of £200. The duration is now 111 days and the direct cost is £38 000 + £200 = £38 200.

12.13.8.3 Second compression
Compress plumbing and heating 2nd fix by 2 days at a cost of 2 × £60 = £120. The duration is now 109 days and the direct cost is £38 200 + £120 = £38 320.

Two paths are now critical in this area and the combined cost slopes are:

Activity	*Max. Compression*	*Cost per day*
Carpentry 2nd fix	1 day	£ 50
Plumbing and Heating 2nd fix	1 day	£ 70
	Total	£120
For further reduction:		
Carpentry 2nd fix	1 day	£ 60
Plumbing and Heating 2nd fix	1 day	£ 70
	Total	£130

2.13.8.4 Third compression
Compress roof construction by 2 days, carpentry 2nd fix by 1 day and plumbing and heating 2nd fix by 1 day at a cost of 2 × £120 + £50 + £70 = £360.
The duration is now 106 days and the direct cost is £38 320 + £360 = £38 680.
The greater cost slope of £160 per day now applies to roof construction.

2.13.8.5 Fourth compression
Compress carpentry 2nd fix and plumbing and heating 2nd fix by a further 1 day at a cost of £130. The duration is now 105 days and direct cost is £38 680 + £130 = £38 810.

2.13.8.6 Fifth compression
Compress internal fittings by 7 days (only 4 are effective) at a cost of £560. The duration is now 101 days and the direct cost is £38 810 + £560 = £39 370.

2.13.8.7 Sixth compression
Compress roof construction a further 2 days at a cost of 2 × £160 = £320. The duration is now 99 days and the direct cost is £39 370 + £320 = £39 690.

2.13.8.8 Seventh compression
Compress erect frame to 3 days at a total cost of £540. The duration is now 96 days and the direct cost is £39 690 + £540 = £40 230.

2.13.8.9 Eighth compression
Reduce excavate foundations a further 3 days at a cost of 3 × £300 = £900. The duration is now 93 days and the direct cost is £40 230 + £900 = £41 130.

Fig. 2.54 shows the network analysed for 93 days duration.

2.13.8.10 Optimising the cost
To determine the optimum cost of the project, indirect costs must be added as shown in Fig. 2.55. The optimum cost is £53 510 and the optimum project duration is 101 days. Note: The cost is £53 510 for a project duration of 105 days but the shortest duration has been selected as this gives more flexibility for controlling the project during progress of the work.

Fig. 2.56 shows the network analysed for 101 days.

Graphs can now be produced which show the direct cost curve, the indirect cost curve and the total cost curve (see Figs. 2.57 and 2.58).

2.13.8.11 Optimum cost at minimum duration
This can be found directly by following the procedure set out in 2.13.4, but taking the additional factors set out in 2.13.8 into account.

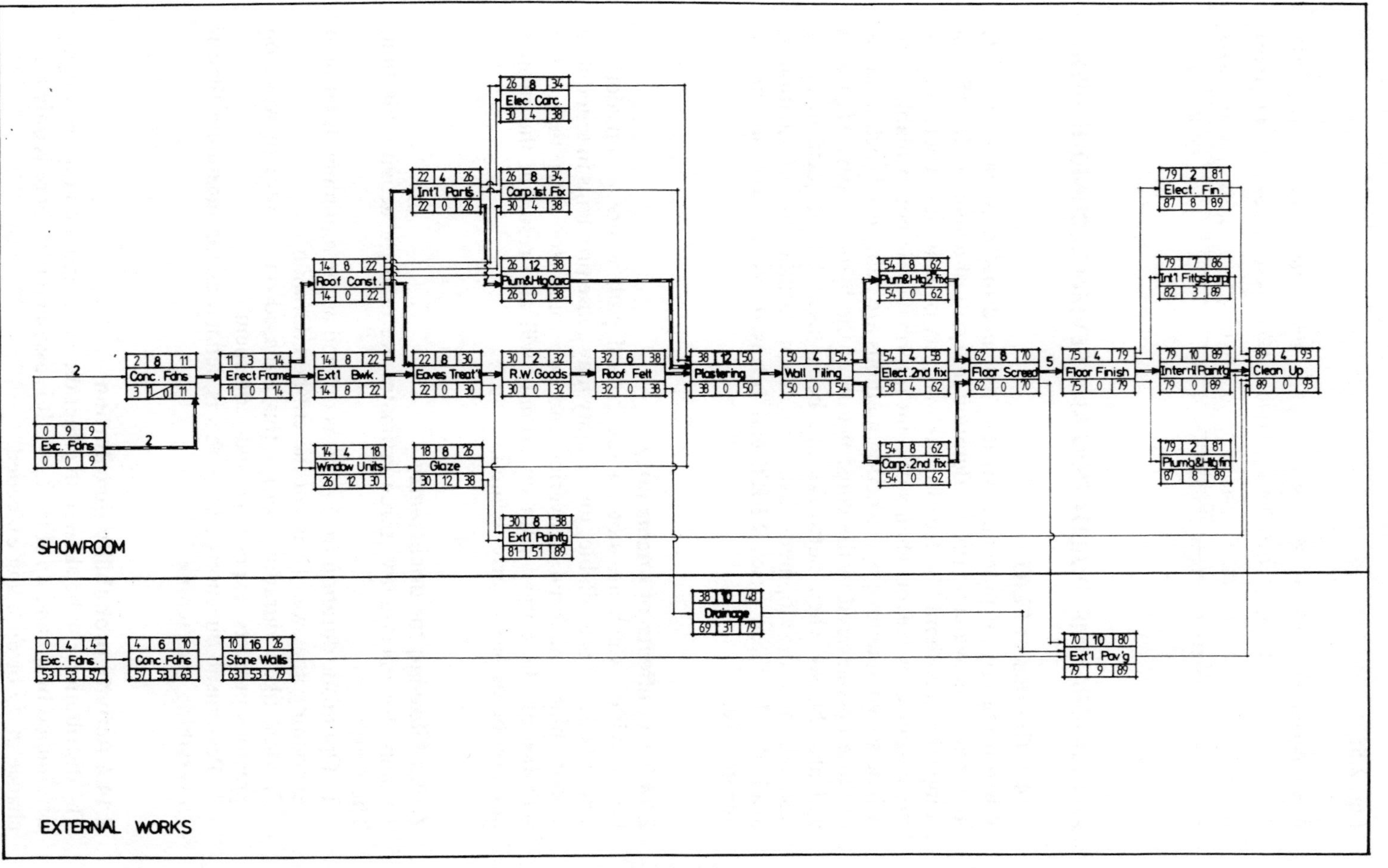
SHOWROOM
EXTERNAL WORKS
Exc. Fdns
Conc. Fdns
Erect Frame
Roof Const.
Ext'l Bwk.
Window Units
Int'l Partis.
Eaves Treat't
Glaze
Elec. Conc.
Carp. 1st Fix
Plum & Htg Conc
R.W. Goods
Ext'l Paintg
Roof Felt
Plastering
Wall Tiling
Plum & Htg 2 fix
Elect 2nd fix
Carp 2nd fix
Floor Screed
Floor Finish
Elect. Fin.
Int'l Fittings
Internl Paint'g
Plumb & Htg fin
Clean Up
Drainage
Ext'l Pav'g
Exc. Fdns.
Conc. Fdns
Stone Walls

Fig. 2.54

Fig. 2.55

Project durations	93	96	99	101	105	106	109	111	115
Indirect costs	13020	13440	13860	14140	14700	14840	15260	15540	16100
Direct costs	41130	40230	39690	39370	38810	38680	38320	38200	38000
Total costs	54150	53670	53550	53510	53510	53520	53580	53740	54100

2.14 PROGRAMME EVALUATION AND REVIEW TECHNIQUE (PERT)

2.14.1 The use of PERT

When using critical path analysis it is assumed that reasonably accurate information is available on the duration and cost of activities. In efficient companies this should be true for the majority of activities. There may be circumstances however when sufficient information is not available. This will be so when unique construction methods are being used such as many of those encountered in the construction of the Sydney Opera House. It will also be so when networks are being used for projects involving research and development work or when insufficient information is available. In these cases PERT can be used, as it takes account of uncertainty.

2.14.2 The effects of uncertainty

For activities which are not on the critical path where a considerable amount of float is available, uncertainty may cause problems in scheduling labour, plant, material deliveries and sub-contractor start times. For activities on the critical path uncertainty will also result in the project duration being less clearly defined.

2.14.3 Allowing for uncertainty

To allow for uncertainty, three estimates are used for activity duration. These are:

1. Optimistic duration (*do*) – this is defined as the minimum duration if everything goes well (it is *not* the crashed duration).
2. Most likely duration (*dm*) – this is based on analysis of work on previous projects, experience and judgement.
3. Pessimistic durations (*dp*) – this is defined as the maximum time if everything goes wrong.

2.14.4 Activity probability distribution

The distribution can be skewed in either direction and the range is roughly determined by *do* and *dp* (Fig. 2.59). It is assumed that there is only a 1% chance of *do* or *dp* being exceeded.

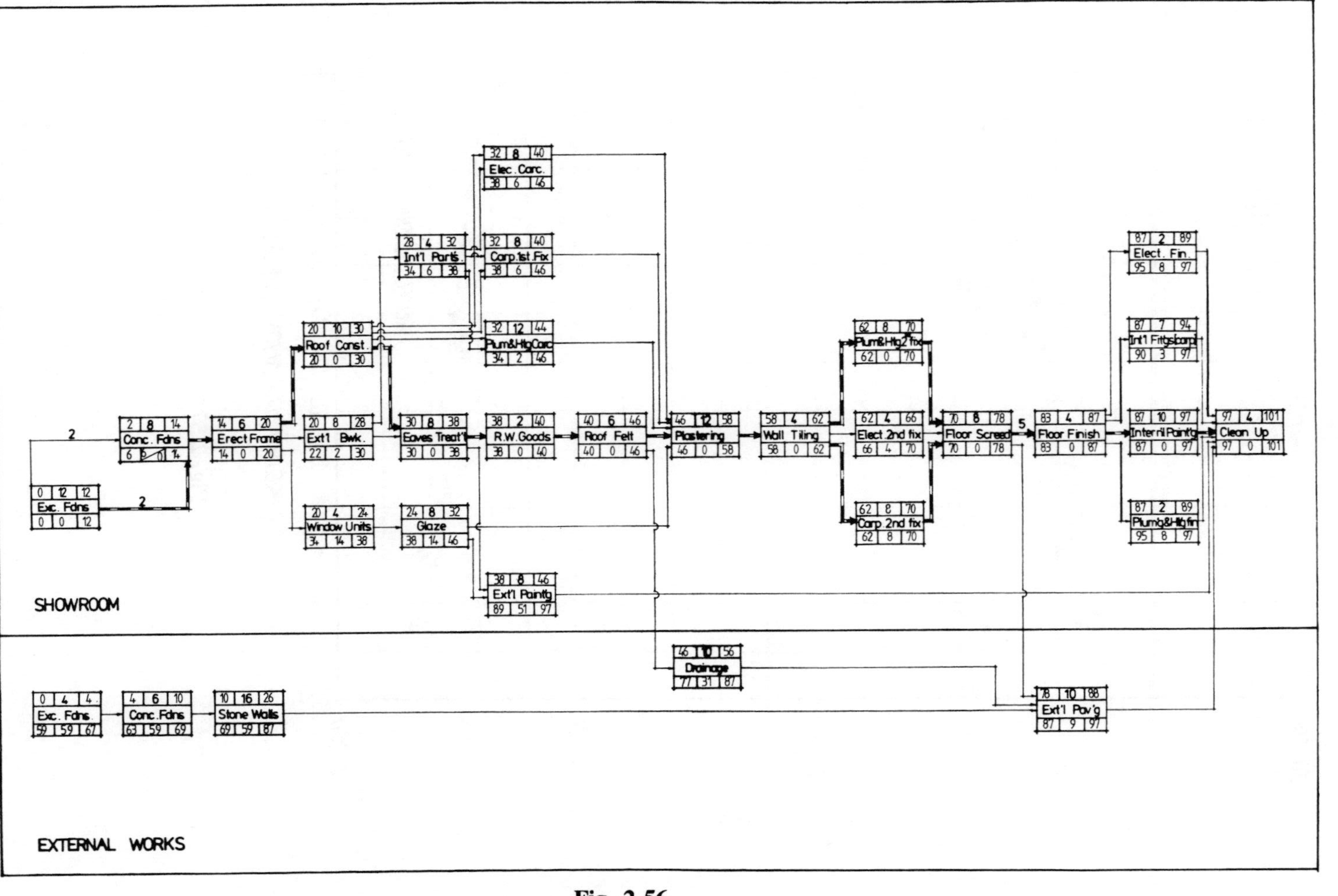
SHOWROOM
Exc. Fdns
Conc. Fdns
Erect Frame
Roof Const.
Ext'l Bwk.
Window Units
Int'l Parts.
Eaves Treat't
Glaze
Elec. Conc.
Carp. 1st Fix
Plum&Htg Conc
R.W. Goods
Ext'l Paint'g
Roof Felt
Plastering
Wall Tiling
Plum&Htg 2nd fix
Elect 2nd fix
Carp 2nd fix
Floor Screed
Floor Finish
Elect. Fin.
Int'l Fittgs(carp)
Intern'l Paint'g
Plumb&Htg fin
Clean Up
EXTERNAL WORKS
Exc. Fdns.
Conc. Fdns
Stone Walls
Drainage
Ext'l Pav'g

Fig. 2.56

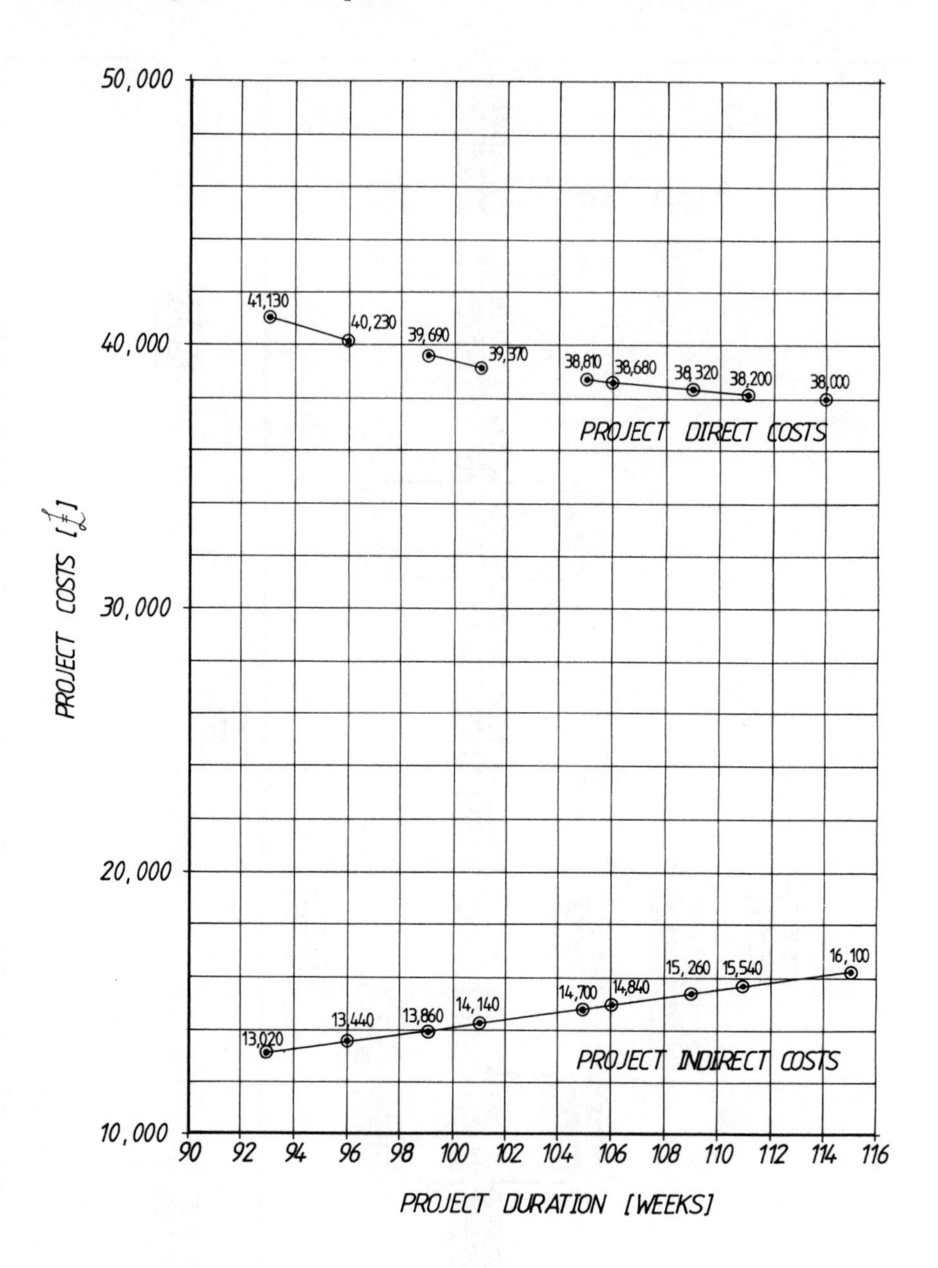
50,000
40,000
30,000
20,000
10,000
PROJECT COSTS [£]
41,130
40,230
39,690
39,370
38,810
38,680
38,320
38,200
38,000
PROJECT DIRECT COSTS
13,020
13,440
13,860
14,140
14,700
14,840
15,260
15,540
16,100
PROJECT INDIRECT COSTS
90 92 94 96 98 100 102 104 106 108 110 112 114 116
PROJECT DURATION [WEEKS]

Fig. 2.57

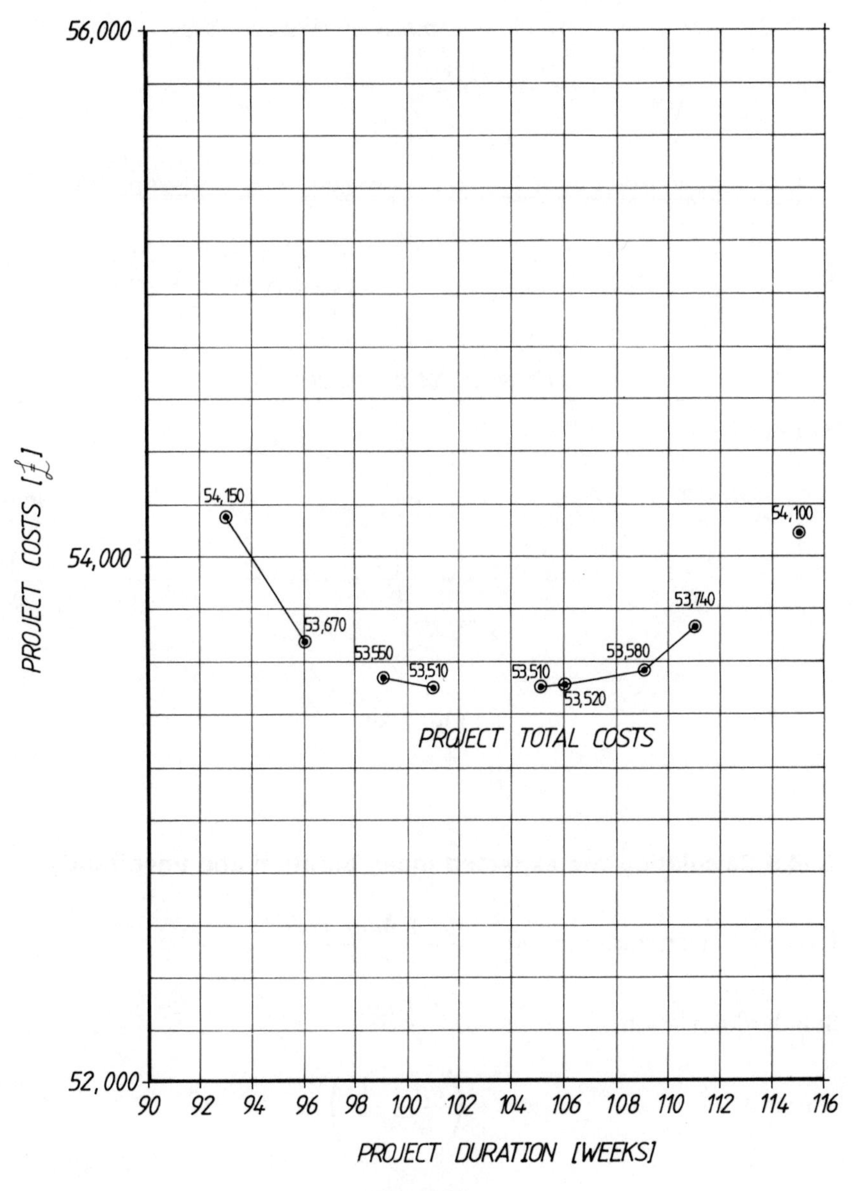
56,000
54,000
52,000
PROJECT COSTS [£]
54,150
53,670
53,550
53,510
53,510
53,520
53,580
53,740
54,100
PROJECT TOTAL COSTS
90 92 94 96 98 100 102 104 106 108 110 112 114 116
PROJECT DURATION [WEEKS]

Fig. 2.58

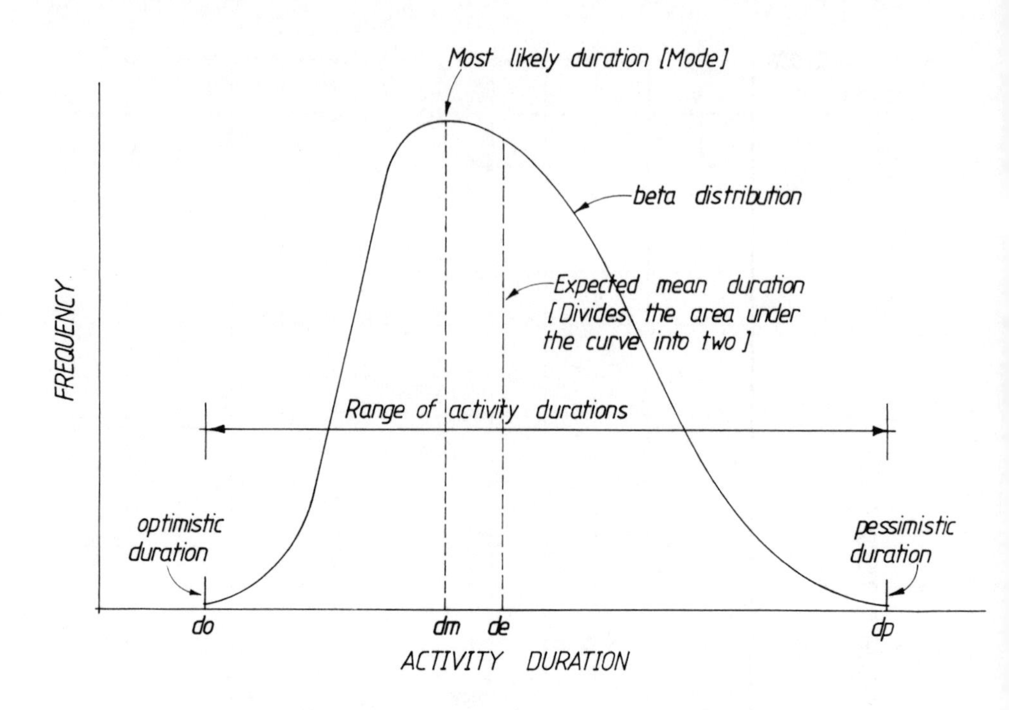

Fig. 2.59

2.14.5 Calculating the expected mean duration and uncertainty

Expected mean duration $de = \dfrac{do + 4dm + dp}{6}$

Standard deviation $\sigma de = \dfrac{dp - do}{6}$

Variance $\nu de = \left(\dfrac{dp - do}{6}\right)^2$

2.14.6 Critical path calculations

Calculations are carried out as set out in Analysis of Arrow Diagrams (p. 108), but the standard deviation and variance are also calculated.

2.14.7 Event variance

1. Based on earliest event times – is the sum of the variances up to the event being considered following the critical path from the start of the project. For non-critical events, it is the sum of the variances at each

event, following the path which determined the earliest event time.
2. Based on latest event times – is the sum of the variances up to the event being considered following the critical path from the expected completion date. This variance gives the uncertainty remaining from the event being considered up to the end of the project measured along the critical path. For non-critical events it is the sum of the variances of each event, following the path which determined the latest event times.

2.14.8 Multiple critical paths

Where there is more than one critical path, the one with the maximum variance is used.

2.14.9 Meeting specific event schedule times

Event completion times are assumed to have a normal probability distribution with a mean value of Tx, the variance of Tx being calculated as stated in 2.14.7, and from this the standard deviation can be calculated. This assumption is strictly correct only for an infinite series but is assumed to be accurate enough in practice for large projects. The event times in the example which follows are assumed to have normal distributions. To calculate the probability of achieving a specific event time, the area cut off by the specified time is calculated as a percentage of the total area beneath the curve (Fig. 2.60). Standard probability tables can be used for normal

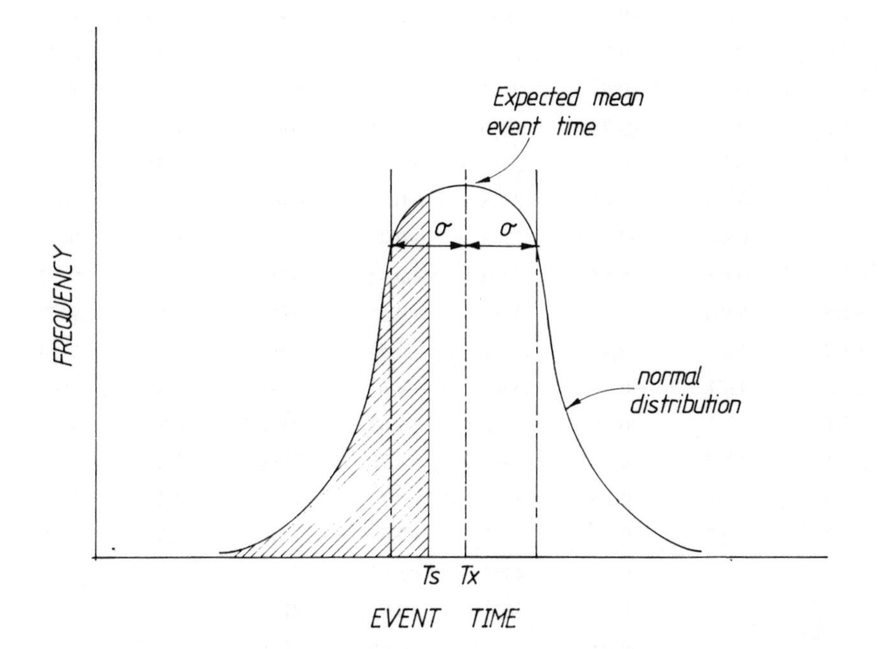

Fig. 2.60

distribution functions and a summarised table is given in Fig. 2.61. This table is accurate enough for construction projects.

The factor Z is calculated thus:

$$Z = \frac{Ts - Tx}{\sigma Tx}$$

Using this value the probability of meeting the specific event scheduled time (Ts) can be read off from the table interpolating as necessary.

2.14.10 Determining a scheduled time based on a specific probability

The scheduled time for a specific probability can be established by transposing the formula thus:

$$Ts = Tx + Z\sigma Tx$$

2.14.11 Applying PERT to a construction project

Fig. 2.62 shows a building project and Fig. 2.63 shows the three estimated

Fig. 2.61 Standard normal distribution

z	*probability*	*z*	*probability*	*z*	*probability*	*z*	*probability*
−2.8	.0062	−0.95	.1711	+0.05	.5799	+1.0	.8413
−2.2	.0139	−0.9	.1841	+0.1	.5398	+1.05	.8531
−2.0	.0223	−0.85	.1977	+0.15	.5596	+1.1	.8643
−1.9	.0287	−0.8	.2119	+0.20	.5793	+1.15	.8749
−1.8	.0359	−0.75	.2266	+0.25	.5987	+1.2	.8849
−1.7	.0446	−0.7	.2412	+0.3	.6179	+1.25	.8944
−1.6	.0548	−0.65	.2578	+0.35	.6768	+1.3	.9032
−1.5	.0668	−0.6	.2743	+0.4	.6554	+1.4	.9192
−1.4	.0808	−0.55	.2912	+0.45	.6736	+1.5	.9332
−1.3	.0968	−0.5	.3085	+0.5	.6915	+1.6	.9452
−1.25	.1056	−0.45	.3264	+0.55	.7088	+1.7	.9554
−1.2	.1151	−0.4	.3446	+0.6	.7257	+1.8	.9641
−1.15	.1251	−0.35	.3632	+0.65	.7422	+1.9	.9713
−1.1	.1357	−0.3	.3821	+0.7	.7588	+2.0	.9777
−1.05	.1469	−0.25	.4013	+0.75	.7734	+2.2	.9861
−1.0	.1587	−0.2	.4207	+0.8	.7881	+2.8	.9938
		−0.15	.4404	+0.85	.8023		
		−0.10	.4602	+0.9	.8159		
		−0.05	.4801	+0.95	.8289		
		−0	.5				

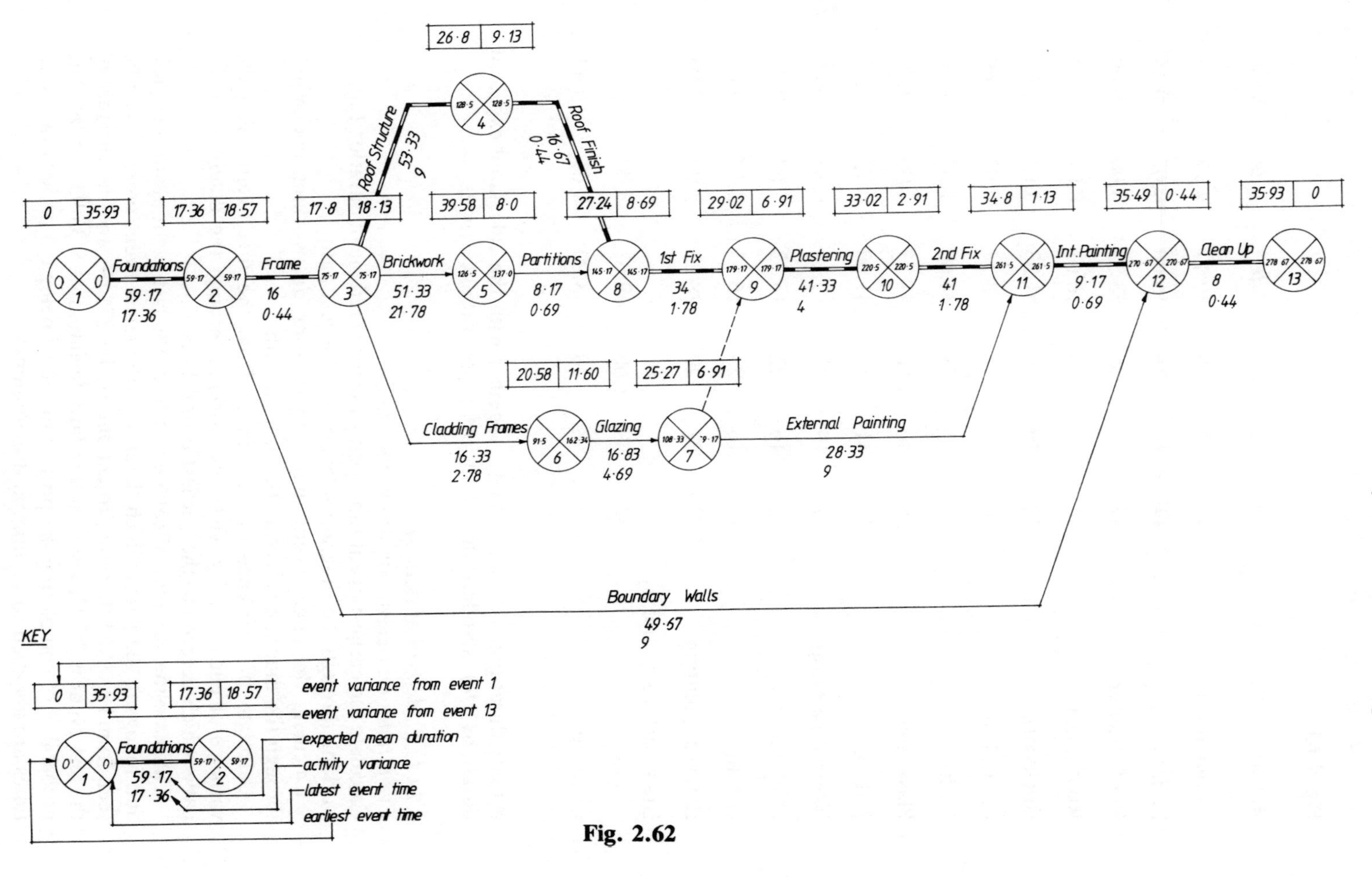
26·8 | 9·13
Roof Structure
53·33
9
Roof Finish
16·67
0·44
0 | 35·93
17·36 | 18·57
17·8 | 18·13
39·58 | 8·0
27·24 | 8·69
29·02 | 6·91
33·02 | 2·91
34·8 | 1·13
35·49 | 0·44
35·93 | 0
Foundations
59·17
17·36
Frame
16
0·44
Brickwork
51·33
21·78
Partitions
8·17
0·69
1st Fix
34
1·78
Plastering
41·33
4
2nd Fix
41
1·78
Int. Painting
9·17
0·69
Clean Up
8
0·44
20·58 | 11·60
25·27 | 6·91
Cladding Frames
16·33
2·78
Glazing
16·83
4·69
External Painting
28·33
9
Boundary Walls
49·57
9
KEY
event variance from event 1
event variance from event 13
expected mean duration
activity variance
latest event time
earliest event time

Fig. 2.62

Fig. 2.63

Activity	*do*	*dm*	*dp*	*de*	*σde*	*√de*
Foundations	45	60	70	59.17	4.17	17.36
Frame	14	16	18	16.00	0.67	0.44
Roof structure	47	52	65	53.33	3.00	9.00
Roof finish	14	17	18	16.67	0.67	0.44
Brickwork	42	49	70	51.33	4.67	21.78
Partitions	6	8	11	8.17	0.83	0.69
1st fix	30	34	38	34.00	1.33	1.78
Plastering	38	40	50	41.33	2.00	4.00
2nd fix	37	41	45	41.00	0.75	1.78
Internal painting	7	9	12	9.17	0.83	0.69
Cladding frames	12	16	22	16.33	1.67	2.78
Glazing	12	16	25	16.83	2.17	4.69
External painting	22	27	40	28.33	3.00	9.00
Brickwork to boundary walls	42	49	60	49.67	3.00	9.00
Clear up	6	8	10	8.00	0.67	0.44

activity durations for each activity, together with the calculated expected durations (*de*), standard deviation (*σde*), and variance (*vde*).

2.14.11.1 Method of analysis
Event times and event variances (Fig. 2.62).
Calculate the earliest event times using the expected mean duration (*de*) by working through the diagram selecting the longest path.

Calculate the latest event times, again using *de*, working backwards through the diagram selecting the longest path.

Calculate the variance of each event from the first event using the variance of each activity. (*vde*) by working through the diagram on the paths which determine the earliest event times.

Calculate the variance of each event from the last event again using *vde* and following the paths which determine the latest event times. It can be seen from Fig. 2.62 that the project duration is 278.67, and the variance of the last event is 35.93, giving a standard deviation of $\sqrt{35.93} = 5.99$. The probability of completing the project in 278.67 being 0.5 or 50% (as event times are assumed to be normal distributions).

2.14.11.2 Probability of achieving other project durations

To calculate the probability of completing the project in 270, first calculate the factor Z

$$Z = \frac{270 - 278.67}{5.99} = -1.45$$

From Fig. 2.61. Probability is .07 or 7%.

2.14.11.3 Project duration for a specific probability

To calculate the project duration if it is required to be 90% certain of completion on time; first look up the factor Z for 90% probability. This equals 1.3

$$\begin{aligned} Ts &= Tx + Z\sigma Tx \\ &= 278.67 + 1.3 \times 5.99 \\ &= 286.46 \end{aligned}$$

to be 90% sure of completion on time
project duration = 286.46

2.14.11.4 Project duration determined by paths other than the critical path

If a particular string of activities has very little float and a substantial standard deviation, this string could become critical. The probability of the string becoming critical can be calculated.

The calculations become very complex when three or more strings could become critical and will not therefore be discussed in this introduction to PERT.

In Fig. 2.62 string 3-5-8 may become critical. To calculate the probability of this, calculations start from the branching event (3).

The difference between two independent normal distributions results is another normal distribution whose variance is determined by adding the variances of the two component distributions. It is assumed that the two strings are approximately normal distributions for the sake of illustration.

Fig. 2.64 shows the two distributions.

The variance from event 3 on 3-5-8 is 22.47 $\therefore \sigma = 4.74$
The variance from event 3 on 3-4-8 is 9.44 $\therefore \sigma = 3.07$
Sum of variance = 22.47 + 9.44 = 31.91

$$\sigma = 5.65$$

$$Z = \frac{59.50 - 70.00}{5.65} = -1.86$$

probability = 0.03 or 3%

There is therefore a 3% probability that path 3-5-8 will affect the project duration.

Fig. 2.65 shows the probability curve for the time difference.

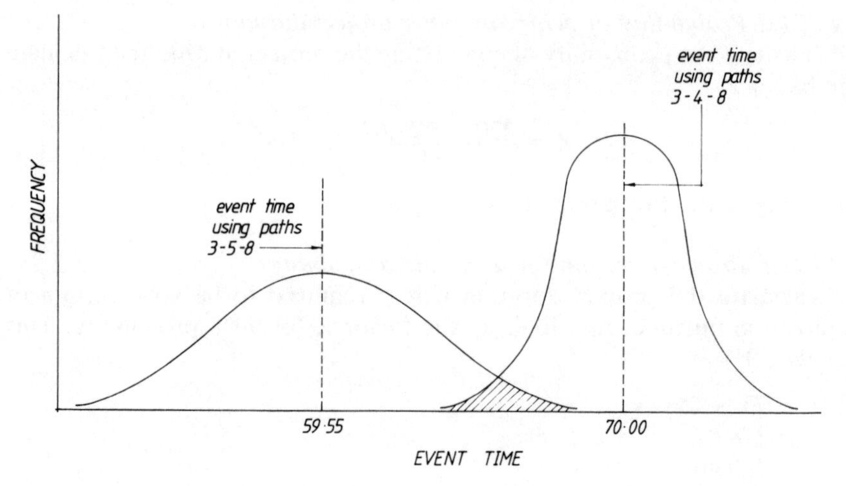

Fig. 2.64

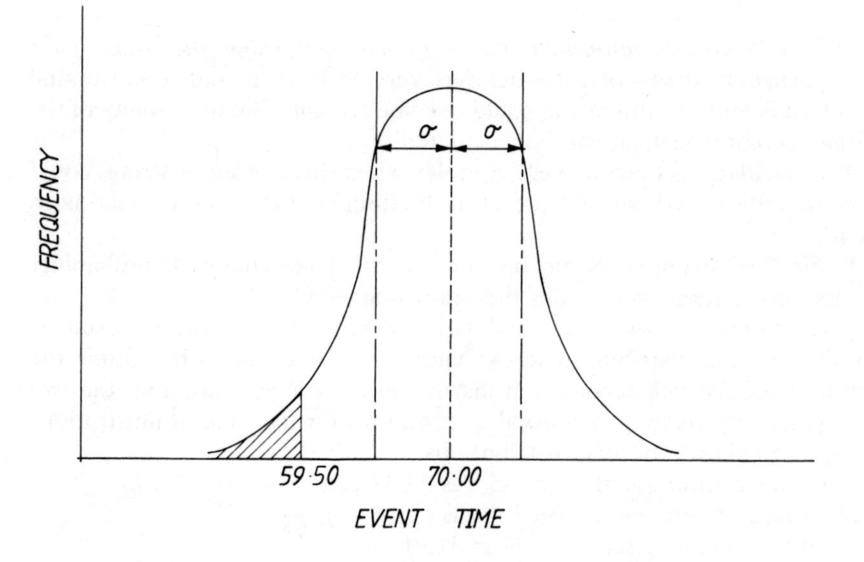

Fig. 2.65

3 Work Study

3.1 INTRODUCTION

The human aspect is a most important part of work study, both on site and at higher management level, and the subject consists just as much of human relations (i.e.problems of communication and eliciting co-operation) as it does of techniques. In this chapter only the application of work study techniques will be considered; for a more detailed consideration of the subject as a whole, other texts listed in the bibliography should be consulted.

3.2 PURPOSE OF WORK STUDY

The purpose of work study is the provision of factual data to assist management in making decisions and to enable them to utilize with the maximum of efficiency all available resources (i.e. labour, plant, materials and management) by applying a systematic approach to problems instead of using intuitive guesswork.

3.3 BREAKDOWN OF WORK STUDY

Work study has two main aspects, method study and work measurement, which are very closely related. For convenience they will be considered separately in this text, but their interdependence must be appreciated at all times.

3.4 METHOD STUDY

The aim of method study is to provide information which will assist management in taking decisions related to the method it is proposed to use by making a systematic analysis of a problem and developing alternative methods, thus determining the optimum layouts and the most effective use of resources.

There are six basic steps involved in carrying out a method study:

1. Select the work to be studied and define the problem.
2. Record the relevant facts using the recording techniques.
3. Examine these facts critically and without bias to ascertain whether each element in the work is necessary, asking a series of questions and examining alternative solutions.
4. Develop the best method from the alternatives and submit this to management for approval.
5. Install the new method. Installation follows a decision by management to accept the method.

6. Maintain the new method. This consists of checking that the new method is adhered to by regular inspection on site or by watching output records which will indicate deviations from the standards set.

3.4.1 Recording techniques

There are three main groups of recording techniques available for setting down a problem. These are:

1. Charts	Outline process charts Flow process charts (man-type, material-type and equipment-type) Two-handed process charts Multiple activity charts
2. Diagrams and models (Two-and three-dimensional)	Flow diagrams String diagrams Cut-out templates (two-dimensional models) Models (three-dimensional)
3. Photographic	Photographs Films

Templates, i.e. silhouettes of plant, equipment, storage areas etc., are very useful when considering layouts, and it is also possible to obtain charting kits, which consist of translucent sheets and stick-on templates that may be used for die-line prints. This is an excellent method of recording when it is desired to compare one method with another.

3.4.1.1 Flow process charts

This technique can be useful in helping to solve problems of layout such as those in site workshop areas where the operations are likely to be repetitive.

All activities are shown by means of symbols with a description against each. This is a very simple method of showing a sequence of work, and employs the following symbols:

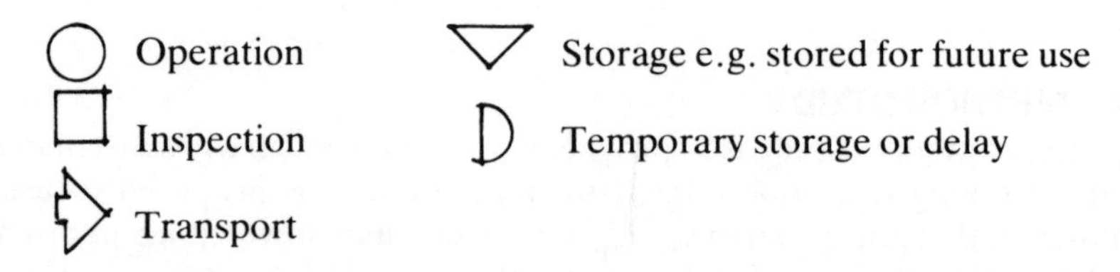

The movements of men, materials or equipment are followed through a process and the symbols are used to indicate what is happening at the various stages. The distance travelled may be shown on the transport activities and the time for each element can be given.

3.4.1.2 Flow diagrams

A flow diagram can be used in conjunction with a flow process chart to show where the activities take place. The same symbols that are employed on flow process charts are used but in this case they are superimposed onto a drawing and the descriptions are not necessary. All movements and distances can be clearly shown.

3.4.1.3 String diagrams

This technique is very useful in solving problems of movement. It is applied to repetitive situations and is therefore most useful in working areas such as factories producing industrialized components, machine shops, pre-casting yards, steel bending areas on site etc. The diagram will show up points of congestion and any excessive distances travelled.

The procedure for improving a layout is to first draw to scale a plan of the area under consideration with the work places or stacking areas etc., and all changes in direction denoted by pins. String or thread is then tied round the starting point and passed from pin to pin showing movement. Men, materials or machines can be denoted by different coloured string. If the string is then measured, any excessive distances travelled will be obvious and any points of congestion will be seen on the diagram.

Routes which are travelled regularly should be kept as short as possible, and alternative methods can be examined to obviate the faults of the first (templates can be very useful here).

3.4.2 Critical examination

The step examine is the key step in a method study, and consists of a detailed examination of every aspect of the work.

The purpose is to:

1. Establish the true facts surrounding the problem.
2. Establish the reasons for these facts and determine whether they are valid.
3. On this foundation to consider all the possible alternatives and hence the optimum solution.

Examination is carried out by a questioning technique for which a critical examination sheet is very useful (Fig. 3.1). There should be liaison with the site management personnel who will carry out the work and they should be encouraged to put their ideas forward.

3.4.3 Site layout problems

When deciding on the relative positions of plant, working areas, storage areas etc. on site, reference has to be made to the various activities to be performed, i.e. the number and types of activity and the minimum transports and storages required. This basic information can be provided from materials flow process charts, which may be used in conjunction with a scale drawing of the site layout and paper templates. String diagrams can then be used either to evaluate or to visually represent the movement intensity for the alternative arrangements of the templates. These methods were used in the selection of the site layout in chapter 1 (Fig. 1.44).

3.4.4 Steel cutting and bending area

Flow charts, flow diagrams and string diagrams are used to assist in considering the layout of the steel bending area in the site for the amenity centre and office block (Fig. 1.44).

The space required for storing steel in the form it is delivered and after it has been prepared will be determined by the amount of steel involved and

CRITICAL EXAMINATION SHEET

DESCRIPTION OF OPERATION:
DESCRIPTION OF ELEMENT:

INVESTIG'N	PRIMARY QUESTION	SECONDARY QUESTION	POSSIBLE ALTERNATIVES	SELECT ALTERNATIVES
purpose	what is acheived	is it necessary if so why	what else could be done	what should be done
place	where is it done	why there	where else could it be done	where should it be done
sequence	when is it done	why then	when else could it be done	when should it be done
person	who does it	why him	who else could do it	who should do it
means	how is it done	why that way	how else could it be done	how should it be done

Fig. 3.1

the timing of deliveries. A power bender will be required, but a hand cutter should be adequate for cutting purposes as large diameter bars and bars for the floor slab are being purchased cut to length.

The site manager has suggested the layout shown in Fig. 3.2 and a material-type flow diagram has been superimposed on it; a flow process chart has also been drawn up for this arrangement (Fig. 3.3).

The chart and diagram are now critically examined to find improvements, and from the flow diagram (Fig. 3.2) it can be seen that a fair amount of movement is involved in transporting steel to and from the various positions where operations would take place. A string diagram is therefore produced to show the movement of materials for one day (Fig. 3.4) and it becomes evident that many long distances are travelled an excessive number of times.

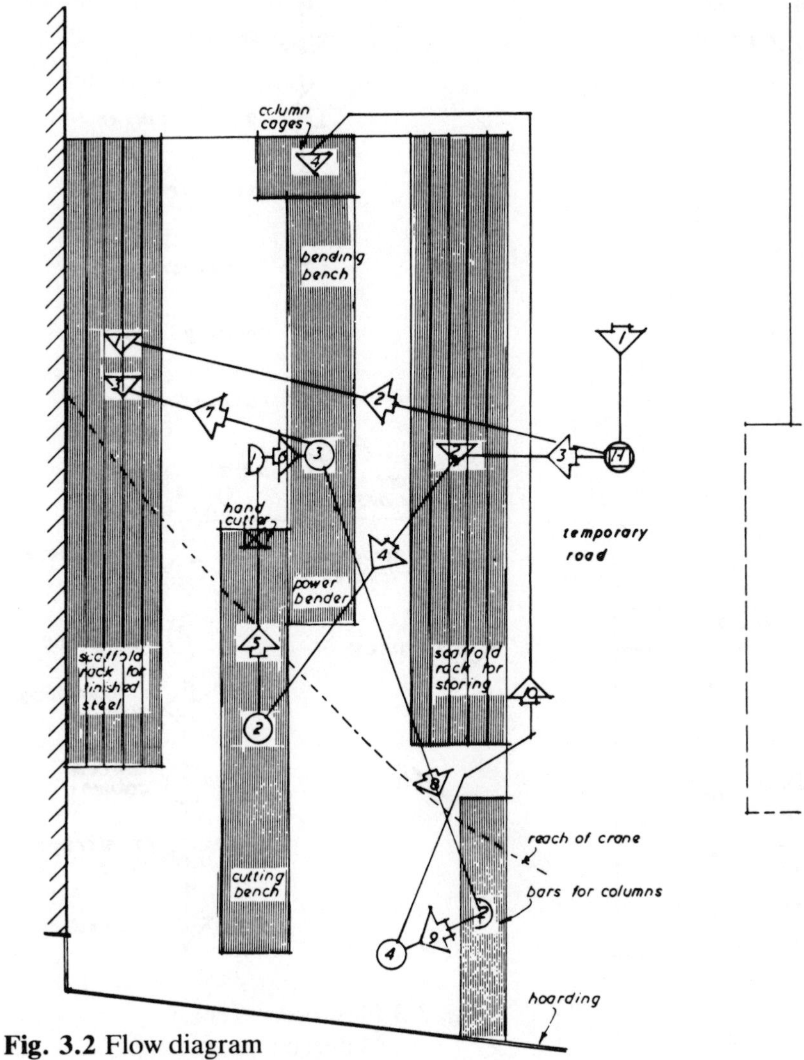

Fig. 3.2 Flow diagram
Cutting and bending area

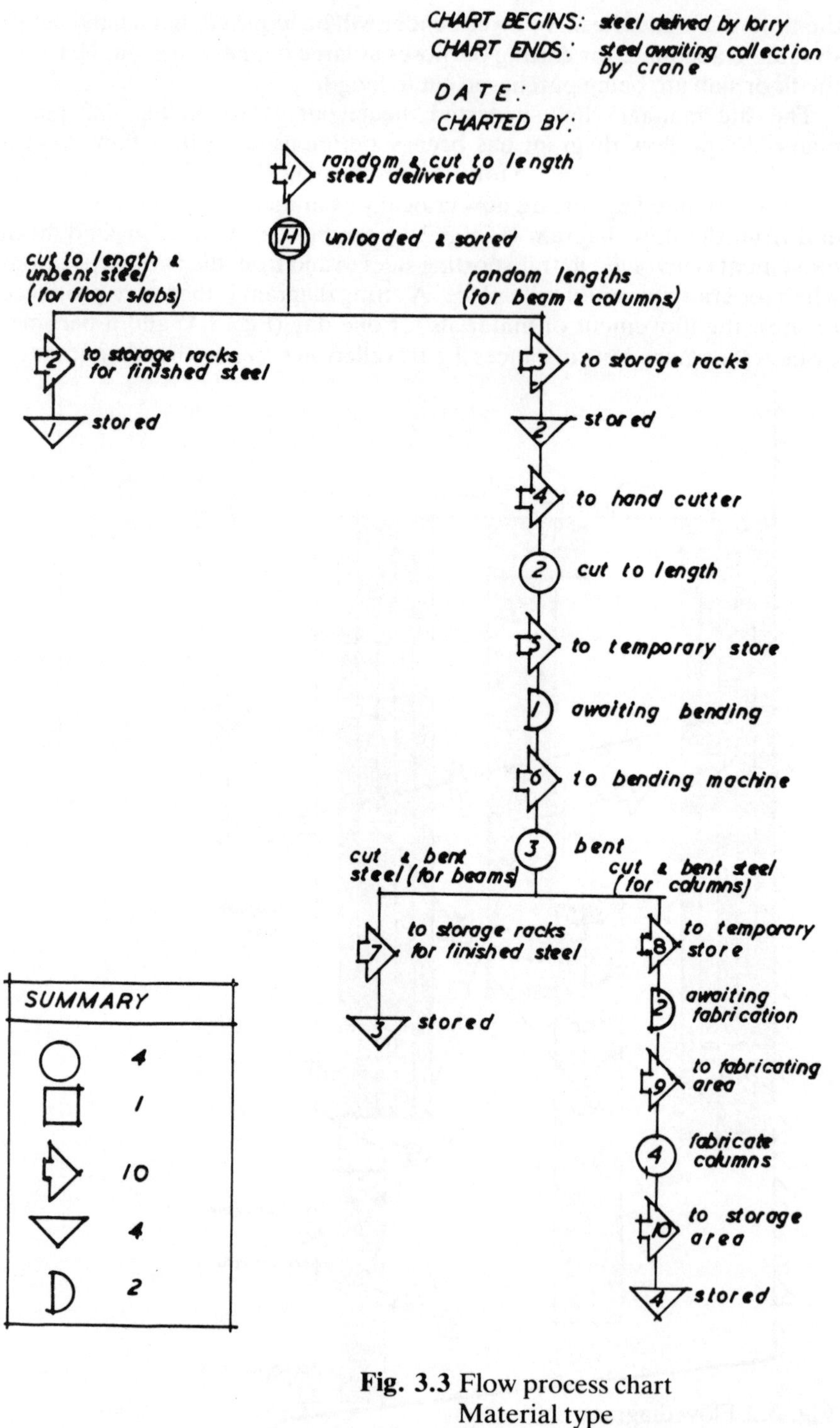

Fig. 3.3 Flow process chart
Material type
Cutting and bending area

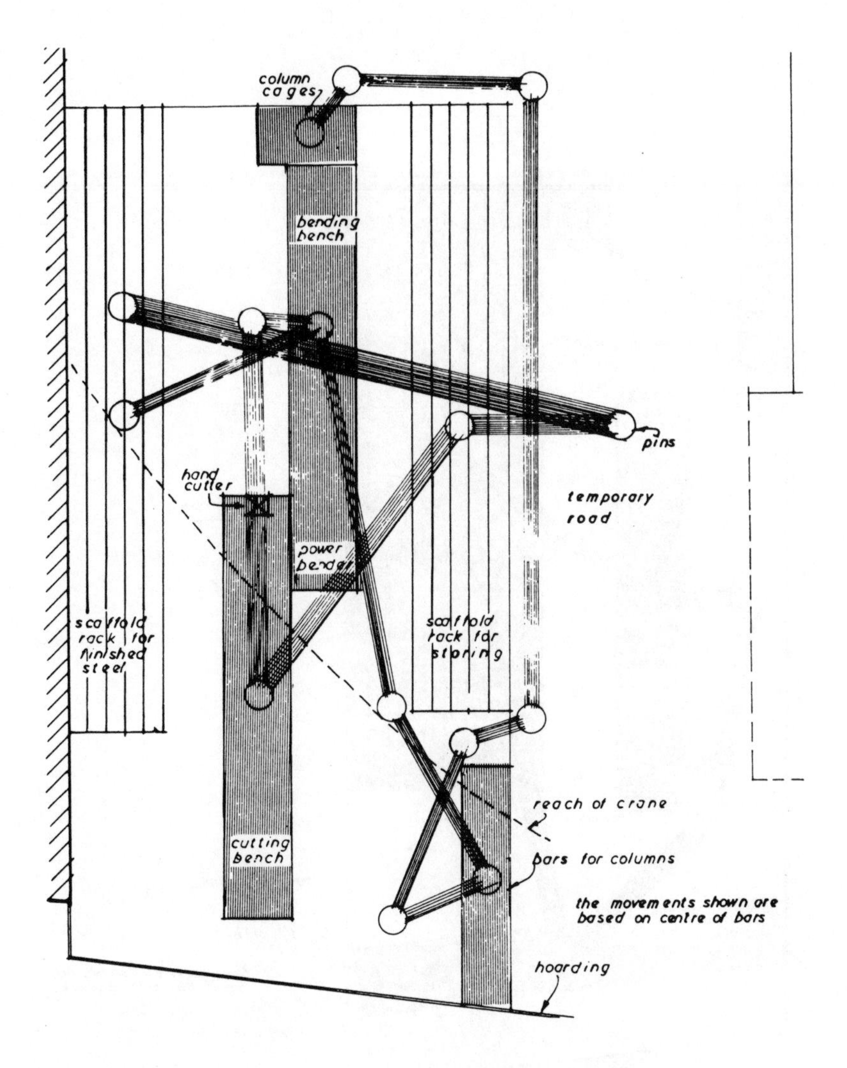

Fig. 3.4 String diagram
Cutting and bending area

Alternative layouts are examined to eliminate long transports on frequently used routes, and Fig. 3.5 shows the string diagram for the selected layout, with shortening of the routes travelled most frequently; the flow diagram for this layout is shown in Fig. 3.6. If required, different coloured string can be used to show the movement of bars cut to length on delivery and those prepared on site etc.

Had the layout suggested by the site manager not been put forward, the first stage would have been to produce an outline process chart, followed by flow process charts, flow diagrams, and string diagrams.

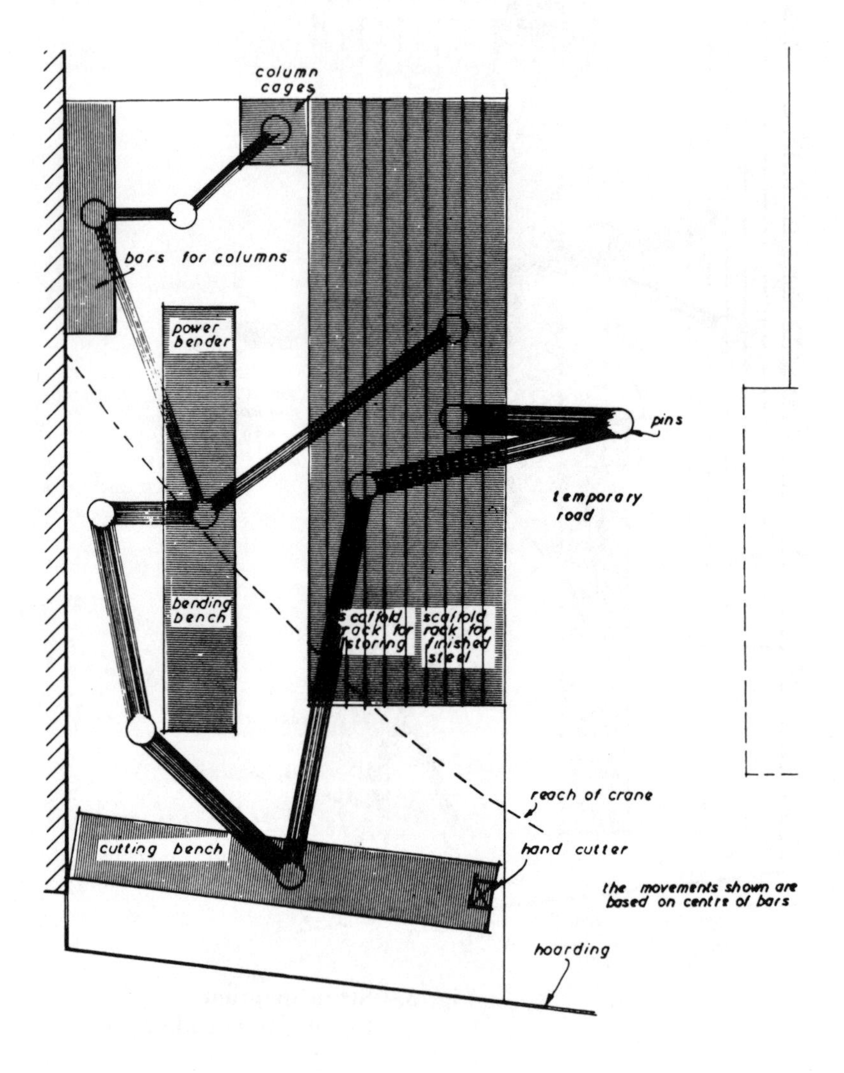

Fig. 3.5 String diagram
Cutting and bending area

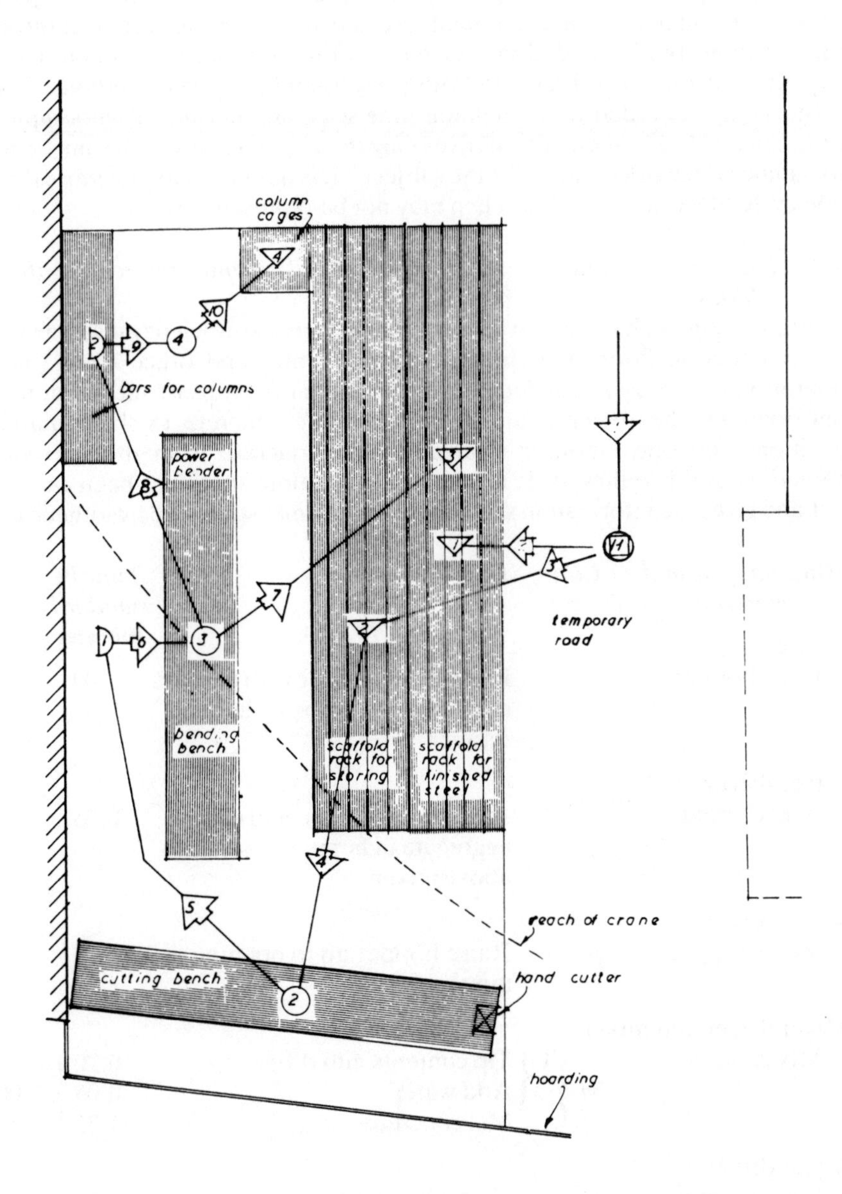

Fig. 3.6 Flow diagram
Cutting and bending area

3.4.5 Multiple activity charts

This technique is used to help solve problems when a number of subjects (operatives, machines or equipment) are dependent on each other. It shows the occupied time (divided into elements of work if necessary) and unoccupied time for the subjects in both present and proposed methods. The subjects are recorded on a common time scale and periods of unoccupied time can be readily seen; alternatives may then be considered, the aim being to balance the work content for the subjects. It is necessary to plot more than one cycle of work as the first taken may not be representative.

3.4.5.1 Laying the concrete upper floor slabs of amenity centre and office block

Multiple activity charts are to be used to help increase efficiency in laying the upper concrete floor slabs in the amenity centre and office block. The operation is at present carried out by two men at the mixer (one driver and one loader), the tower crane transporting the concrete to the required position, and one labourer discharging the concrete from the skip and spreading and levelling it. It is assumed that a time study has been carried out and gives the information shown below (the times stated are hypothetical).

Man/machine and element	*Code*	*Description*	*Time in standard minutes*	
Loader				
Load hopper	L	Load mixer hopper with fine and coarse aggregate using hand scraper	2.00	
Mixer driver				
Add cement	ct	Add cement to fine and coarse aggregate in hopper (from silo above mixer)	0.25	
Mixer driver				
Raise hopper	R	Raise hopper up to opening in mixer drum	0.25	
Mixer driver and mixer				
Mix concrete	M { T {	Tip contents into drum	0.20	2.00
		Add water	0.05	
		Mix concrete	1.75	
Mixer driver				
Lower hopper	H	Lower hopper from opening in mixer drum down to ground level	0.25	
Mixer driver				
Discharge concrete	D	Position crane skip and discharge contents of mixer drum into it	0.50	

Crane and driver Deliver concrete	DC	Lift concrete in skip, slew and deliver concrete to placing position	1.50
Spreader Discharge concrete	DL	Discharge concrete from crane skip	0.25
Spreader Spread concrete	SC	Spread and vibrate concrete between reinforcing steel and tamp	5.50
Crane and driver Return empty skip	RE	Return empty skip from placing position to concrete mixer	1.50

3.4.5.2 Method 1

The multiple activity chart (Fig. 3.7) shows the operation as carried out at present. One man loads the hopper on the mixer with sand and coarse aggregate. The mixer driver checks the weight on a dial and discharges the cement from the silo above the hopper; he next elevates the hopper and discharges the contents into the mixer drum, feeds water into the drum, and lowers the hopper. When the concrete is mixed, the mixer driver places the skip in position and then discharges the concrete from the mixer into it; the crane then lifts the concrete in the skip and transports it to the desired position. Finally a spreader discharges the skip and spreads the concrete, the crane returning with empty skip to the mixer.

The chart shows that the time cycle of 5.75 standard minutes is determined by the spreader, and this results in unoccupied time for other operatives and machines.

3.4.5.3 Method 2

The multiple activity chart (Fig. 3.8) shows an alternative method. Two spreaders are used here to discharge and spread concrete from the crane skip, so that the crane now determines the time cycle of 3.75 standard minutes – a marked increase in efficiency.

3.4 5.4 Method 3

The multiple activity chart (Fig. 3.9) shows another alternative. This procedure is similar to method 2 but the mixer driver also does the loading. The cycle time remains the same, and this results in a further increase in efficiency.

3.4.5.5 Other methods

Additional alternatives could be tried and one possibility is to use two crane skips, which might cut down some of the time at present wasted whilst the crane waits for the skip to be filled at the mixer. There would of course be extra time involved in attaching and detaching the skip, etc.

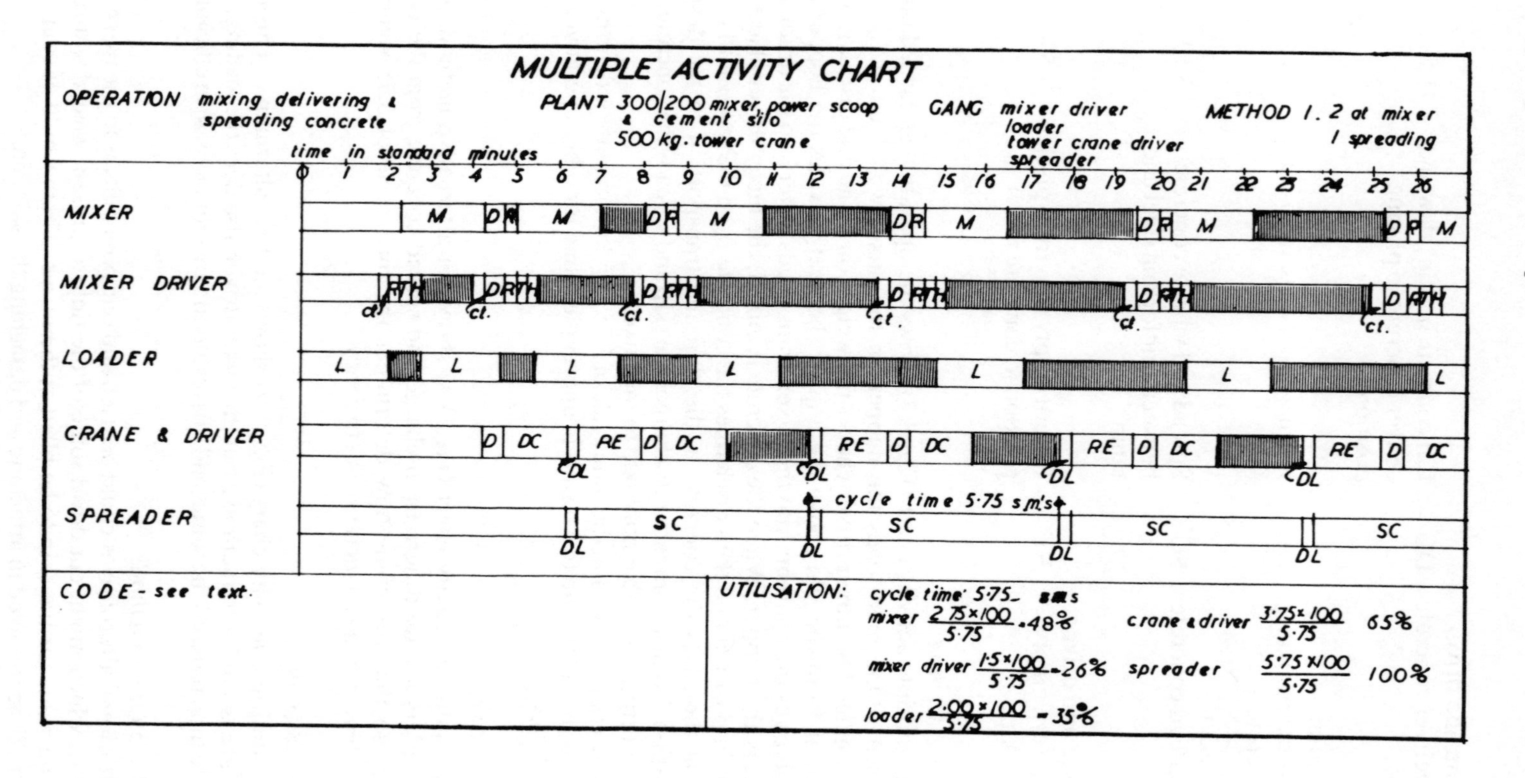

MULTIPLE ACTIVITY CHART
OPERATION mixing delivering & spreading concrete
PLANT 300/200 mixer, power scoop & cement silo
500 kg. tower crane
GANG mixer driver
loader
tower crane driver
spreader
METHOD 1. 2 at mixer
1 spreading
time in standard minutes
MIXER
MIXER DRIVER
LOADER
CRANE & DRIVER
SPREADER
cycle time 5·75 s m's
CODE - see text.
UTILISATION: cycle time 5·75
mixer 2·75×100/5·75 = 48%
crane & driver 3·75×100/5·75 65%
mixer driver 1·5×100/5·75 = 26%
spreader 5·75×100/5·75 100%
loader 2·00×100/5·75 = 35%

Fig. 3.7

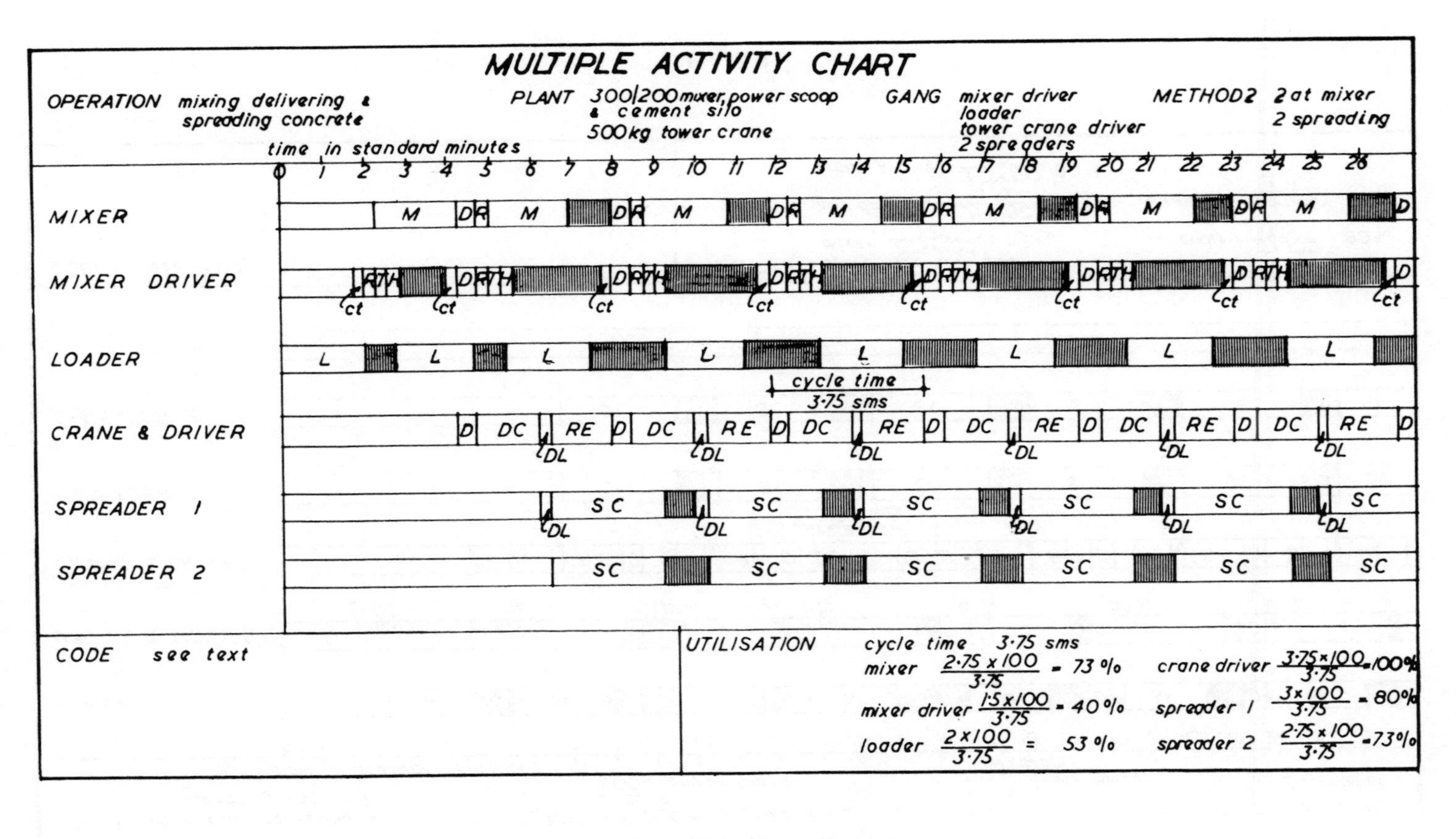
MULTIPLE ACTIVITY CHART
OPERATION mixing delivering & spreading concrete
PLANT 300/200 mixer, power scoop & cement silo
500kg tower crane
GANG mixer driver
loader
tower crane driver
2 spreaders
METHOD2 2 at mixer
2 spreading
time in standard minutes
MIXER
MIXER DRIVER
LOADER
cycle time 3·75 sms
CRANE & DRIVER
SPREADER 1
SPREADER 2
CODE see text
UTILISATION
cycle time 3·75 sms
mixer 2·75 x 100 / 3·75 = 73 %
mixer driver 1·5 x 100 / 3·75 = 40 %
loader 2 x 100 / 3·75 = 53 %
crane driver 3·75 x 100 / 3·75 = 100%
spreader 1 3 x 100 / 3·75 = 80%
spreader 2 2·75 x 100 / 3·75 = 73%

Fig. 3.8

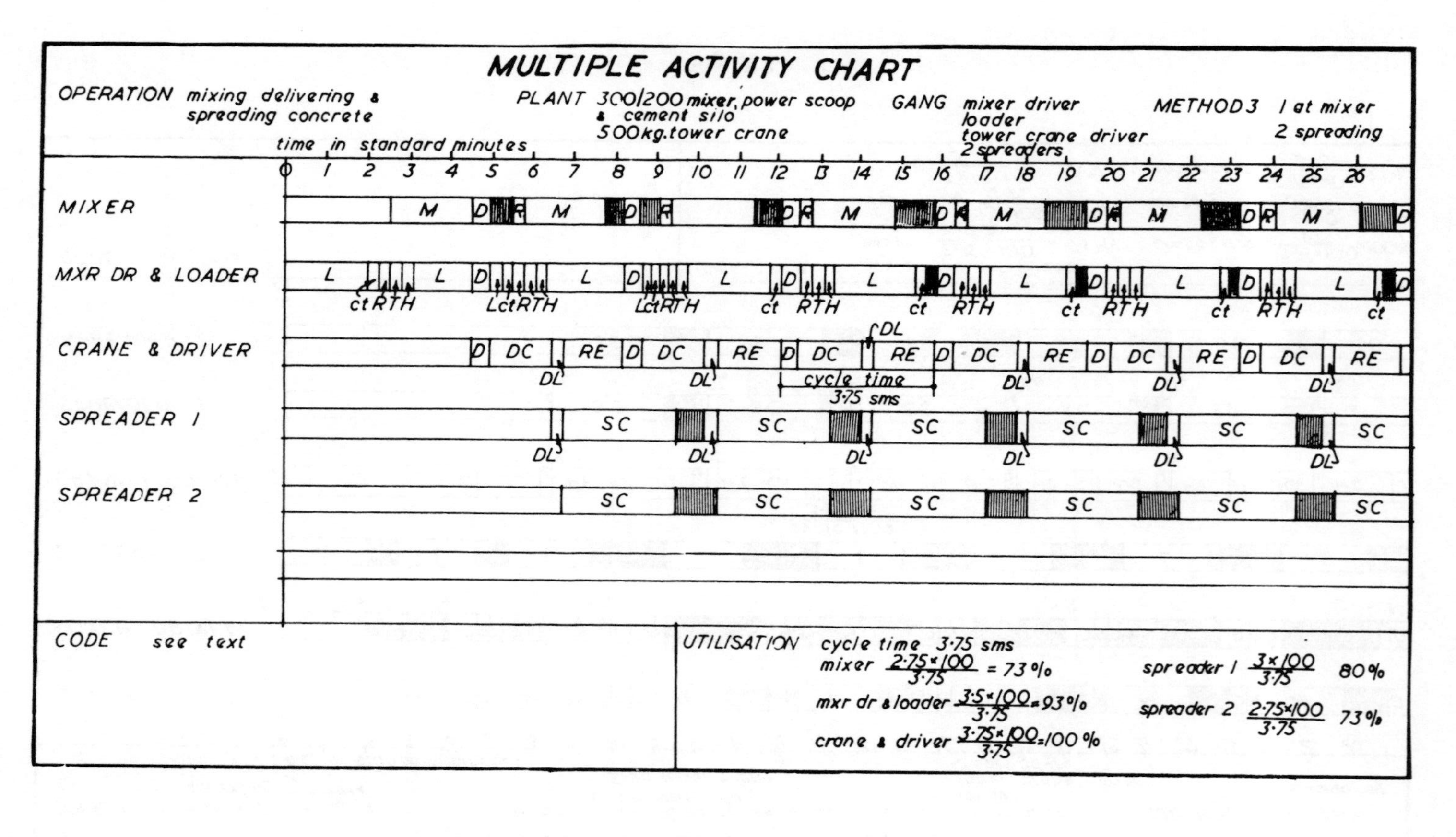
MULTIPLE ACTIVITY CHART
OPERATION mixing delivering & spreading concrete
PLANT 300/200 mixer, power scoop & cement silo 500kg. tower crane
GANG mixer driver loader tower crane driver 2 spreaders
METHOD 3 1 at mixer 2 spreading
time in standard minutes
0 1 2 3 4 5 6 7 8 9 10 11 12 13 14 15 16 17 18 19 20 21 22 23 24 25 26
MIXER
MXR DR & LOADER
CRANE & DRIVER
SPREADER 1
SPREADER 2
M D R
L D ct RTH LctRTH
D DC RE DL
cycle time 3·75 sms
SC
CODE see text
UTILISATION cycle time 3·75 sms
mixer 2·75×100 / 3·75 = 73%
mxr dr & loader 3·5×100 / 3·75 = 93%
crane & driver 3·75×100 / 3·75 = 100%
spreader 1 3×100 / 3·75 80%
spreader 2 2·75×100 / 3·75 73%

Fig. 3.9

3.5 WORK MEASUREMENT

The aim of work measurement is to determine the time it takes for a qualified worker to carry out a specific job at a defined level of performance and to eliminate ineffective elements of work. It seeks to provide the standard times for jobs and thus supplies basic, essential data for management.

3.5.1 Measurement techniques common in the construction industry

The work measurement techniques common in the construction industry are:

1. Time study
2. Activity sampling
3. Synthesis from standard and synthetic data
4. Analytical estimating.

3.5.2 Time study

The stages involved in carrying out a time study are:

1. Selecting the work to be measured
2. Analysing and breaking the work down into elements
3. Rating and timing each element
4. Extending the observed time to basic time
5. Selecting basic times, allocating allowances and building up the final standard.

Timing and rating are obviously carried out in the field and consequently in the example given it is assumed that the times and ratings have already been obtained.

3.5.2.1 Methods of timing

The methods of timing commonly used in the construction industry are cumulative timing and flyback timing.

Cumulative timing is the more common as it is better for observing a number of operatives in a gang and requires only an accurate wrist watch. The cumulative time is recorded after each element.

Flyback timing is carried out with a flyback stopwatch, the observer recording the time for each element as work proceeds. The watch has a flyback button on it which returns the hands to zero when pressed; on releasing the button the watch re-commences timing.

To check the accuracy, the start and finish times are taken and the difference between them is compared with the total of the readings.

3.5.2.2 Working up the standard time for an operation

To determine a standard time, a number of time studies of the operation will have to be completed; it is assumed that this has been done using cumulative timing.

In the example the following terms are used:

Element	A distinct part of an operation which can be easily defined.

Rating	The method used to take into account variations in working pace.
Cumulative time	The time recorded on the study sheet.
Observed time	The time taken to perform an element of work.
Standard rating	The average rate at which a qualified man will work given sufficient motivation and instruction.
Basic time	The time required for carrying out an element of work at standard rating.
Relaxation allowance	The time given to allow a worker to recover from the effects of fatigue and to attend to personal needs.
Contingency allowance	The time given to allow for occurrences throughout the day, e.g. a painter cleaning brushes.

3.5.2.3 Method used for calculating standard time

1. Calculate the observed time for each element by subtracting the previous cumulative time from the cumulative time for the element in question. Check that the total observed time equals the last cumulative time recorded (Fig. 3.10).

2. Extend the observed times to basic times, i.e. show all times at the standard rating by taking the observed rating into account. The formula for this calculation (Fig. 3.11) is:

$$\frac{\text{Observed time} \times \text{observed rating}}{\text{Standard rating}}$$

3. Calculate the average of the basic times over the series of studies (Fig. 3.12).

4. Add the appropriate relaxation allowances and contingency allowances (Fig. 3.13), remembering that unoccupied time allowance and interference allowances must also sometimes be made.

This procedure will give the standard time for each element; to calculate the standard time for an operation, add the elements in the operation together.

NOTE: This example is very much simplified and the times are hypothetical.

Fig. 3.10 AMENITY CENTRE AND OFFICE BLOCK

OPERATION Concreting upper floor slabs

Element	*Rating*	*Cumulative time*	*Observed time*
Check time		2.75	2.75
A. Discharge crane skip	85	2.99	0.24

Fig. 3.10 *continued*

Element	*Rating*	*Cumulative time*	*Observed time*
B. Spread concrete	75 80 70	4.89	1.90
C. Vibrate concrete	100	5.33	0.44
D. Level and tamp concrete	95 90 85 90	7.63	2.30
			7.63

Fig. 3.11

Element	*Rating*	*Cumulative time*	*Observed time*	*Basic time*
Check time	85	2.75	2.75	
A. Discharge crane skip	85	2.99	0.24	$\frac{0.24 \times 85}{100} = 0.20$
B. Spread concrete	75 80 70	4.89	1.90	$\frac{1.90 \times 75}{100} = 1.43$
C. Vibrate concrete	100	5.33	0.44	$\frac{0.44 \times 100}{100} = 0.44$
D. Level and tamp concrete	95 90 85 90	7.63	2.30	$\frac{2.30 \times 90}{100} = 2.07$
			7.63	

Fig. 3.12

Element	*Basic time*	*Total basic times*	*Frequency*	*Average*
A. Discharge crane skip	0.20, 0.18, 0.19, 0.18, 0.21, 0.20, 0.18, 0.19, 0.19, 0.18	1.90	10	0.19
B. Spread concrete	1.43, 1.27, 1.31, 1.49, 1.45, 1.45, 1.72, 1.48, 1.80, 1.60	15.00	10	1.50
C. Vibrate concrete	0.44, 0.46, 0.44, 0.42, 0.44, 0.45, 0.45, 0.44, 0.43, 0.43	4.40	10	0.44
D. Level and tamp concrete	2.07, 2.13, 2.35, 2.35, 2.23, 2.27, 2.24, 2.00, 2.16, 2.20	22.00	10	2.20

Assume that the total relaxation allowance and contingency allowance are as follows:

Fig. 3.13

Element	*Basic time*	*Total relaxation allowance*	*Contingency allowance*	*Total allowances*	*Standard time (minutes)*
A. Discharge crane skip	0.19	25%	5%	30%	$\frac{0.19 \times 130}{100} = 0.25$
B. Spread concrete	1.50	30%	5%	35%	$\frac{1.50 \times 135}{100} = 2.03$
C. Vibrate concrete	0.44	26%	5%	31%	$\frac{0.44 \times 131}{100} = 0.58$
D. Level and tamp concrete	2.20	30%	5%	35%	$\frac{2.20 \times 135}{100} = 2.97$

The standard time for the operation is 0.25 + 2.03 + 0.58 + 2.97 = 5.83 standard minutes.

3.6 ACTIVITY SAMPLING

Activity sampling is also known as snap observation studies or random observation studies. It is a very useful technique in the construction industry, and is used to determine the activity levels of machines and operatives. Using this method a number of subjects can be observed concurrently.

3.6.1 Basis of the method

In this procedure the principles of the statistical method of random samples are used. Snap readings are taken at intervals and the percentage of readings taken per element will give a result very near to the actual percentage on each providing the sample is big enough. From this it is possible to calculate the time taken on each element if necessary.

One observer can record the activities of a whole gang and readings can be almost continuous. The timing may be by either the fixed interval method (easier to use on site), or the random interval method, which is better for work of a cyclic nature.

It essential that a recording is made the instant a subject is observed.

3.6.2 Uses of activity sampling

On site, activity sampling can be used:

1. To assess unoccupied time as a basis for analysing cause.

2. To find the percentage of time spent on each element of work in an operation by each operative and/or machine.
3. To find the percentage utilization of machines or operatives as a basis for cutting down unoccupied time (rated activity sampling can be used for this, which is an extension of the method).

3.6.3 An example on activity sampling

The following example (Fig. 3.14) shows the application of activity sampling to determine the proportions of productive and non-productive time for three excavators owned by a contractor.

A total of 999 observations were made on the three excavators and at each observation the machines were recorded as being productive, suffering from major delays, or suffering from minor delays.

Weather delays, opening up working areas, and machine repairs were defined as a major delay, whilst insufficient lorries, spotting lorries, trimming and clearing up, moving the machine, routine maintenance, and operator delays were held to constitute minor delays.

If it is felt that the delays are excessive, a further analysis can be carried out to ascertain the actual cause of delay under the more detailed headings listed above. As stated previously, the accuracy of the result is determined to some extent by the size of the sample and formulae available which will give:

(a) the number of observations to produce a given accuracy, and
(b) the limit of accuracy obtained from the observations made.

These formulae are:

$$\text{number of observations required} = \frac{4P\,(100 - P)}{L^2}$$

where P is the expected percentage occurrence of the activity and L is the limit of accuracy required (percentage $\pm$)

$$\text{limit of accuracy} = 2\sqrt{\left(\frac{P\,(100 - P)}{N}\right)}$$

where N is the number of observations and P is the percentage of observations recorded for a particular activity.

There is a 95 per cent certainty that the value of P will be somewhere between $P + L$ and $P - L$.

These formulae will now be applied to the example to ascertain:

(a) the number of observations necessary to give a limit of accuracy of $\pm$ 3 per cent on minor delays, and
(b) the accuracy achieved with the number of observations taken.

Before starting the calculations it is necessary to estimate what the percentage on the activity will be, either from a short preliminary study or from other information. In this example, an estimate of 33 per cent has been made.

(a) Number of observations required

$$= \frac{4P\,(100 - P)}{L^2}$$

$$= \frac{4 \times 33\,(100 - 33)}{3^2}$$

$$= \frac{4 \times 33 \times 67}{9}$$

$$= 983 \text{ observations}$$

999 observations were actually taken

(b) Limit of accuracy of result obtained (Fig. 3.14)

$$= 2\sqrt{\left(\frac{P\,(100 - P)}{N}\right)}$$

$$= 2\sqrt{\left(\frac{21 \times 79}{999}\right)}$$

$$= 2\sqrt{1.6}$$

$$= 2 \times 1.26$$

$$= \pm\, 2.52 \text{ per cent}$$

The formulae can also be applied to major delays and productive time if required.

Fig. 3.14

Number of tours of site		*Excavator 1*	*Excavator 2*	*Excavator 3*	
1		P	SD	P	
2		SD	P	LD	
3		LD	LD	P	
etc		etc	etc	etc	
333		P	SD	LD	
					Totals
Totals for	P	133	137	140	410
999	SD	73	70	67	210
Observations	LD	127	126	126	379

Productive time = P   Minor delays = SD   Major delays = LD

Calculations:

$$\text{productive time } \frac{410 \times 100}{999} = 40 \text{ per cent}$$

$$\text{minor delays } \frac{210 \times 100}{999} = 21 \text{ per cent}$$

$$\text{major delays } \frac{379 \times 100}{999} = 38 \text{ per cent}$$

3.6.4 Carrying out an activity sampling study

Assume that the activities of a carpenter over a two day period are as shown in Fig. 3.15 (obviously this information is not available when the study is carried out because the work has not yet been started). Activity sampling is to be used to determine the percentage of time spent on studying drawings, working on beam formwork, working on slab formwork, unoccupied and absent from the job. It has been decided to take 30 observations in total, initially. A table of random numbers could be used as shown in Fig. 3.16 to determine the timing of each observation. Assume we start the study with random No. 01, row 3 column 2 and read off along the rows. The numbers can now be converted to time as shown in Fig. 3.17. By applying these times to Fig. 3.15, the percentage of observations on each activity can be calculated and the limit of accuracy obtained as described in 3.6.3(b) (see Fig. 3.18). Clearly, the accuracy of the percentages on each activity would probably be unacceptable and further calculations would be carried out to determine the number of additional observations necessary to reach the accuracy required.

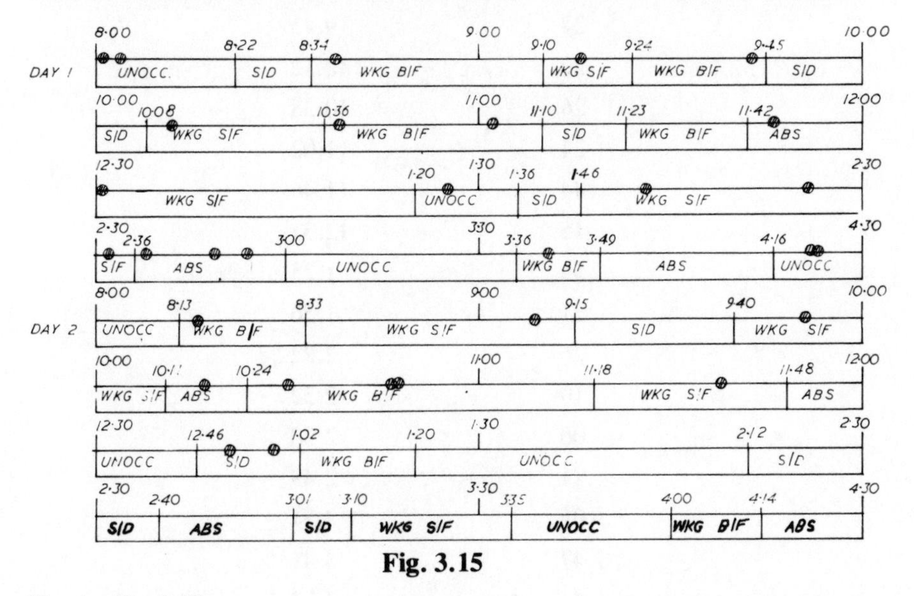

Fig. 3.15

KEY (*to Fig. 3.15*)

S/D = Studying drawings; WKG B/F = Working on beam formwork;

WKG S/F = Working on slab formwork; UNOCC = Unoccupied;

ABS = Absent from job

Fig. 3.16

02	49	59	26	39	45	12	32	18	56	10	03
22	02	43	19	23	55	39	05	09	17	44	52
00	01	03	34	38	27	29	26	24	44	15	54
31	26	10	06	11	05	47	41	01	23	53	42
26	13	16	01	51	43	07	42	21	05	24	57
34	56	36	31	41	17	44	50	25	53	02	03
45	07	16	18	42	51	33	55	43	20	46	54
23	09	21	01	04	23	08	04	23	52	38	01
17	43	07	28	19	54	40	40	36	54	07	58
14	04	50	10	11	25	36	00	39	37	59	59
34	16	05	29	25	28	30	32	35	58	02	51
35	08	52	47	52	10	05	36	02	48	17	31

Fig. 3.17

Random number	*Time*
1	8.01
3	8.04
34	8.38
38	9.16
27	9.43
29	10.12
26	10.38
24	11.02
44	11.46
15	12.31
54	1.25
31	1.56
26	2.22
10	2.32
06	2.38
11	2.49
05	2.54
47	3.41
41	4.22
01	4.23
23	8.16
53	9.09

Fig. 3.17 *continued*

42	9.51
26	10.17
13	10.30
16	10.46
01	10.47
51	11.38
43	12.51
07	12.58
748	

Fig. 3.18

Activity	*Tally*	*Total*	*%*	*Limit of accuracy* $L = 2\sqrt{\left(\frac{P(100-P)}{N}\right)}$
1. Study drawings	||	2	6.67	9.12%
2. Working on beam form-work	~~||||~~ ||||	9	30.00	16.74%
3. Working on slab formwork	~~||||~~ ||||	9	30.00	16.74%
4. Unoccupied	~~||||~~	5	16.67	13.60%
5. Absent	~~||||~~	5	16.67	13.60%
TOTAL		30	100.01	

3.7 SYNTHESIS AND SYNTHETIC DATA

The purpose of synthetic data is to enable time values to be compiled for jobs where direct measurement is unnecessary or impracticable. Many operations are very similar to ones which have been carried out before, perhaps with variation due to physical dimensions etc. (examples are: sawing different lengths of timber and paving different areas); these various jobs have many common elements and a suitably referenced stock of data relating to these elements can be built up over a period of time and used for quickly compiling the work content of various jobs.

3.8 ANALYTICAL ESTIMATING

To calculate the work content of non-repetitive jobs, the work value can be compiled by analytical estimating, using whatever information is available from past time studies or standard data and estimating the time for the remaining elements.

3.9 FINANCIAL INCENTIVES

3.9.1 Aims of financial incentive schemes

The aim of a sound financial incentive scheme is to increase productivity by increasing the efficiency of the individual, and enable him to increase his earnings.

A good scheme will achieve these objectives, but a poor one will cause nothing but trouble. Incentive schemes should be very closely related to short term planning and cost control. Careful supervision is necessary to ensure quality is maintained.

3.9.2 Features of a sound scheme

The main features of a sound financial incentive scheme are:

1. The amount of bonus paid to the operatives should be a direct proportion of the time saved and there should be no limit to the amount that can be earned.
2. Targets should be issued for all operations before work commences. The extent and nature of the operation should also be clearly understood. The length of time for an operation should ideally be kept to about two days.
3. Targets should be set for small gangs (there are exceptions to this in repetitive work similar to flow production when group bonuses may be more appropriate).
4. Targets should not be altered during an operation without agreement of both parties.
5. The system used for calculating bonus earnings should be easily understandable by all persons concerned.
6. Operatives must not be penalised for lost time which is caused by delays outside their control.
7. Bonus should be paid weekly.
8. Arrangement should be made for dealing with time lost due to reasons outside the operatives control.

3.9.3 Basis for setting targets

3.9.3.1 The estimate

For large projects where the estimate is based on a bill of quantities the rates used are not really suitable for direct use as bonus targets. The figures used in the bill are average, e.g. brickwork operations will vary in difficulty and, therefore, speed of completion depending on their situation. Brickwork in small panels takes longer per unit area to build, than a large expanse of uninterrupted brickwork. This is often not clear from the traditional Bill of Quantities. In addition it is a laborious job to collect bill items together which are relevant to a particular operation.

For small projects where the builder has to prepare the estimate from drawings it is easier to extract the labour content of operations and set targets based on this. By this means a check can easily be made on profits. The labour content in the estimate should of course be based on data from past projects.

3.9.3.2 Data from past projects

This is probably the most reliable information reasonably available to most firms where work measurement has not yet been used. The data will be built up over a period of years and can be used as a basis for estimating and as a basis for setting bonus targets, etc. Great care is necessary in analysing this data as it must be appreciated that the level of productivity, i.e. *the rating* will very considerably. The conditions under which operations are carried out should also be analysed.

3.9.3.3 Work measurement

This is by far the best method but, as stated above, is not at present available to the vast majority of firms. The targets based on work measurement are more accurate and the method to be used is taken into account. Targets based on work measurement are less likely to be questioned by the operatives, but where targets are questioned a check should be made before making adjustments.

3.9.3.4 Other methods

Experience, bargaining and other methods are used but are not very reliable generally and will not be considered here.

3.9.4 Expected earnings when using bonus schemes

The general level of earnings may well be fixed by competition from local firms and this may determine the basis for setting targets.

It is generally felt, however, that an operative should be paid $\frac{1}{3}$ of his basic weekly earnings, as a bonus when he works at a rating of 100, and the following discussion is based on this.

3.9.5 Rate of payment for time saved

Operatives are paid differing rates of payment for time saved depending on circumstances. This will affect the targets set in order to allow the $\frac{1}{3}$ bonus at 100 rating.

Examples of rates of payment which could be used are:

1. Payment of full rate for each hour saved (100% system)
2. Payment of 75% of hourly rate for each hour saved (75% system)
3. Payment of 50% of hourly rate for each hour saved (50% system)
4. Payment of 25% of hourly rate for each hour saved (25% system).

When the payments made for time saved are less than the full hourly rate the systems are known as *Geared Systems*.

3.9.6 Effects of different rates of payment on earnings and direct costs (see Fig. 3.19)

3.9.6.1 Effects on earnings

Assume an operation takes x hours to perform at 100 rating and at this $\frac{1}{3}$ bonus will be earned.

1. *Payment of all time saved (100% system)*

$$\text{Target} = x + \frac{x}{3} = 1\tfrac{1}{3}x$$

2. *Payment of 75% of time saved (75% system)*

$$\text{Target} = x + \frac{\frac{4}{3}}{3}x = 1\tfrac{4}{9}x$$

3. *Payment of 50% of time saved (50% system)*

$$\text{Target} = x + \tfrac{2}{3}x = 1\tfrac{2}{3}x$$

4. *Payment of 25% of time saved (25% system)*

$$\text{Target} = x + \tfrac{4}{3}x = 2\tfrac{1}{3}x$$

Rating at which bonus will start to be earned

100% system:

target is $\frac{4}{3}$ of operations time at 100 rating

$\therefore$ bonus starts at $\frac{3}{4} \times 100 = 75$ rating.

75% system:

target is $\frac{13}{9}$ of operations time at 100 rating

$\therefore$ bonus starts at $\frac{9}{13} \times 100 = 69.2$ rating

50% system:

target is $\frac{5}{3}$ of operations time at 100 rating

$\therefore$ bonus starts at $\frac{3}{5} \times 100 = 60$ rating

25% system:

target is $\frac{7}{3}$ of operations time at 100 rating

$\therefore$ bonus starts at $\frac{3}{7} \times 100 = 42.9$ rating

3.9.6.2 Effects on direct costs

It is assumed that the labour content of estimated prices are based on outputs which approximate to a rating of 75. The following comparisons show the effects of different systems of payment on actual direct costs.

1. *100% system*

At $37\frac{1}{2}$ rating, cost

$$= \frac{75}{37\frac{1}{2}} \times 100 = 200\%$$

At 40 rating, cost

$$= \frac{75}{40} \times 100 = 187\tfrac{1}{2}\%$$

At 60 rating, cost

$$= \frac{75}{60} \times 100 = 125\%$$

At 75 rating and above cost is constant at 100%.

75% system – target $1\tfrac{4}{9}x$

Cost as above until bonus is earned at 42.9 rating.

At 75 rating time required $= \tfrac{4}{3}x$

$$\text{time saved} = 1\tfrac{4}{9}x - \tfrac{4}{3}x = \tfrac{1}{9}x$$

$$\text{bonus} = \tfrac{3}{4} \times \tfrac{1}{9}x = \tfrac{1}{12}x$$

$$\text{cost} = \frac{\tfrac{4}{3}x + \tfrac{1}{12}x}{\tfrac{4}{3}x} = 106\tfrac{1}{4}\%$$

At 100 rating time required $= x$

$$\text{time saved} = 1\tfrac{4}{9}x - x = \tfrac{4}{9}x$$

$$\text{bonus} = \tfrac{1}{3}x$$

$$\text{cost} = \frac{x + \tfrac{1}{3}x}{\tfrac{4}{3}x} = 100\%$$

At 125 rating time required $= \tfrac{4}{5}x$

$$\text{time saved} = 1\tfrac{4}{9}x - \tfrac{4}{5}x = \tfrac{29}{45}x$$

$$\text{bonus} = \tfrac{29}{60}x$$

$$\text{cost} = \frac{\tfrac{4}{5}x + \tfrac{29}{60}x}{\tfrac{4}{3}x} = 96\tfrac{1}{4}\%$$

At 150 rating time required $= \tfrac{2}{3}x$

$$\text{time saved} = 1\tfrac{4}{9}x - \tfrac{2}{3}x = \tfrac{7}{9}x$$

$$\text{bonus} = \tfrac{7}{12}x$$

$$\text{cost} = \frac{\tfrac{2}{3}x + \tfrac{7}{12}x}{\tfrac{4}{3}x} = 93\tfrac{3}{4}\%$$

At 175 rating time required $= \tfrac{4}{7}x$

$$\text{time saved} = 1\tfrac{4}{9}x - \tfrac{4}{7}x = \tfrac{55}{63}x$$

$$\text{bonus} = \tfrac{55}{84}x$$

$$\text{cost} = \frac{\tfrac{4}{7}x + \tfrac{55}{84}x}{\tfrac{4}{3}x} = 92\%$$

3. *50% system – target* = $1\frac{2}{3}x$

Cost as above until bonus is earned at 60 rating.

At 75 rating

$$\text{time required} = \tfrac{100}{75}\,x = 1\tfrac{1}{3}x$$

$$\text{target} = 1\tfrac{2}{3}x$$

$$\text{time saved} = 1\tfrac{2}{3}x - 1\tfrac{1}{3}x = \tfrac{1}{3}x$$

$$\text{bonus} = \tfrac{1}{6}x$$

$$\text{cost} = \frac{1\frac{1}{3}x + \frac{1}{6}x}{1\frac{1}{3}x} = 112\tfrac{1}{2}\%$$

At 100 rating

$$\text{time required} = x$$

$$\text{time saved} = 1\tfrac{2}{3}x - x = \tfrac{2}{3}x$$

$$\text{bonus} = \tfrac{1}{3}x$$

$$\text{cost} = \frac{x + \frac{1}{3}x}{1\frac{1}{3}x} = 100\%$$

At 125 rating

$$\text{time required} = \tfrac{100}{125}x = \tfrac{4}{5}x$$

$$\text{time saved} = 1\tfrac{2}{3}x - \tfrac{4}{5}x = \tfrac{13}{15}x$$

$$\text{bonus} = \tfrac{13}{30}x$$

$$\text{cost} = \frac{\frac{4}{5}x + \frac{13}{30}x}{1\frac{1}{3}x} = 92\tfrac{1}{2}\%$$

At 150 rating

$$\text{time required} = \tfrac{100}{150}x = \tfrac{2}{3}x$$

$$\text{time saved} = 1\tfrac{2}{3}x - \tfrac{2}{3}x = x$$

$$\text{bonus} = \tfrac{1}{2}x$$

$$\text{cost} = \frac{\frac{2}{3}x + \frac{1}{2}x}{1\frac{1}{3}x} = 87\tfrac{1}{2}\%$$

At 175 rating

$$\text{time required} = \tfrac{100}{175}x = \tfrac{4}{7}x$$

$$\text{time saved} = 1\tfrac{2}{3}x - \tfrac{4}{7}x = \tfrac{23}{21}x$$

$$\text{bonus} = \tfrac{23}{42}x$$

$$\text{cost} = \frac{\frac{4}{7}x + \frac{23}{42}x}{1\frac{1}{3}x} = 84\%$$

4. *25% system* – target = $2\frac{1}{3}x$
   Cost as above until bonus is earned at 42.9 rating

At 50 rating

$$\text{time required} = \tfrac{100}{50}\,x = 2x$$

$$\text{time saved} = 2\tfrac{1}{3}x - 2x = 1\tfrac{1}{3}x$$

$$\text{bonus} = \tfrac{1}{12}x$$

$$\text{cost} = \frac{2x + \frac{1}{12}x}{1\frac{1}{3}x} = 156\tfrac{1}{4}\%$$

At 75 rating

$$\text{time required} = \tfrac{100}{75}\,x = \tfrac{4}{3}x$$

$$\text{time saved} = 2\tfrac{1}{3}x - 1\tfrac{1}{3}x = x$$

$$\text{bonus} = \tfrac{1}{4}x$$

$$\text{cost} = \frac{\frac{4}{3}x + \frac{1}{4}x}{1\frac{1}{3}x} = 118\tfrac{3}{4}\%$$

At 100 rating

$$\text{time required} = x$$

$$\text{time saved} = 2\tfrac{1}{3}x - x = 1\tfrac{1}{3}x$$

$$\text{bonus} = \tfrac{1}{3}x$$

$$\text{cost} = \frac{x + \frac{1}{3}x}{1\frac{1}{3}x} = 100\%$$

At 125 rating

$$\text{time required} = \tfrac{100}{125}x = \tfrac{4}{5}x$$

$$\text{time saved} = 2\tfrac{1}{3}x - \tfrac{4}{5}x = 1\tfrac{8}{15}x$$

$$\text{bonus} = \tfrac{23}{60}x$$

$$\text{cost} = \frac{\frac{4}{5}x + \frac{23}{60}x}{1\frac{1}{3}x} = 88\tfrac{3}{4}\%$$

At 150 rating

$$\text{time required} = \tfrac{100}{150}x = \tfrac{2}{3}x$$

$$\text{time saved} = 2\tfrac{1}{3}x - \tfrac{2}{3}x = 1\tfrac{2}{3}x$$

$$\text{bonus} = \tfrac{5}{12}x$$

$$\text{cost} = \frac{\frac{2}{3}x + \frac{5}{12}x}{1\frac{1}{3}x} = 81\tfrac{1}{4}\%$$

At 175 rating

$$\text{time required} = \tfrac{100}{175}x = \tfrac{4}{7}x$$

$$\text{time saved} = 2\tfrac{1}{3}x - \tfrac{4}{7}x = \tfrac{37}{21}x$$

$$\text{bonus} = \tfrac{37}{84}x$$

$$\text{cost} = \frac{\tfrac{4}{7}x + \tfrac{37}{84}x}{1\tfrac{1}{3}x} = 76\%$$

3.9.7 General comments

3.9.7.1 The 100% system

As shown in Fig. 3.19 the cost is constant when the rating exceeds 75. The operatives earn a greater bonus when the rating exceeds 100 when this system is used.

3.9.7.2 Geared systems

Direct costs are higher than the 100% system below 100 rating down to the level at which bonus earnings start. Direct costs break even at a rating of 100.

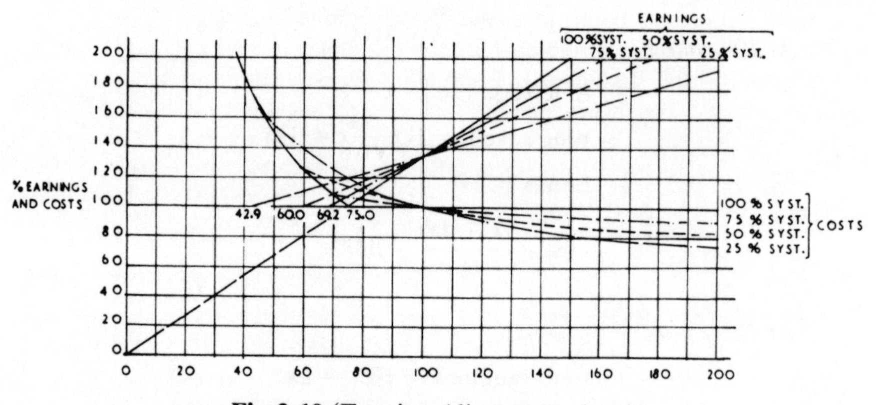

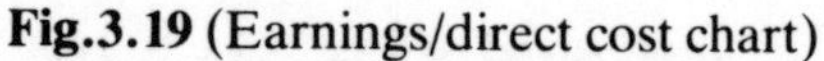
Fig.3.19 (Earnings/direct cost chart)

3.9.8 Effects of indirect costs on overall costs

In practice, savings for the employer are greater than those resulting from direct costs. The all-in labour rate used in the estimate includes indirect costs which cover Sick Pay Allowances, Public Holiday Pay, Annual Holiday Pay and death benefit, National Insurance, C.I.T.B. Levy, Severence Pay Allowance, Employers Liability Insurance, inclement weather and other NWR allowances.

These items vary in cost but at present add approximately one third to the basic weekly wage. For illustration purposes it is assumed that the basic hourly rate is £3.00 and the all-in rate for estimating purposes is £4.00.

The operation used in the examples which follow is assumed to take 45 man hours at 100 rating. The 100%, 50% and 25% systems only, will be considered in this section.

3.9.8.1 Effects on cost when the estimate is based on a rating of 75

EARNINGS AND COST CALCULATIONS The results shown in Fig. 3.20 to 3.23 are calculated as follows:

$$\text{Time taken } (T_t) = \frac{\text{Time at 100 rating} \times 100}{\text{rating required}}$$

$$\text{Time saved } (T_s) = \text{Target} - \text{Time taken}$$

$$\text{Bonus hours } (B_h) = \text{Time saved} \times \%\ \text{payback}$$

$$\text{Bonus pay} = \text{Bonus hours} \times £3.00$$

$$\text{Percentage cost} = \frac{100\,(T_t \times £4.00 + B_p)}{60 \times £4.00}$$

$$\text{Percentage earnings} = \frac{100\,(T_t \times £3.00 + B_p)}{T_t \times £3.00}$$

3.9.9 General comments

3.9.9.1 100% System or direct system (Figs 3.20 and 3.23)
The rate of increase in earnings rises rapidly above 75 rating.

The 100% system is suitable where the targets set are very accurate, i.e. in factory conditions, particularly when work measurement is used as a basis for the targets. As shown in Fig. 3.23 savings are made when the rating exceeds 75 and these savings increase as the rating increases.

For site work it is very difficult to set really accurate targets, particularly in view of the fact that the vast majority of firms do not use work measurement as a basis for setting targets.

Fig. 3.20

Direct system (Pay back all time saved). Target = 45 + 15 = 60 hours

	Rating							
	37½	50	60	75	100	125	150	175
Time taken	120	90	75	60	45	36	30	25.71
Time saved	Nil	Nil	Nil	Nil	15	24	30	34.29
Bonus hours	Nil	Nil	Nil	Nil	15	24	30	34.29
Bonus pay in £	Nil	Nil	Nil	Nil	45	72	90	102.87
Percentage cost	200	150	125	100	94	90	87	86
Percentage earnings	100	100	100	100	133	167	200	233

If tight targets are set, the operatives want them adjusted. If the targets are too easy, the operatives earn high wages, and the company can loose money even at ratings above 75. When productivity is below 75 rating costs are high, but when productivity is high gains for the company are lower than in geared systems.

Fig. 3.21

Geared system (pay back 50% of time saved). Target = 45 + 30 = 75 hours

	Rating							
	37½	50	60	75	100	125	150	175
Time taken	120	90	75	60	45	36	30	25.71
Time saved	Nil	Nil	Nil	15	30	39	45	49.29
Bonus hours	Nil	Nil	Nil	7.5	15	19.5	22.5	24.65
Bonus pay in £	Nil	Nil	Nil	22.5	45	58.5	67.5	73.95
Percentage cost	200	150	125	109	94	84	78	74
Percentage earnings	100	100	100	113	133	154	175	196

Fig. 3.22

Geared system (pay back 25% of time saved). Target = 45 + 60 = 105 hours

	Rating							
	37½	50	60	75	100	125	150	175
Time taken	120	90	75	60	45	36	30	25.71
Time saved	Nil	15	30	45	60	69	75	79.29
Bonus hours	Nil	3.75	7.5	11.25	15	17.25	18.75	19.82
Bonus pay in £	Nil	11.25	22.5	33.75	45	51.75	56.25	59.46
Percentage cost	200	155	134	116	94	82	73	68
Percentage earnings	100	104	110	119	133	148	163	177

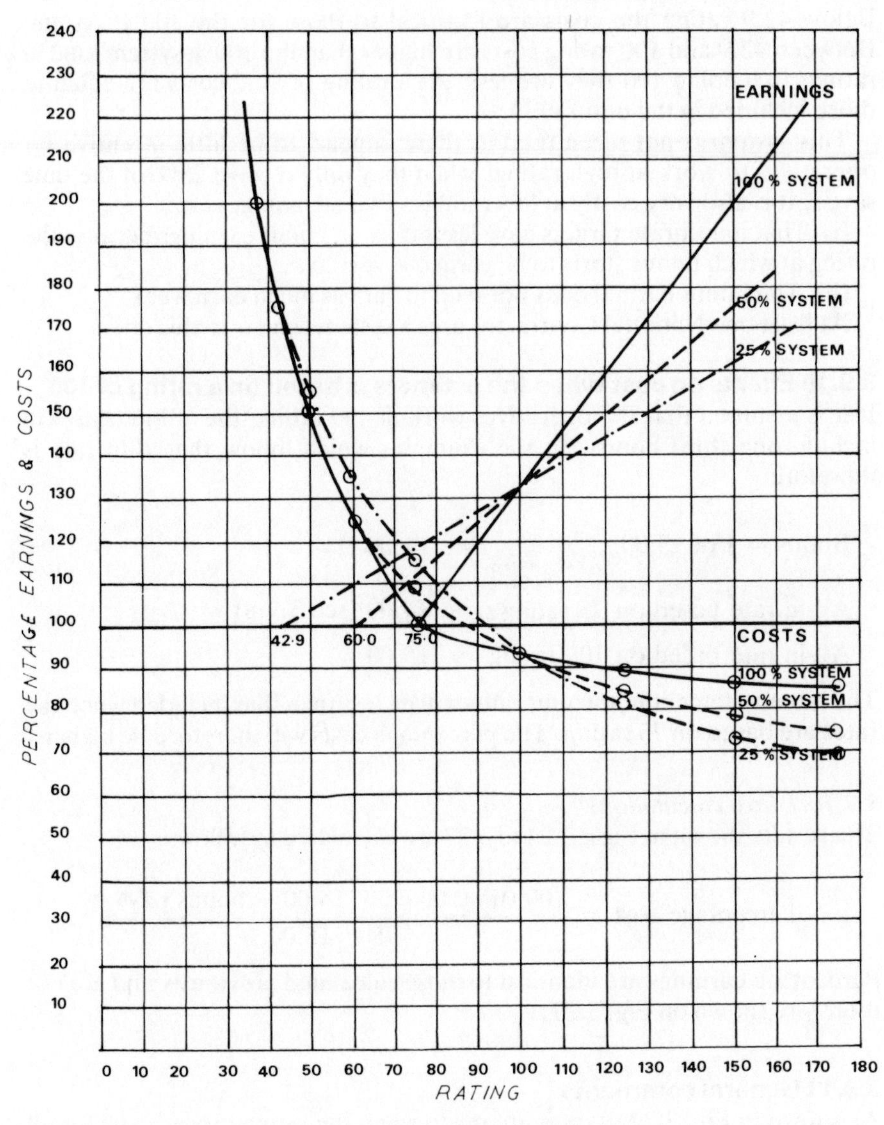

Fig. 3.23

3.9.9.2 50% System (Figs 3.21 and 3.23)
Below 60 rating the costs are identical to those for the 100% system. Between 60 and 100 rating costs are higher than the 100% system, and at ratings exceeding 100 they are less. At a rate of 88.2, the costs are equal to those included in the unit rates.

The 50% system is the one most used in the construction industry.

3.9.9.3 25% System (Figs 3.22 and 3.23)

Below 42.9 rating the costs are identical to those for the 100% system. Between 42.9 and 100 rating costs are higher than the 100% system, and at ratings exceeding 100 they are less. At a rating of 90.7 costs are equal to those included in the unit rates.

This system is not often used as there appears to be little incentive for operatives to work at high ratings when they only receive 25% of the time saved. It is also very costly at low ratings. Two advantages are:

(i) That inaccurate targets have less effect on bonus earnings because the rating at which bonus starts to be earned is very low.

(ii) The bonus earned does not tend to vary as much each week.

At least one National Contractor uses a system similar to this one.

3.9.10 Effects on cost when the estimate is based on a rating of 100

If it is assumed that the operatives work at 100 rating, the 'all-in' rate will include one third bonus. In the examples which follow the all-in rate is therefore:

Bonus = $\frac{1}{3}$ of £3.00 = £1.00

All-in rate based on 75 rating = £4.00 (see 3.9.8)

All-in rate based on 100 rating = £5.00

The labour element in the unit rates will be less than that included when the rates are based on 75 rating. The percentage costs will therefore be higher.

3.9.10.1 Cost calculations

The results shown in Figs. 3.24 to 3.27 are calculated as follows:

$$\text{Percentage cost} = \frac{100\ (\text{time taken} \times £4.00 + \text{bonus pay})}{45 \times £5.00}$$

Percentage earnings are identical to those calculated previously and are not therefore shown on Fig. 3.27.

3.9.11 General comments

As shown in Fig. 3.27 savings are made when the rating exceeds 100 for all the systems and these savings increase gradually as the rating increases. At low ratings, costs are considerably higher than in the previous method.

3.9.12 Conclusions

Considerable increases in productivity and wages can result from a good scheme. Savings in excess of those illustrated are achieved in practice because any reduction in project duration results in savings in overheads. Good supervision is necessary to ensure that quality is maintained and careful control of material usage is essential. Care must also be taken to ensure that safety is not overlooked.

Fig. 3.24

Direct system (pay back all time saved)								
	Rating							
	37½	50	60	75	100	125	150	175
Percentage cost	214	160	134	107	100	96	93	92

Fig. 3.25

Geared system (pay back 50% of time saved)								
	Rating							
	37½	50	60	75	100	125	150	175
Percentage cost	214	160	134	117	100	90	83	79

Fig. 3.26

Geared system (pay back 25% of time saved)								
	Rating							
	37½	50	60	75	100	125	150	175
Percentage cost	214	165	144	124	100	87	78	72

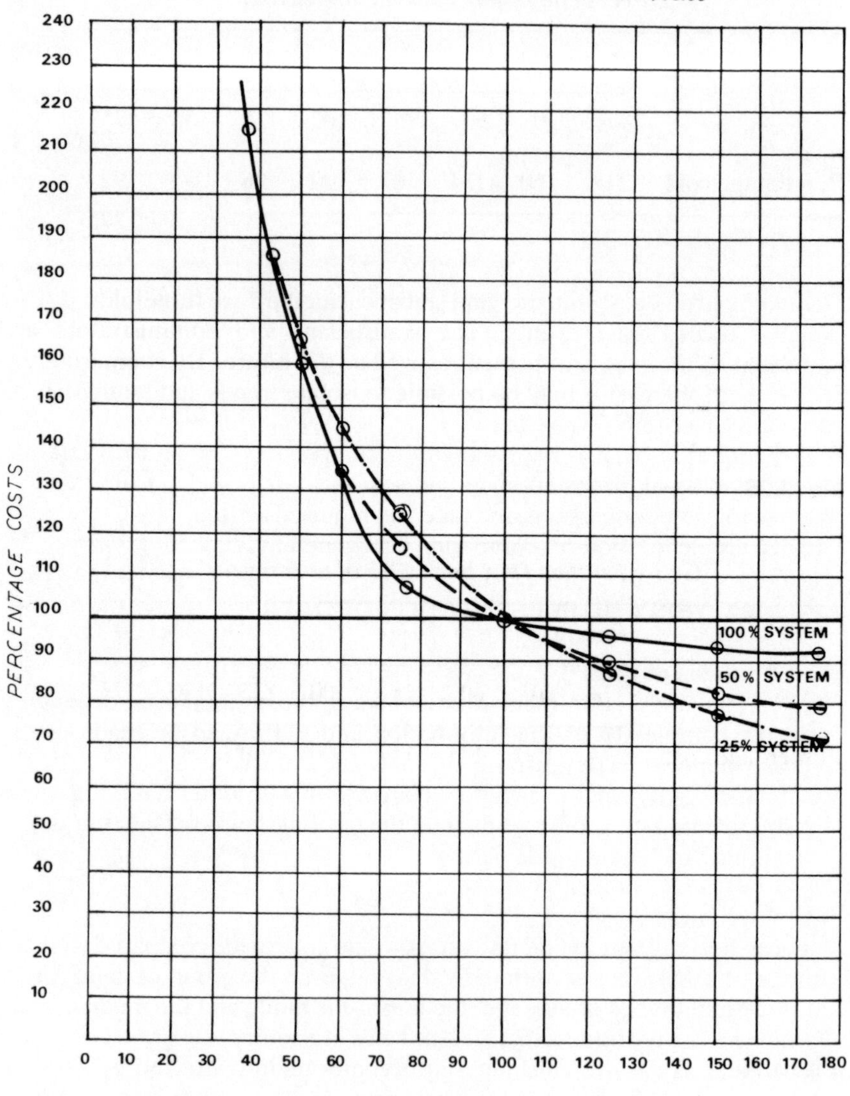
COSTS/ ALL IN LABOUR RATE £5.00
240
230
220
210
200
190
180
170
160
150
140
130
120
110
100
90
80
70
60
50
40
30
20
10
PERCENTAGE COSTS
100 % SYSTEM
50 % SYSTEM
25% SYSTEM
0
10
20
30
40
50
60
70
80
90
100
110
120
130
140
150
160
170
180

Fig. 3.27

4 Statistics

4.1 INTRODUCTION

The main purposes of statistics and statistical method are threefold:

1. To record past events. The statisticians and administrators are interested in the past and present as well as the future. By comparing the present with the past it may be possible to isolate trends and tendencies so that a forecast can be arrived at.
2. To identify what is happening in what appears to be an irregular and unrelated mass of information and present the data is such a form that it is possible to draw conclusions and take any required action.
3. To create methods of comparing data generally.

4.2 ANALYSIS AND PRESENTATION OF DATA

4.2.1 Analysis of data

Data can be analysed in many different ways, the method used depending upon the complexity of the information and the use to be made of the resulting analysis. The purpose of analysing data is to achieve a better understanding of it and to facilitate comparison with other data; it is usual practice to compare groups of data on the basis of their average, frequency distribution, and standard deviation.

4.2.1.1 Averages

Data can be analysed to find its average value and can be compared with the average of other data or with individual values in the group of data. There are three main types of average, the mean, the mode and the median, each of which has its usefulness in different fields of enquiry.

The mean is easy to calculate but becomes unduly affected by extreme values and consequently can be misleading. It is the arithmetical average – the sum of all values in a group divided by the number of values concerned.

The mode is unaffected by abnormal values in the group, and unlike the mean represents an actual value – the most frequently occurring one in a group.

The median is the middle value in a group and also eliminates the effect of extreme values. Half the observations are greater than it and half the observations are smaller; when the number of values is even, it is usual to take the mean of the two middle values to find the median.

4.2.1.2 Frequency distribution

Another important method of analysis is to sort data into predetermined groups having certain limits so that one group can be compared with another on the basis of frequency of occurrence in each group. Using this method of analysis it is also possible to compare one list of results with another by analysing each list into a number of groups: comparison is then carried out by considering the number of occurrences in each group in each list. For example, in the section on quality control 25 test cube results have been taken and sorted into groups having various predetermined strengths (Fig. 4.15.2), the analysis being represented by the use of a histogram. It would be possible to analyse another 25 test cube results in the same manner using the same group limits, and then to compare the first 25 results with the second 25 results on the basis of frequency distibution, i.e. the manner in which the results fall into the predetermined groups.

A complete example of this method of analysis is shown in the section on quality control (section 4.3.4).

4.2.1.3 Standard deviation

This is another useful way of comparing one set of data with another. Like the mean, mode and median, it expresses the characteristics of a group of data in one numerical expression (although with greater precision), and is used to measure the deviation of items in a set of data from the mean value of that set of data.

By using these measures of the central tendency, two sets of data could appear to be identical because they possess the same average, despite the fact that the range and distribution of results is very different. This contrast would be brought out by finding the standard deviation (Fig. 4.15.5) for each set of data and comparing them on this basis.

4.2.2 Presentation of data

The analyst may discover that he can achieve a clearer understanding of the meaning of the analysed data if he illustrates the results by some suitable graphical technique, a procedure which is often also useful for presenting the data to others who are less familiar with it.

Data must always be presented in a form that is easily understood by the recipient, and only data that is pertinent to the recipient's particular interest should be presented to him. For example, the managing director of a large construction firm would require to know weekly whether each of his projects was operating at a profit or a loss; information about variation in unit costs would consequently be unimportant to him, and it would be pointless to present him with a weekly cost control sheet for each project as he would only require information on total value and total cost from this sheet.

4.2.2.1 Tabulation (Fig. 4.1)

When using this form of presentation it must be borne in mind that however well presented they may be, columns of figures have very little meaning to some recipients. Tables should always be kept as simple as possible so that they do not become unintelligible due to the mass of data presented, and a useful addition in complex tables is the percentage (which has been used in Fig. 4.1 and makes it far easier to appreciate the proportions). Various types and weights of print can be used to attract attention to important features in a table, and a system of bold type for totals is shown in Fig. 4.1.

4.2.2.2 Pictograms

This form of presentation has a limited use, since it stimulates interest in a topic rather than presenting detailed facts. In Fig. 4.2.1 comparisons are made by counting the number of units shown for each group being compared. In Fig. 4.2.2 (top) the respective quantities are represented by the height of each unit, but in this form of presentation there is a danger of areas being compared and creating a false impression. This problem has been overcome in Fig. 4.2.2 (bottom) by standardizing on the width of each unit's base.

4.2.2.3 Pie chart (Fig. 4.3)

This is a very simple form of presentation that creates an immediate impression. These diagrams can vary from ones which have three or four slices to ones which have many slices but it must be remembered that a large number of slices make the chart more difficult to assimilate at a glance. This can be overcome to some extent by the use of colours or shading and the use of percentages. In the pie chart, each slice is proportional to the quantity represented.

4.2.2.4 Bar charts (Figs 4.4.1, 4.4.2 and 4.4.3)

This technique is used to compare items that have a common characteristic (usually time). The principle is to compare items by area of bar, so for convenience it is better to standardize on width so that the bars can be compared by their length.

Bar charts take many forms with various horizontal or vertical arrangements, and compound bars are also frequently used. Fig. 4.4.1 (top) shows horizontal compound bars, whereas Fig. 4.4.1 (bottom) demonstrates the same information broken down into individual bars, and Fig. 4.4.2 gives the same information in one compound bar. To make the chart more impressive, the major divisions of the compound bar have been shaded differently, thus drawing attention to the major divisions.

Figure 4.4.3 shows information abstracted from Fig. 4.1. This figure again uses the compound bar chart to present the information, Fig. 4.4.3 (top) presenting the information on a numerical bases whilst Fig. 4.4.3 (bottom) presents the same information on a percentage basis. It is far easier to gain an impression of the relative proportions involved from the latter than from the former.

It is interesting to compare Figs 4.3, 4.4.1 and 4.4.2 to ascertain which form of presentation creates the greatest impression, since these three figures present the same information in different ways.

Fig. 4.1

Year	All dwellings					Houses					Flats				
	1 bed	*2 beds*	*3 beds*	*4 beds +*	*Total*	*1 bed*	*2 beds*	*3 beds*	*4 beds +*	*Total*	*1 bed*	*2 beds*	*3 beds*	*4 beds +*	*Total*
1	14864	48523	73876	2702	**139965**	4606	28614	67016	2331	**102567**	10258	19909	6860	371	**37398**
	10.62	34.67	52.78	1.93	100%	4.49	27.90	65.34	2.27	100%	27.43	53.24	18.34	0.99	100%
2	18287	49361	67017	2919	**137584**	5557	27725	59319	2566	**95167**	12730	21636	7698	353	**42417**
	13.29	35.88	48.71	2.12	100%	5.84	29.13	62.33	2.70	100%	30.01	51.01	18.15	0.83	100%
3	20132	40863	49681	2470	**113146**	6121	21217	43415	2214	**72967**	14011	19646	6266	256	**40179**
	17.79	36.12	43.91	2.18	100%	8.39	29.08	59.50	3.04	100%	34.87	48.90	15.60	0.64	100%
4	21840	34339	41347	1930	**99456**	6832	16133	34445	1681	**59091**	15008	18206	6902	249	**40365**
	21.96	34.53	41.57	1.94	100%	11.56	27.30	58.29	2.84	100%	37.18	45.10	17.10	0.62	100%
5	27057	35463	39050	1665	**103235**	8941	16708	32323	1513	**59485**	18116	18755	6727	152	**43750**
	26.21	34.35	37.83	1.61	100%	15.03	28.09	54.34	2.54	100%	41.41	42.87	15.38	0.35	100%
6	24257	29690	37018	1915	**92880**	7885	13589	30755	1430	**53659**	16372	16101	6263	485	**39221**
	26.17	31.97	39.86	2.06	100%	14.69	25.32	57.31	2.66	100%	41.74	41.05	15.97	1.24	100%
7	28971	34012	40275	2044	**105302**	8460	14163	33707	1740	**58070**	20511	19849	6568	304	**47232**
	27.51	32.30	38.25	1.94	100%	14.57	24.39	58.05	3.00	100%	43.43	42.02	13.91	0.64	100%
8	27079	31347	36539	2050	**97015**	7041	12279	29434	1845	**50599**	20038	19068	7105	205	**46416**
	27.91	32.31	37.66	2.11	100%	13.92	24.27	58.17	3.65	100%	43.17	41.08	15.31	0.44	100%
9	32952	41291	42740	2485	**119468**	8385	15263	34825	2151	**60624**	24567	26028	7915	334	**58844**
	27.58	34.56	35.77	2.08	100%	13.83	25.18	57.44	3.55	100%	41.75	44.23	13.45	0.57	100%
10	36351	47723	46226	2724	**133024**	7474	16671	36847	2365	**63357**	28877	31052	9379	359	**69667**
	27.33	35.88	34.75	2.05	100%	11.80	26.31	58.16	3.73	100%	41.45	44.57	13.46	0.52	100%

NEW DWELLINGS COMPLETED FOR LOCAL AUTHORITIES
Analysis by bedroom type in England and Wales (based on the *Annual Digest of Statistics* [H.M.S.O.])

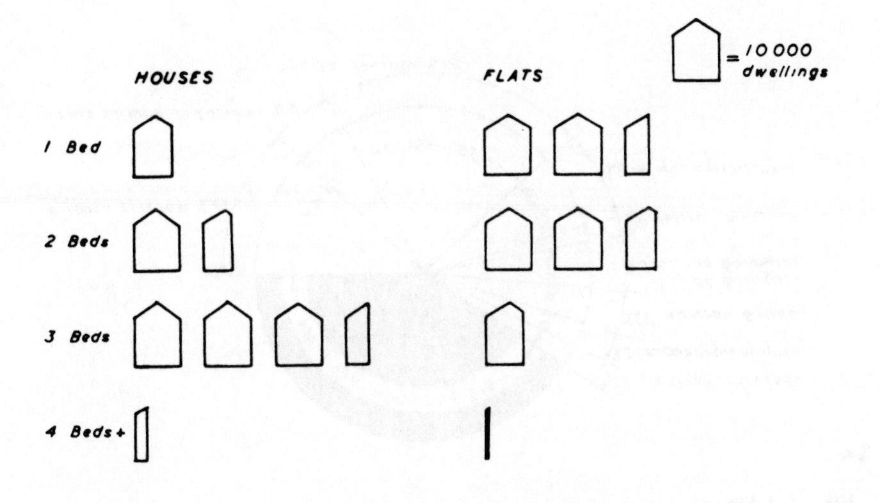

Fig. 4.2.1 (*based on Fig. 4.1*)

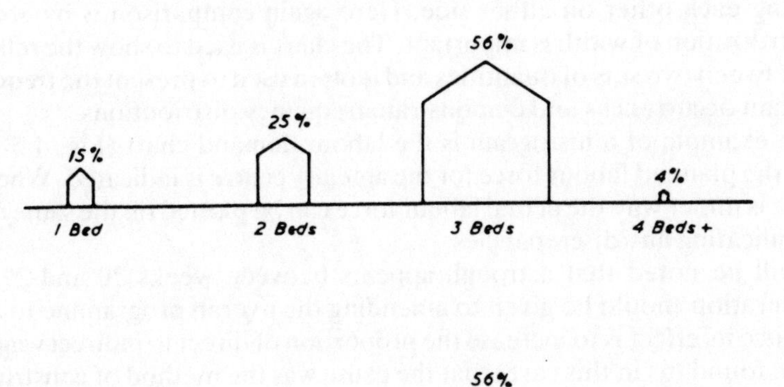

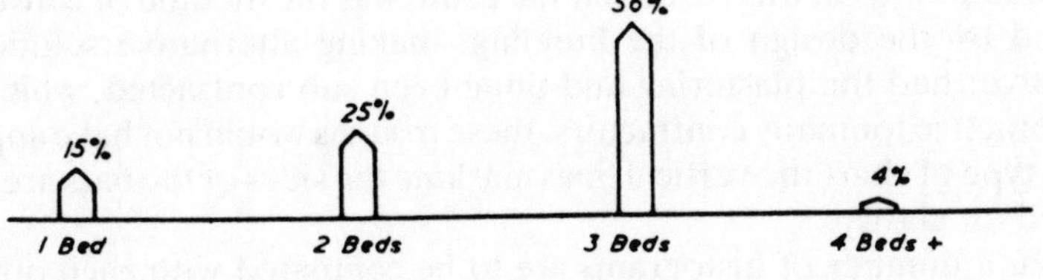

Fig. 4.2.2 (*based on Fig. 4.1*)

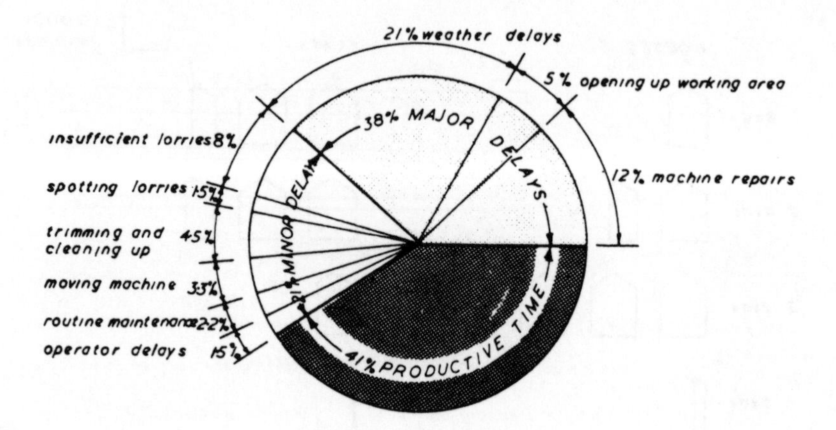

Fig. 4.3 Breakdown of productive and non-productive time for excavators owned by A.B.C. Builders Ltd.

4.2.2.5 Histogram

This is a type of vertical bar chart in which it is common to have the bars touching each other on either side. Here again comparison is by area, so standardization of width is important. The chart is used to show the relationship between two sets of quantities and is often used to present the frequency of certain occurrences and demonstrate frequency distributions.

One example of a histogram is the labour demand chart (Fig. 4.5.1), in which the planned labour force for the amenity centre is indicated. When the project is under way the actual labour force can be plotted on the same chart, thus indicating any discrepancies

It will be noted that a trough appears between weeks 20 and 25, and consideration should be given to amending the overall programme to avoid this, since its effect is to increase the proportion of direct to indirect wages. It may be found (as in this case) that the cause was the method of construction dictated by the design of the building, making alternative solutions too expensive: had the plastering and tiling been sub-contracted, which is the usual practice for many contractors, these troughs would not have appeared. In this type of chart the vertical lines marking the sides of the bars are usually omitted for clarity.

When a number of histograms are to be compared with each other, the three dimensional technique is useful (Fig. 4.5.2).

4.2.2.6 Array

This chart is similar to the bar chart but quantities are represented by the length of a line rather than the area of a bar. The chart is used to represent a large number of items which are arranged in order of magnitude, with the smallest item at the top (Fig. 4.6).

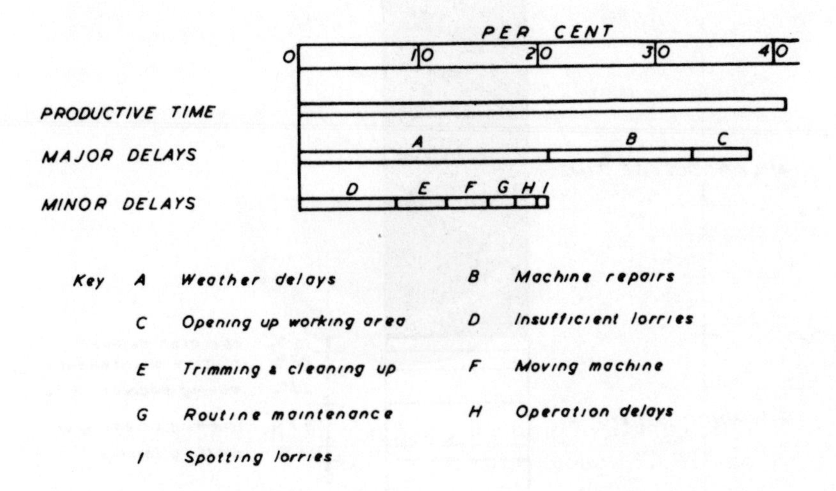

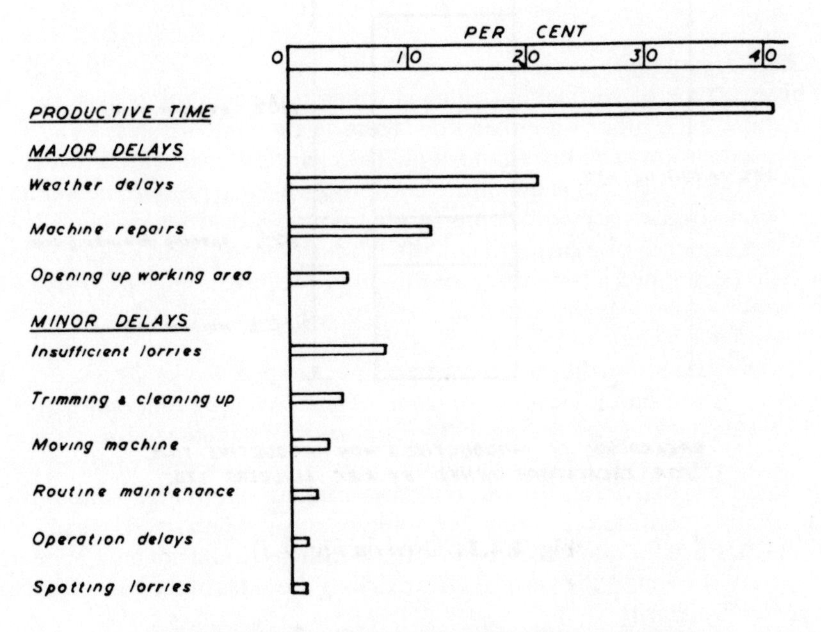

Fig. 4.4.1 (*based on Fig. 4.3*)

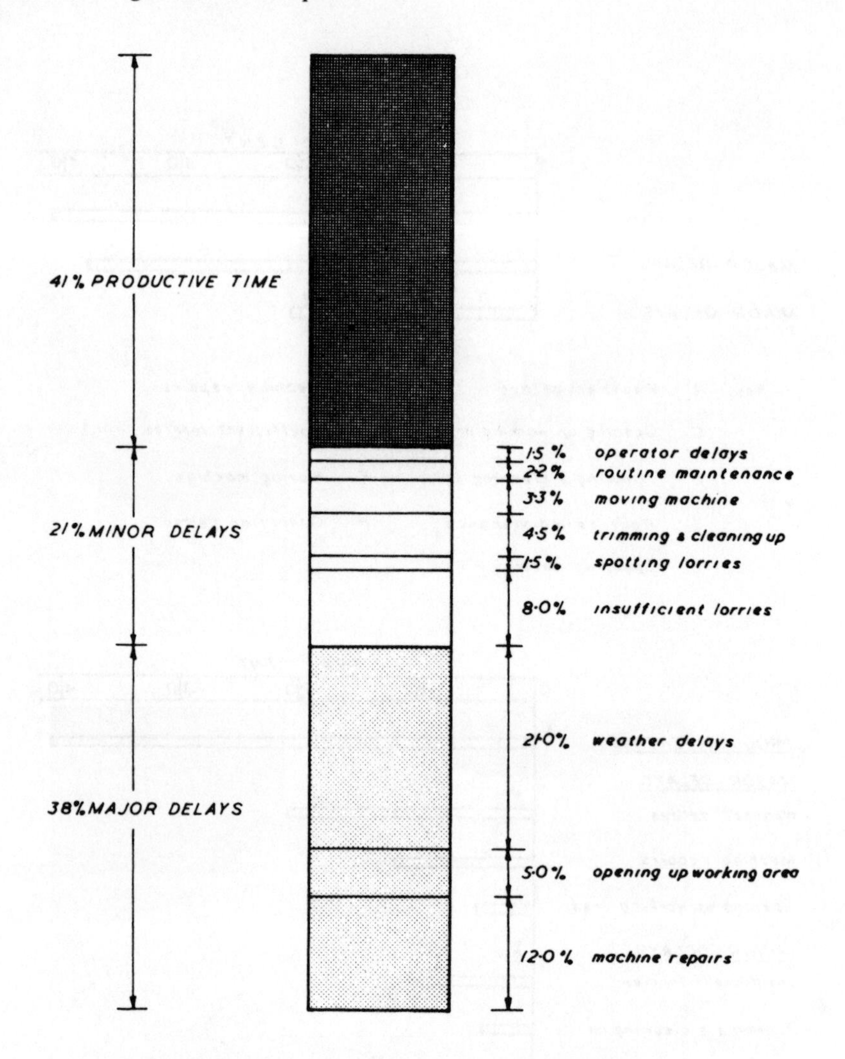

Fig. 4.4.2 (*based on Fig. 4.3*)

4.2.2.7 Graphs

A graph is a line which is plotted on a chart field. The chart field is generally divided into equal vertical divisions and equal horizontal divisions (an arithmetic chart field), but other chart fields are sometimes used which are divided into equal horizontal divisions but have the vertical scale calibrated by logarithms (semi-log chart field).

The following general rules should be observed when constructing chart fields and drawing graphs.

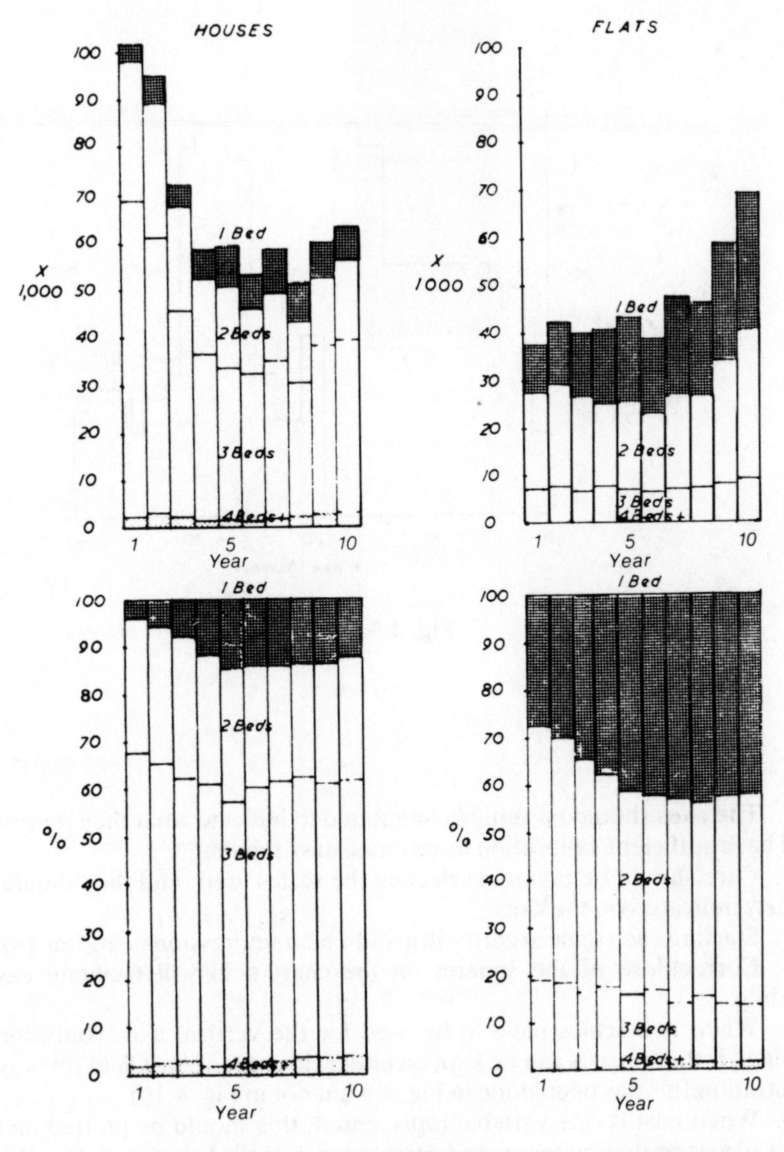

Fig. 4.4.3 (*based on Fig. 4.1*)

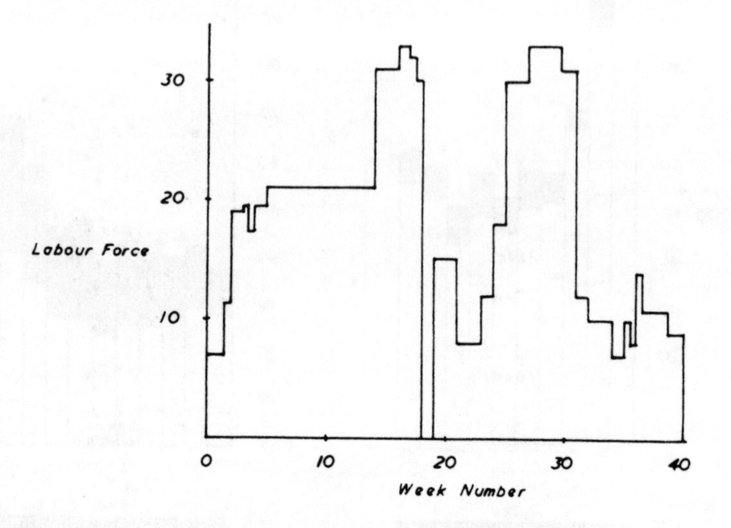

Fig. 4.5.1

1. The axes should be suitably annotated to indicate what they represent and have sufficient calibration to facilitate easy reading.
2. Care should be taken in selecting the scales used, and they should be clearly indicated on the axes.
3. Starting each scale at zero will avoid a false impression being imparted.
4. Correct use of the squares on the chart field will facilitate easier reading.
5. When two scales have to be used for the vertical axis, confusion is avoided if these scales can be kept on either side of the chart field (by way of illustration this has been done in Fig. 4.9 but not in Fig. 4.10).
6. When cost is one variable represented, this should be plotted on the vertical axis so that increases and decreases can easily be appreciated. When time is a variable this is usually plotted on the horizontal axis.
7. The figures from which the chart is made up should be clearly indicated on the chart by dots or crosses to ensure that values are not missed when the graph is being drawn.
8. When a number of graphs are being compared on the same chart field, advantage should be taken of the use of colours or different line codes. However, the number of lines used must be restricted when they fall in the same part of the chart field.

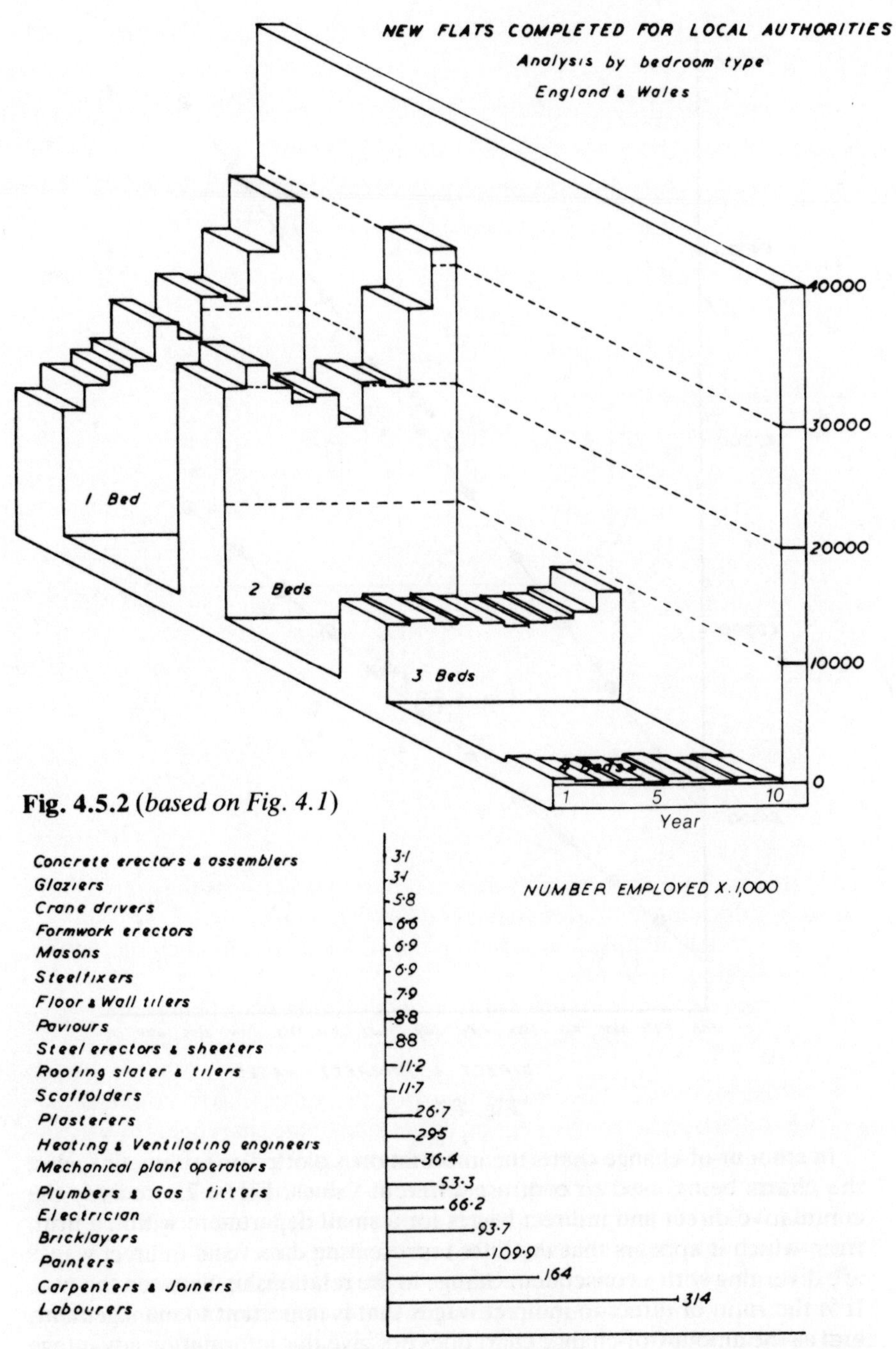

Fig. 4.5.2 (*based on Fig. 4.1*)

N.B. All other building & civil engineering crafts not shown above total 61·8 × 1000
All other occupations not shown above total 54·5 × 1000

Fig. 4.6

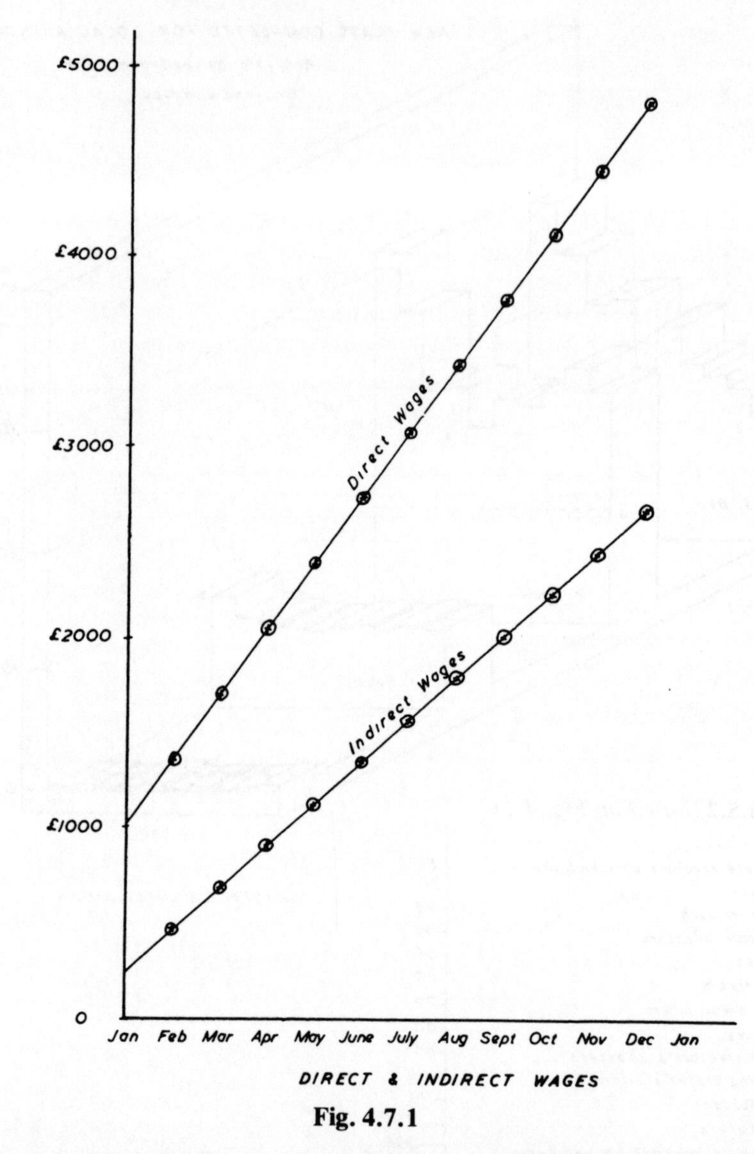

Fig. 4.7.1

In amount-of-change charts the information is plotted on arithmetic scales, the charts being used to contrast different values. Fig. 4.7.1 records the cumulative direct and indirect wages for a small department within a firm, from which it appears that the lines representing direct and indirect wages are diverging with a consequent change in the relationship between the two. It is the ratio of direct to indirect wages that is important to management, and as the amount-of-change chart does not give this information advantage should be taken of the rate-of-change chart, which plots the information on a semi-log chart field.

The rate-of-change chart is used to show whether two or more sets of values are maintaining a constant ratio with each other. If the rate of change is the same in all sets of values, then the lines will remain a constant distance apart, but the lines will either converge or diverge if the rate of change alters. An example of this technique is to be found in the control of the direct to indirect wages ratio (Fig. 4.7.2): it can be seen that the lines representing direct and indirect wages are parallel, which indicates a constant relationship between the two. Another use of the rate-of-change chart is in the comparison of small values with very large values on one chart. In Fig. 4.8 a comparison is being made between a maximum of about 31000 two-bedroomed flats and a minimum of about 350 four-bedroomed flats. Clearly if this presentation had been attempted on arithmetic paper of the same size or any other reasonable size the minimum value of 350 would not have been noticeable.

Other examples of graphs are given in the following section.

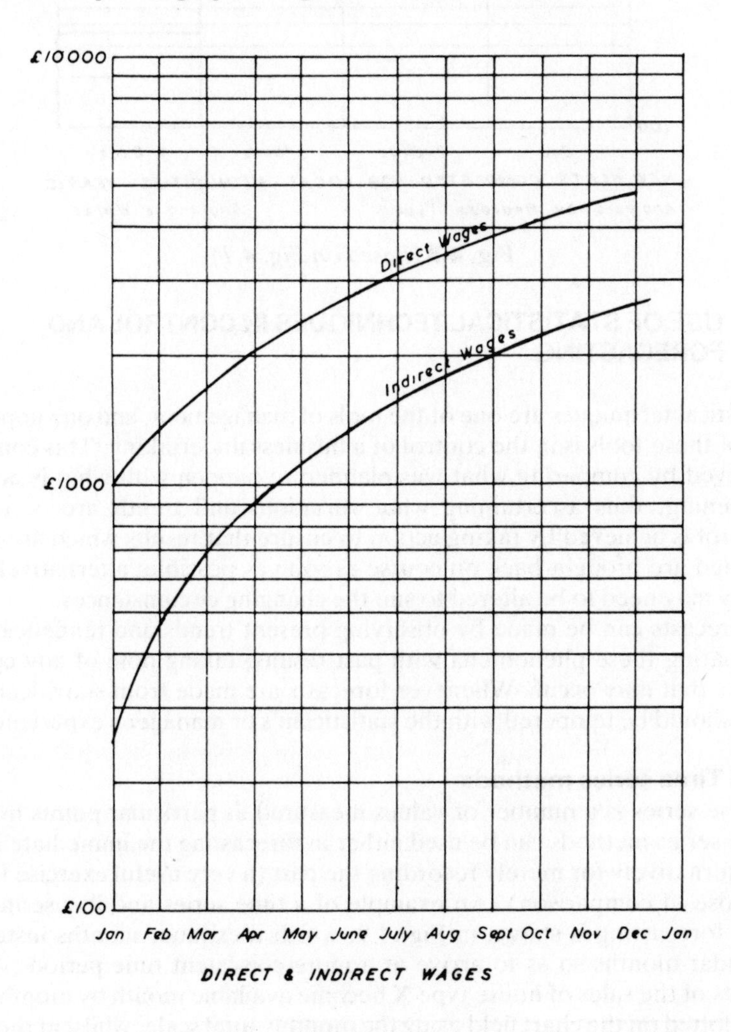

Fig. 4.7.2

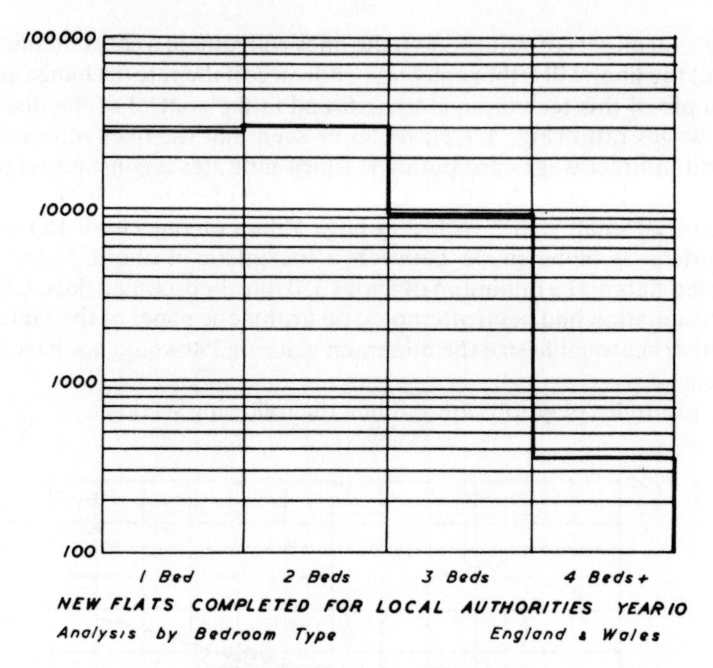

Fig. 4.8 (*based on Fig. 4.1*)

4.3 USE OF STATISTICAL TECHNIQUES IN CONTROL AND FORECASTING

Statistical techniques are one of the tools of management, and one important use of these tools is in the control of a business undertaking. This control is achieved by comparing what was planned to happen with what is actually happening, thus ascertaining what variations and trends are occurring. Control is achieved by taking action to ensure that results which are not as planned are brought back on course as soon as possible; alternatively, the policy may need to be altered to suit the changing circumstances.

Forecasts can be made by observing present trends and tendencies and comparing these phenomena with past results, taking note of any cyclical trends that may occur. Whenever forecasts are made from statistical data, they should be tempered with the statistician's or manager's experience.

4.3.1 Time series methods

A time series is a number of values measured at particular points in time. Time series methods can be used either in forecasting the immediate future or alternatively for merely recording the past (a very useful exercise for the purpose of comparison). An example of a time series and its use in short term forecasting is shown in Fig. 4.9, which uses lunar months instead of calendar months so as to arrive at a more consistent time period. As the results of the sales of house type X become available month by month, they are plotted on the chart field using the monthly total scale, whilst at the same time the moving annual total can be plotted on the chart field using the

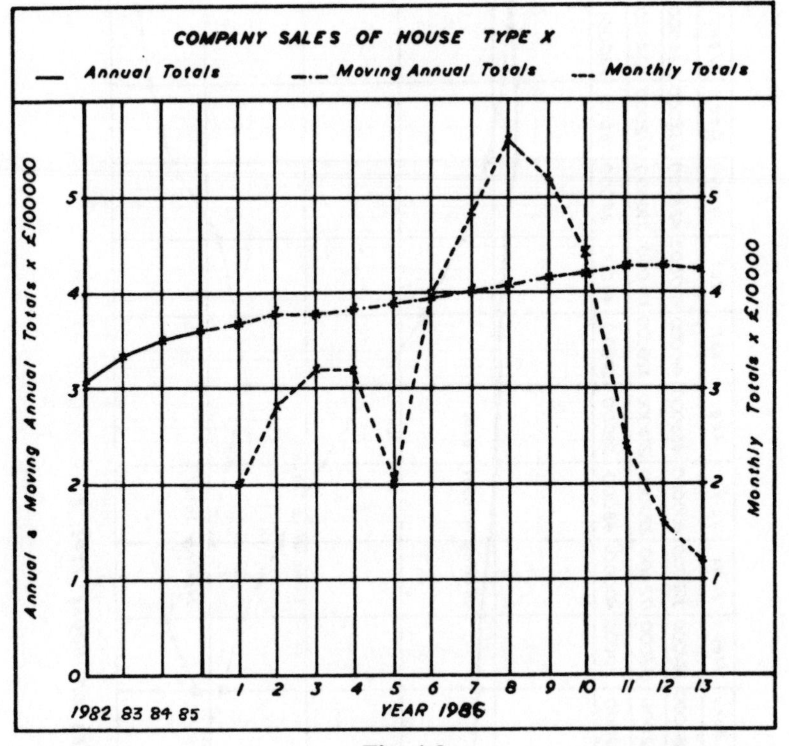

Fig. 4.9

moving annual total scale. The moving annual total is found by adding the present month's result to the previous twelve results (note that if this example had been based on calendar months the present month's result would be added to the previous eleven results). The advantage of the moving annual total is that it smooths out the variations that are occurring monthly, and indicates the trend of events. An example of this can be seen in Fig. 4.9, where although at the end of time period 5 there has been a considerable drop in sales it can nonetheless be seen from the moving annual total that the general trend of sales is still upwards.

From the trends indicated by the moving annual total it should be possible to forecast the result at the end of the year.

4.3.2 Z-chart

This chart is called the Z-chart because of the characteristic shape set up by the three graphs (monthly total, cumulative monthly total and moving annual total) plotted on the chart field.

This type of chart is particularly useful for recording sales statistics, but can be used to advantage with any data where trends are important. By plotting the budgeted sales in the form of a Z-chart and then plotting actual results on the chart as they become available, it is possible to depict discrepancies so that control can take place.

An example of a Z-chart is given for the sales of house type Y (Fig. 4.10).

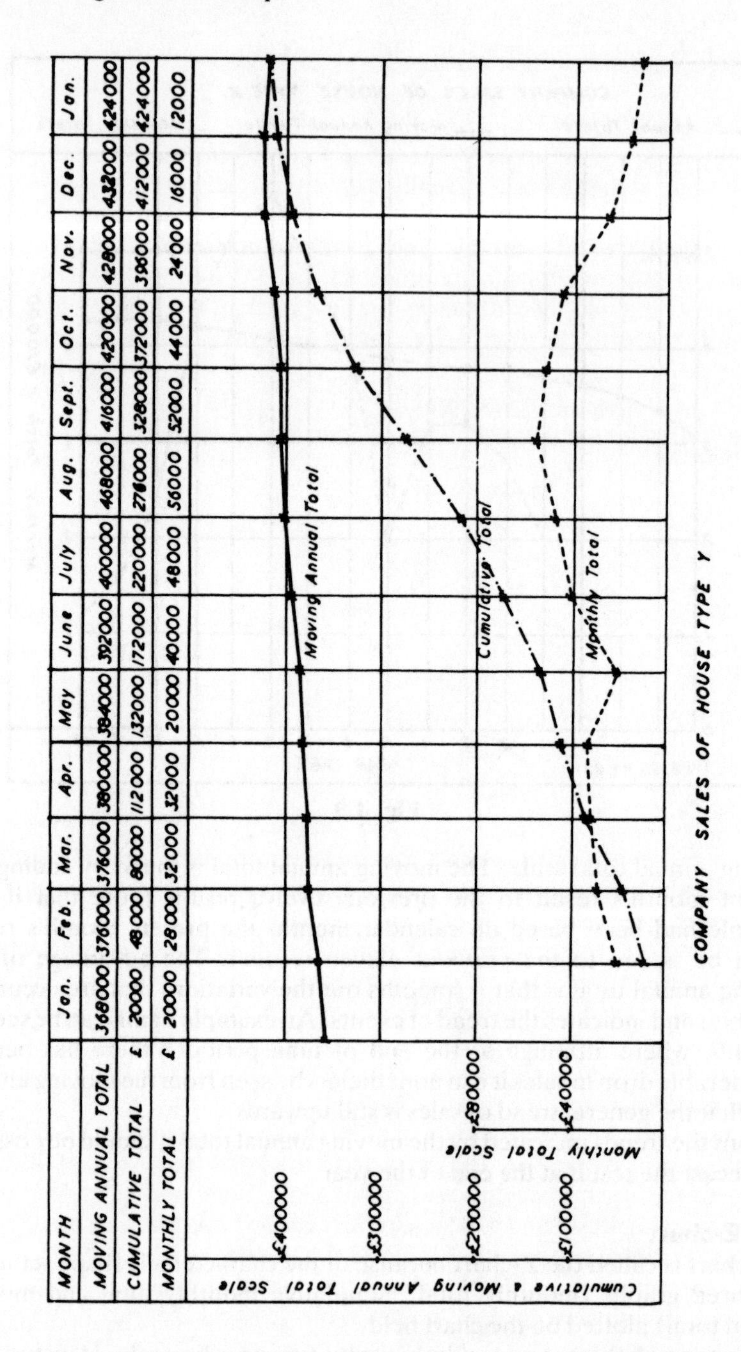

MONTH	Jan.	Feb.	Mar.	Apr.	May	June	July	Aug.	Sept.	Oct.	Nov.	Dec.	Jan.
MOVING ANNUAL TOTAL £	368000	376000	376000	380000	384000	392000	400000	468000	416000	420000	428000	432000	424000
CUMULATIVE TOTAL £	20000	48000	80000	112000	132000	172000	220000	276000	328000	372000	396000	412000	424000
MONTHLY TOTAL £	20000	28000	32000	32000	20000	40000	48000	56000	52000	44000	24000	16000	12000

Fig. 4.10

In this chart, as with the previous chart, it is possible to arrive at the trend in sales by plotting the moving annual total and thus smoothing out the monthly variations. It is common practice to plot the monthly figures to a larger scale than the cumulative and annual figures.

4.3.3 Relatives and index numbers

4.3.3.1 Relatives

The relative is a ratio between the value for any one year's result and the value for the standard year (usually called the base year), expressed as a percentage of the standard year. Relatives can be used for comparing changes in value of any variable, such as the volume of production or wage rates. As an example a list of relatives has been drawn up for the value of new orders obtained by contractors for new work in the housing field (Fig. 4.11); note that only the totals arrived at for each year are being considered.

Fig. 4.11 New housing orders obtained by contractors for new work in Great Britain (*Based on the Annual Digest of Statistics* [*H.M.S.O.*])

	Public sector		*Private sector*	
Year	*£ million*	*Relative*	*£ million*	*Relative*
1	273	100	469	100
2	343	126	465	99
3	454	166	526	112
4	465	170	739	158
5	510	187	693	148
6	564	207	590	126

It can be seen from this chart that between the base year (1) and (2), housing in the public sector increased to 126 per cent, whereas housing in the private sector decreased to 99 per cent. In year 6 housing in the public sector had reached 207 per cent whereas housing in the private sector had only attained 126 per cent. It is difficult to compare the increases in work done in each sector without using relatives as both these series start at different levels.

4.3.3.2 Index numbers

Index numbers are a means by which the related movements of statistical variables can be estimated. They are used in cases where it would be difficult to measure actual movements and provide in a single term an indication of the variation against time of a group of related values. Index numbers are particularly useful because they indicate the trend.

If a contractor wanted to know how the cost of all the materials he used was varying from year to year, he could arrive at the answer by compiling index numbers for the cost of all these materials for each year and comparing them with some base year. The first step would be to ascertain the price of these materials in the base year and the relative importance of each of these materials (weight) in the contractor's work.

The index for the base year is given the value of 100, and a table is constructed (Fig. 4.12.1) to find the index number for the next year.

Fig. 4.12.1

	Base year + 1		
Material	*Percentage rise in price over base year*	*Weight*	*Producer (weight × percentage rise)*
A	5	25	125
B	17	14	238
C	13	10	130
D	20	12	240
E	11	8	88
F	9	5	45
G	10	20	200
		94	1066

Therefore the increase $= \frac{1066}{94} =$ 11 per cent approximately.

Hence the index number for year $x + 1 = 111$.

This process would be repeated for each year as shown in Figs 4.12.2, 4.12.3 and 4.12.4 so as to build up a list of index numbers.

Fig. 4.12.2

	Year $x + 2$		
Material	*Percentage rise in price over base year*	*Weight*	*Producer (weight × percentage rise)*
A	10	25	250
B	23	14	322
C	17	10	170
D	30	12	360
E	20	8	160
F	16	5	80
G	24	20	480
		94	1822

Therefore the increase $= \frac{1822}{94} =$ 19 per cent approximately.

Hence the index number for year $x + 2 = 119$.

Fig. 4.12.3

Year x +3

Material	*Percentage rise in price over base year*	*Weight*	*Producer (weight × percentage rise)*
A	11	25	275
B	24	14	336
C	18	10	180
D	25	12	300
E	25	8	200
F	15	5	75
G	20	20	400
		94	1766

Therefore the increase $= \frac{1766}{94} =$ 19 per cent approximately.

Hence the index number for year $x + 3 = 119$.

Fig. 4.12.4

Year $x + 4$

Material	*Percentage rise in price over base year*	*Weight*	*Producer (weight × percentage rise)*
A	12	20	240
B	24	14	336
C	18	10	180
D	26	12	312
E	27	8	216
F	17	5	85
G	21	20	420
		89	1789

Therefore the increase $= \frac{1789}{89} =$ 20 per cent approximately.

Hence the index number for year $x + 4 = 120$.

(Note that the weight of material A has changed, denoting a change in usage.)

The index numbers for the materials employed are therefore as follows:

Year x 100
Year $x + 1 = 111$
Year $x + 2 = 119$
Year $x + 3 = 119$
Year $x + 4 = 120$

By building up these index numbers over a period of years it is possible to work out the trends in the cost of the materials used and thus to forecast the possible cost at some point in the future.

4.3.4 Quality control

4.3.4.1 Introduction

The aim of quality control is to ensure that all items produced can be used for their intended purpose by facilitating the elimination of defects and variations from the standards established for the production process. Quality control also seeks to avoid waste of time, materials and money by highlighting the point at which a production process is becoming defective, because if defective work did leave the factory and reach the site the cost of the consequent delays could be many times more than the cost of scrapping the original defective in the factory and providing a replacement. For example, if precast column units had to be altered on site, this would cause a hold-up in the production of the concrete frame which in turn would delay the frame. It may even be impossible to alter some units on site, which would necessitate remanufacture.

4.3.4.2 Statistical quality control

This is one of the most important control techniques based on the application of statistical theory. It may be applied to most processes that are of a repetitive nature, examples in the construction industry being the control of concrete and pre-cast concrete production, and the control of standard joinery in a joiners shop.

Statistical quality control is based upon the technique of statistical sampling.

When a production process is running normally, random samples of the items produced will be distributed within the limits of a normal frequency curve possessing the following properties

1. A mean value corresponding to the peak of the curve.
2. Three standard deviations either side of the mean containing 99.7 per cent of all the results.
3. Two standard deviations either side of the mean containing 95.4 per cent of all the results.
4. One standard deviation either side of the mean containing 68.2 per cent of all the results. (Fig. 4.13).

Any particular process can be represented by drawing up a frequency curve from random samples, and from this it is then possible to ascertain by plotting on any specific tolerance what proportion of the items produced have to be rejected. For example, if the tolerance coincided with three standard deviations to the lower side of the mean, then 0.15 per cent of all manufactured articles would be rejected. Consequently it should be possible, knowing the standards of performance obtainable in the manufacturing process and knowing the proportion of rejects that would be acceptable, to arrive at a combination of the two. This would then be the standard, i.e. the accepted process having a particular specification with an allowed number of rejects (one in *x* tested).

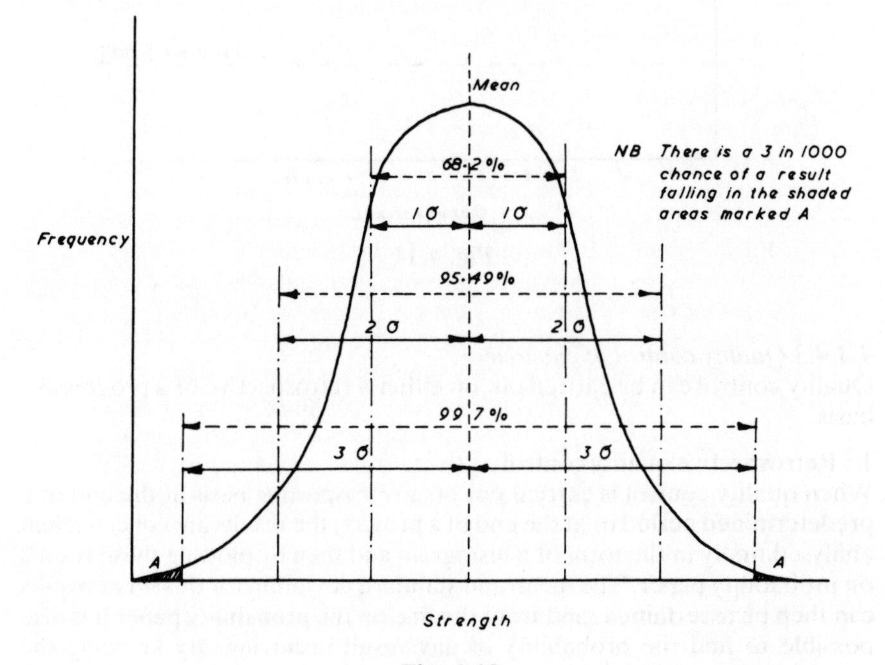

Fig. 4.13

Once this information has been acquired, it is possible to set up a control chart (Fig. 4.14) to carry out progressive control of the process. Customarily the possible values are shown on the vertical axis and the sample number on the horizontal axis, whilst both the mean value and the control limits should also be indicated. A common setting for the upper and lower control limits is $\pm 3\sigma$, but limits can be set at any level providing it is realized that the closer the limits are together, the greater will be the number of rejects.

As results become available they are plotted on the control chart, which throws into relief significant variations from the normal performance that require further investigation. Consistently high or low readings indicate faulty setting of the machine (e.g. if concrete was produced having a consistently high crushing strength, this may indicate that too much cement was being used in the batches, possibly as a result of some fault in the cement weighing mechanism).

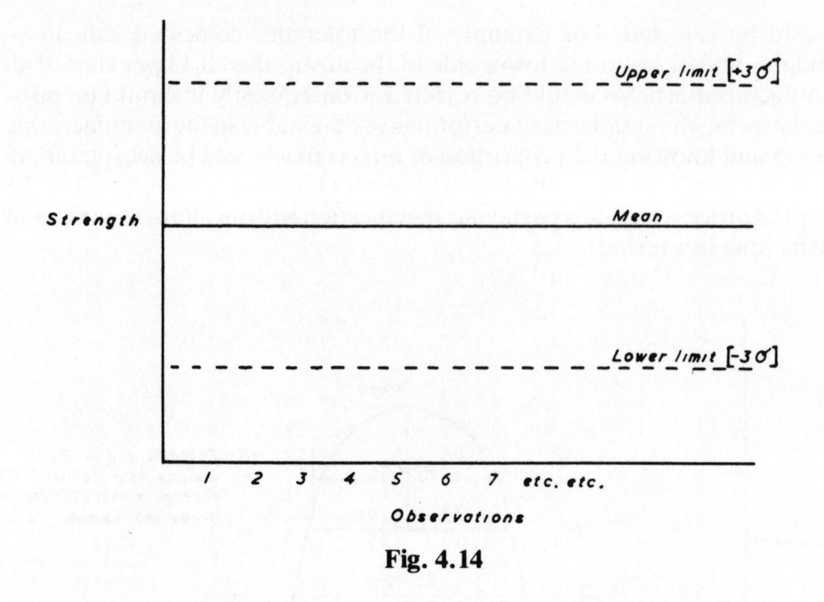

Fig. 4.14

4.3.4.3 Quality control of concrete

Quality control can be carried out on either a retrospective or a progressive basis.

1. Retrospective quality control

When quality control is carried out on a retrospective basis at the end of a predetermined period or at the end of a project, the results are collected and analysed firstly in the form of a histogram and then by plotting these results on probability paper. The mean and standard deviation for this set of results can then be ascertained, and from the line on the probability paper it is also possible to find the probability of any result occurring. By knowing the standard of control for this set of results, it is therefore possible to undertake another project requiring the same standard of control with prior knowledge of the probability that a particular tolerance can be met.

As an example, assume that the results shown in Fig. 4.15.1 are the results obtained on a particular project. The steps to be taken in analysing these results are as follows:

A. Compile a frequency distribution chart (Fig. 4.15.2), which will indicate the number of results occurring in a particular class.

B. Plot these results in the form of a histogram (Fig. 4.15.3), which gives a clearer impression of the distribution of results (the histogram is based on results recorded in the frequency distribution chart). It is helpful for the next stage if a table of the cumulative frequency and cumulative frequency percentage is recorded with the histogram.

C. Construct a probability graph. This is a straight line graph drawn on probability paper (Fig. 4.15.4), the horizontal scale being divided into the class intervals and the vertical scale into percentages. The cumulative percentages are plotted on the chart field in the appropriate class intervals, and

Cube Crushing Strength at 28 days in MN/m²							
1	20.52	26	18.92	51	26.32	76	22.45
2	29.15	27	14.98	52	28.37	77	26.53
3	24.16	28	20.33	53	20.17	78	24.59
4	28.25	29	18.66	54	30.92	79	25.13
5	18.32	30	17.84	55	27.63	80	27.51
6	29.22	31	20.78	56	25.78	81	23.77
7	26.35	32	19.43	57	26.77	82	32.51
8	29.11	33	15.66	58	31.67	83	24.22
9	22.22	34	21.77	59	22.88	84	28.39
10	30.12	35	18.82	60	31.53	85	22.32
11	27.47	36	16.97	61	24.12	86	33.42
12	19.54	37	21.96	62	29.88	87	25.26
13	30.87	38	19.75	63	27.41	88	34.66
14	33.68	39	22.72	64	31.44	89	24.37
15	23.66	40	20.94	65	22.73	90	29.75
16	30.66	41	24.81	66	30.81	91	35.51
17	33.75	42	17.23	67	20.98	92	22.56
18	19.80	43	26.22	68	28.49	93	28.95
19	32.16	44	20.87	69	30.67	94	34.72
20	27.90	45	27.81	70	22.61	95	25.77
21	35.22	46	17.49	71	29.48	96	35.42
22	25.38	47	26.96	72	32.62	97	23.52
23	37.37	48	23.67	73	26.67	98	36.78
24	25.26	49	24.75	74	33.87	99	37.67
25	26.86	50	21.58	75	27.92	100	24.11

Fig. 4.15.1

<table>
<tr><th rowspan="2">CLASS</th><th colspan="4">FREQUENCY</th></tr>
<tr><th colspan="2">1st 25 results</th><th colspan="2">Total 100 results</th></tr>
<tr><td>1400-15.98</td><td></td><td></td><td>||</td><td>2</td></tr>
<tr><td>1600-17.98</td><td></td><td></td><td>||||</td><td>4</td></tr>
<tr><td>1800-19.98</td><td>|||</td><td>3</td><td>𝍸 |||</td><td>8</td></tr>
<tr><td>2000-21.98</td><td>|</td><td>1</td><td>𝍸 𝍸</td><td>10</td></tr>
<tr><td>2200-23.98</td><td>||</td><td>2</td><td>𝍸 𝍸 ||</td><td>12</td></tr>
<tr><td>2400-25.98</td><td>|||</td><td>3</td><td>𝍸 𝍸 ||||</td><td>14</td></tr>
<tr><td>2600-27.98</td><td>||||</td><td>4</td><td>𝍸 𝍸 𝍸</td><td>15</td></tr>
<tr><td>2800-29.98</td><td>||||</td><td>4</td><td>𝍸 𝍸 |</td><td>11</td></tr>
<tr><td>3000-31.98</td><td>|||</td><td>3</td><td>𝍸 ||||</td><td>9</td></tr>
<tr><td>3200-33.98</td><td>|||</td><td>3</td><td>𝍸 ||</td><td>7</td></tr>
<tr><td>3400-35.98</td><td>|</td><td>1</td><td>𝍸</td><td>5</td></tr>
<tr><td>3600-37.98</td><td>|</td><td>1</td><td>|||</td><td>3</td></tr>
<tr><td></td><td colspan="2">TOTAL 25</td><td colspan="2">TOTAL 100</td></tr>
</table>

Fig. 4.15.2

a straight line is drawn through these points. If a perpendicular is dropped from where this line crosses the 50 per cent mark down to the horizontal axis, it will cut the horizontal axis at the mean value, and if lines are dropped from where the inclined line cuts the 16 per cent and 84 per cent lines, these lines will cut the horizontal axis at one standard deviation at either side of the mean (1σ either side of the mean contains approximately 68 per cent of all the results). Consequently we have now obtained the mean and the standard deviation for this series, and from the chart it is also possible to ascertain the probability of achieving a particular result: if the desired minimum crushing strength is 18.00 MN/m², for example, then by drawing a perpendicular line up from the horizontal axis at this point until it cuts the inclined line it can be seen that there is a 9 per cent chance of falling below this limit or alternatively a 91 per cent chance of exceeding the limit. If this rate of failure is acceptable in the new process, then the standard of control exercised before will suffice; if not, then tighter control will be necessary.

As an alternative to finding the standard deviation and the mean by this method, it is possible to calculate them. First the information shown in the frequency distribution table (Fig. 4.15.2) is analysed into the following form:

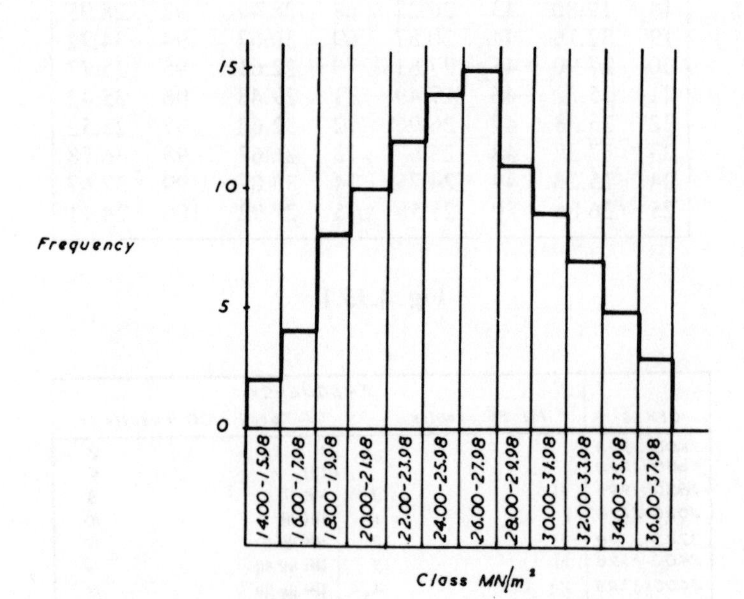

Cumulative frequency %	2	6	14	24	36	50	65	76	85	92	97	100
Cumulative frequency	2	6	14	24	36	50	65	76	85	92	97	100
Frequency	2	4	8	10	12	14	15	11	9	7	5	3

Fig. 4.15.3

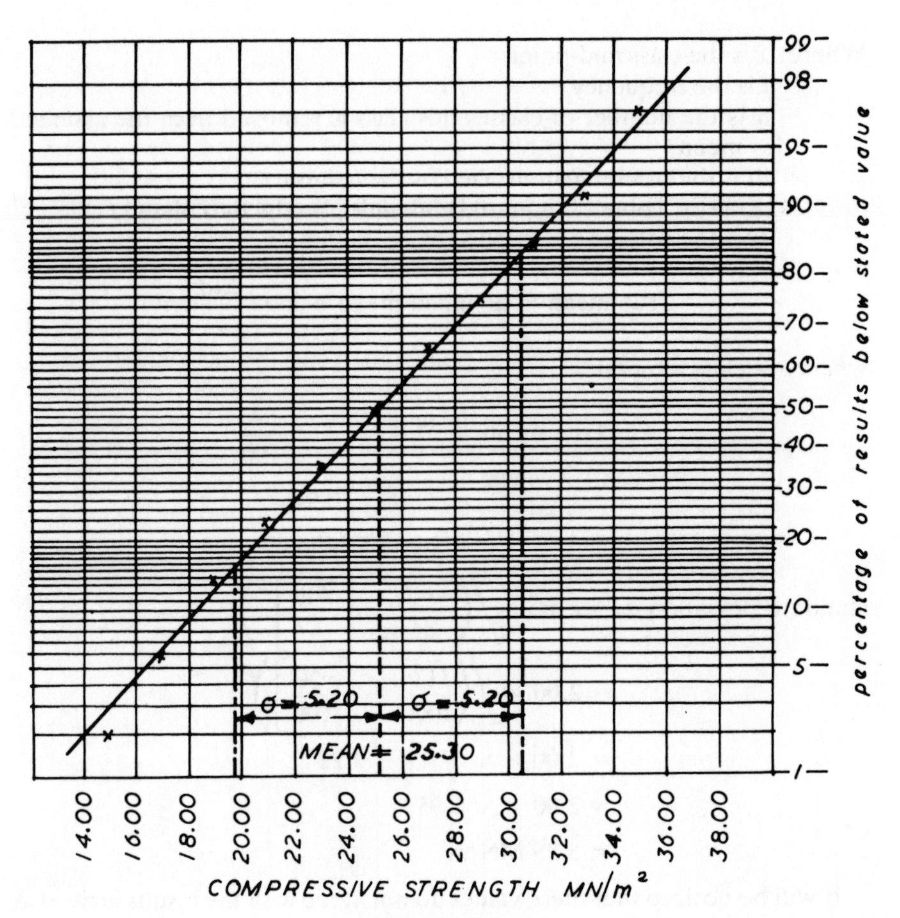

Fig. 4.15.4

Fig. 4.15.5

x	f	u	uf	u^2f	fx
15.00	2	−6	−12	72	30
17.00	4	−5	−20	100	68
19.00	8	−4	−32	128	152
21.00	10	−3	−30	90	210
23.00	12	−2	−24	48	276
25.00	14	−1	−14	14	350
assumed mean 27.00	15	0	0	0	405
29.00	11	1	11	11	319
31.00	9	2	18	36	279
33.00	7	3	21	63	231
35.00	5	4	20	80	175
37.00	3	5	15	75	111
	100		−47	717	2606

Where: x is the class mid-point
f is the frequency
u is the number of classes this class is removed from the assumed mean
uf is the column u multiplied by the column f
u^2f is the column u squared and multiplied by the column f
fx is the column f multiplied by the column x

$$\text{True mean} = \text{assumed mean} + c \times \frac{\Sigma uf}{\Sigma f}$$

where c = class interval

$$\therefore \text{True mean} = 27.00 + 2 \times \frac{-47}{100}$$
$$= 27.00 - 0.94$$
$$= 26.06 \text{ MN/m}^2$$

$$\text{standard deviation } \sigma = c \times \sqrt{\left(\frac{\Sigma u^2 f}{\Sigma f} - \left[\frac{\Sigma uf}{\Sigma f}\right]^2\right)}$$
$$= 2.00 \times \sqrt{\left(\frac{717}{100} - \left[\frac{-47}{100}\right]\right)^2}$$
$$= 2.00 \times \sqrt{7.17 - 0.22}$$
$$= 2.00 \times \sqrt{6.95}$$
$$= 5.25 \text{ MN/m}^2$$

It will be noticed that these values do not agree with the results arrived at by graphical methods, but this is only to be expected.

Fig. 4.16.1

Proportion falling below lower control level	*K*
1 in 10	1.28
1 in 20	1.64
1 in 25	1.75
1 in 33	1.88
1 in 40	1.96
1 in 50	2.05
1 in 100	2.33

2. Progressive quality control

If quality is controlled on a progressive basis, then it is possible to distinguish adverse trends at an early stage and control them by taking corrective action.

To perform this type of control, a chart is drawn up (Fig. 4.14) with concrete strength on the vertical axis and the result number on the horizontal axis. The mean strength and the control levels must now be plotted on the chart, these values being obtained in two different ways.

A. It is quite possible that the designer of the concrete mix made a number of assumptions about the value of the standard deviation and the mean strength, knowing the standard of control that would be possible on site. He will have specified the minimum strength on this basis, and in addition he will know what proportion of the results can be allowed to fall below a certain level. Consequently the anticipated mean strength may be plotted on the chart. The control levels can also be plotted on at $\pm K\sigma$ from the mean where K is a constant from a table (Fig. 4.16.1) and σ is the anticipated standard deviation.

B. The other possibility is that the standard of control is not known. In this case the standard deviation is ascertained after approximately 25 cubes have been taken at random and tested. To find the mean strength necessary to ensure that no more than a specified number of results fall below the specified minimum, the standard deviation σ must be found and $K\sigma$ added to the minimum strength where K would have some desired value (i.e. 1 in 100, $K = 2.33$; 1 in 25, $K = 1.75$). This value is then plotted on the chart, after which the control levels are also entered by using the same method. It is common practice to repeat this process when more results become available. As an example in which the mean and standard deviations are not known, assume that the data given in Fig. 4.15.1 relates to a project that is being controlled progressively, so that the first 25 results can be considered as the results for the trial period. The mean and standard deviation must first be found for these results by following the method used for retrospective control (Figs 4.16.2 and 4.16.3). From Fig. 4.16.3 it can be seen that the mean is 26.30 MN/m^2 and the standard deviation is 5.20 MN/m^2. These results are now used to construct the control chart (Fig. 4.16.4) as follows:

(a) Draw out the axes, with the horizontal representing the test cube number and the vertical and compressive strength.

(b) Insert the minimum strength on the chart (assume this to be 16.50 MN/m^2).

(c) Assuming an allowable failure rate of 1 in 100, the required mean strength can be calculated from mean strength $= K\sigma +$ minimum strength, where σ is the standard deviation of the first 25 results (Fig. 4.16.3).

$$\begin{aligned}\text{Hence mean strength} &= 2.33 \times 5.20 + 16.50 \\ &= 12.11 + 16.50 \\ &\ 28.61\ \text{MN/m}^2\end{aligned}$$

(d) Plot the calculated mean strength on the control chart. If the calculated mean strength is compared with the mean strength for the first 25 results it can be seen that the achieved mean strength is less than that required, so it can be assumed that the control is not good enough. Therefore either the control would have to be improved or the mix design would have to be altered to ensure that only one result in 100 exceeded the limits set.

(e) As more test results become available they are plotted on the chart (Fig. 4.16.5) and if the number of results occurring on either side of the mean is counted it soon becomes evident should the average strength of the concrete being produced fall below the required average. It is also possible to get an impression of the variability after approximately 10 results have been plotted by checking the number of results falling outside the control limits for 1 in 10. When another 10 results are available they can also be checked against the 1 in 10 control limits, and if they are added to the previous 10 results they can then be checked against the 1 in 20 control limits. If more results than planned fall outside these limits then it is necessary to tighten up control to produce the standard deviation.

This example is based on the concept of using individual cube samples as test data. A more sophisticated concept is the use of groups of cubes as samples rather than individual cubes. In the latter case it is necessary to use two control charts – one to control the range and the other to control the mean of the cube sample.

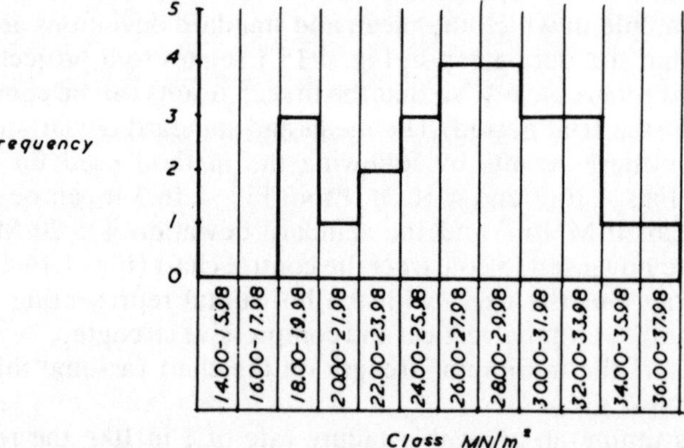

Cumulative frequency %			12	16	24	36	52	68	80	92	96	100
Cumulative frequency			3	4	6	9	13	17	20	23	24	25
Frequency			3	1	2	3	4	4	3	3	1	1

Fig. 4.16.2

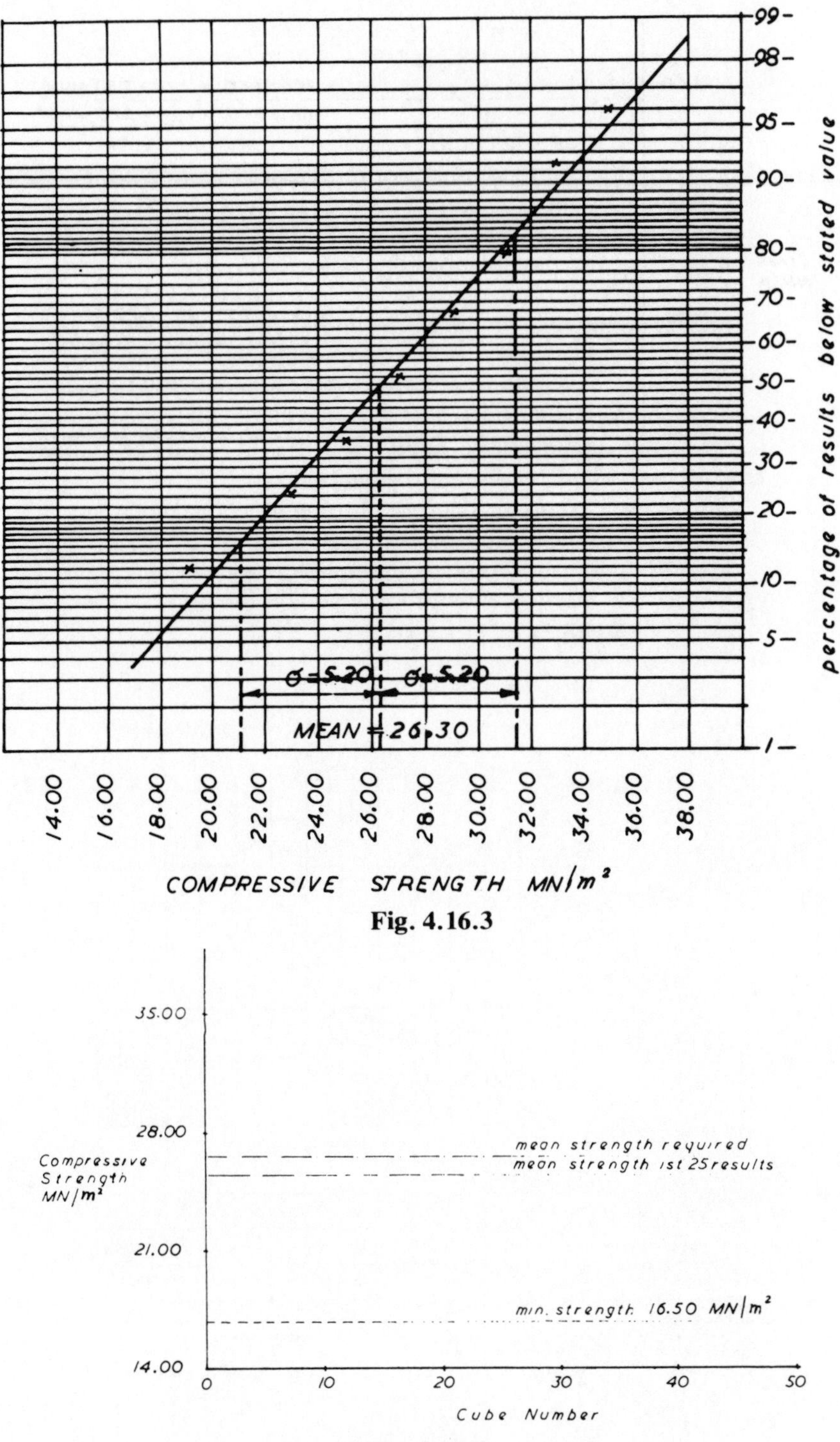

Fig. 4.16.3

Fig. 4.16.4

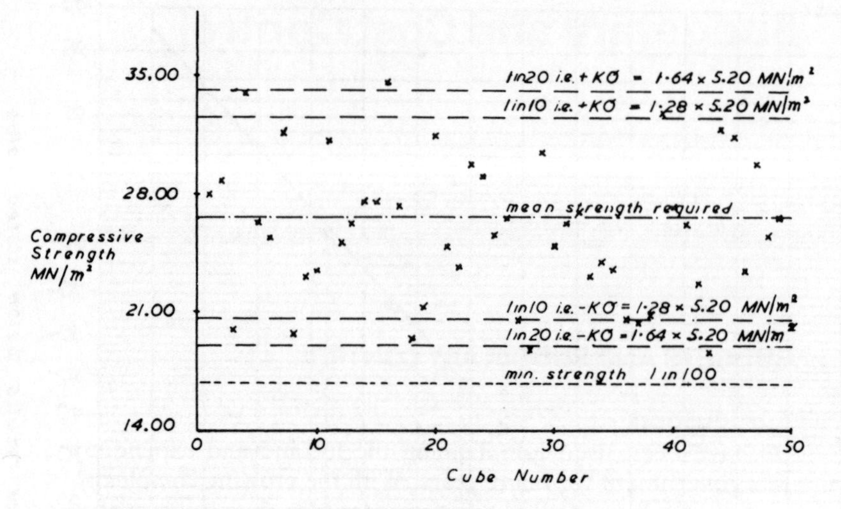
35.00
28.00
21.00
14.00
Compressive Strength MN/m²
1 in 20 i.e. + Kσ = 1·64 × 5.20 MN/m²
1 in 10 i.e. + Kσ = 1·28 × 5.20 MN/m²
mean strength required
1 in 10 i.e. − Kσ = 1·28 × 5.20 MN/m²
1 in 20 i.e. − Kσ = 1·64 × 5.20 MN/m²
min. strength 1 in 100
0
10
20
30
40
50
Cube Number

Fig. 4.16.5

5 Budgetary and Cost Control

5.1 BUDGETS AND BUDGETARY CONTROL

5.1.1 Introduction

Budgets have been used for planning the income and expenditure of a country's government for many years. With the growing complexity of the building industry it is essential that firms also plan their policy well into the future, and it is through budgets that such plans can be converted into the quantitive and monetary terms which are a company's fundamental objectives.

5.1.2 Policy budgets in the construction industry

By compiling budgets a firm can arrive at a selling or tendering price for its projects, having considered what resources are available to carry out the work. The firm will need to know, for example, how much capital is necessary, and having acquired this information it can then decide whether to expand or contract any particular aspect of the business, bearing in mind competition and demand. From past results it may also be possible to determine a change in demand that could give an indication of possible new expansion plans.

5.1.3 Project budgets

It is important for any contractor to know what capital is going to be needed for a project and when it will be required. In order to ascertain this, the contractor will have to draw up a programme for the project and from this find the rate of expenditure and rate of income, based on a particular time period. The difference between these two will give the amount of capital required in this time period (examples are given later in this chapter).

If required, budgets can be prepared for the labour to be used (based on the labour graph), for materials and plant (based on schedules and the programme), and for the site overheads (based on the programme).

5.1.4 Achievement of budgets

Once a budget has been established, the next objective is to achieve the target set. This is attained through a system of budgetary control. Like any other control technique, budgetary control is a continuous comparison of the actual achievement with that planned, which highlights any deviations from the plan so that action can be taken either to bring things back on course or to change the plan.

5.1.5 Examples of budgets and budgetary control

Three examples will be given in the following pages, the first of which is for non-repetitive construction based on the amenity centre (chapter 1, section 1.3.3). The second is an example of repetitive construction based on the shells (chapter 1, section 1.10.3). These examples are very much simplified and are put forward for illustrative purposes only. A more detailed example of non-repetitive construction is given in section 5.1.5.3 in which the cash flow for the project is also ascertained.

5.1.5.1 Non-repetitive construction

It is assumed that a contract has been entered into with the client and that payment for work carried out will be made 14 days after the end of the month in which the work is completed. Retention will be 10 per cent initially with a maximum retention of 5 per cent of the contract sum (£807 900). Half the retention will be released 14 days following the completion of the project, the other half will be released six months after completion.

The cost per operation has been ascertained and allocated to the master construction programme. From this programme it is now possible to arrive at the expenditure for each month (Fig. 5.1.1).

Fig. 5.1.1

	Own expenditure (£)	*Payments to sub-contractors (£)*	*Cumulative expenditure (£)*
Week 4	42200		42200
Week 8	63400		105600
Week 12	64400		170000
Week 16	73800		243800
Week 20	51600		295400
Week 22		5400	300800
Week 24	32300		333100
Week 26		27000	360100
Week 28	69800		429900
Week 30		25740	455640
Week 32	98700		554340
Week 34		64620	618960
Week 36	31400		650360
Week 38		32220	682580
Week 40	27300		709880
Week 42		15820	725700

Using a similar method, it is possible to arrive at the programme of income for the project, the value of each operation having been ascertained from the bill of quantities. The expected amounts of the monthly valuations are shown in Fig. 5.1.2.

Fig. 5.1.2

	Own income (£)	*Retention (£)*	*Income less retention (£)*	*Payments for sub-contractor (£)*	*Cumulative income (£)*
Week 6	45900	4590	41310		41310
Week 10	68900	6890	62010		103320
Week 14	70000	7000	63000		163320
Week 18	80200	8020	72180		238500
Week 22	56600	5660	50940	6000	295440
Week 26	37800	3780	34020	30000	359460
Week 30	78400	4060	74340	28600	462400
Week 34	113200	40000	113200	71800	647400
Week 38	36200	limit of retention	36200	35800	719400
Week 42	30900		30900	17600	767900
		20000 retention repaid			787900
Week 46					
Week 50					
Week 54					
Week 58					
Week 62					
Week 66		20000 retention repaid			807900

(N.B. Payment takes place 14 days after the end of the month in which the work is carried out.)

These figures for income and expenditure have been recorded graphically in Fig. 5.1.3. The capital required at any time can now be ascertained by finding the difference between income and expenditure.

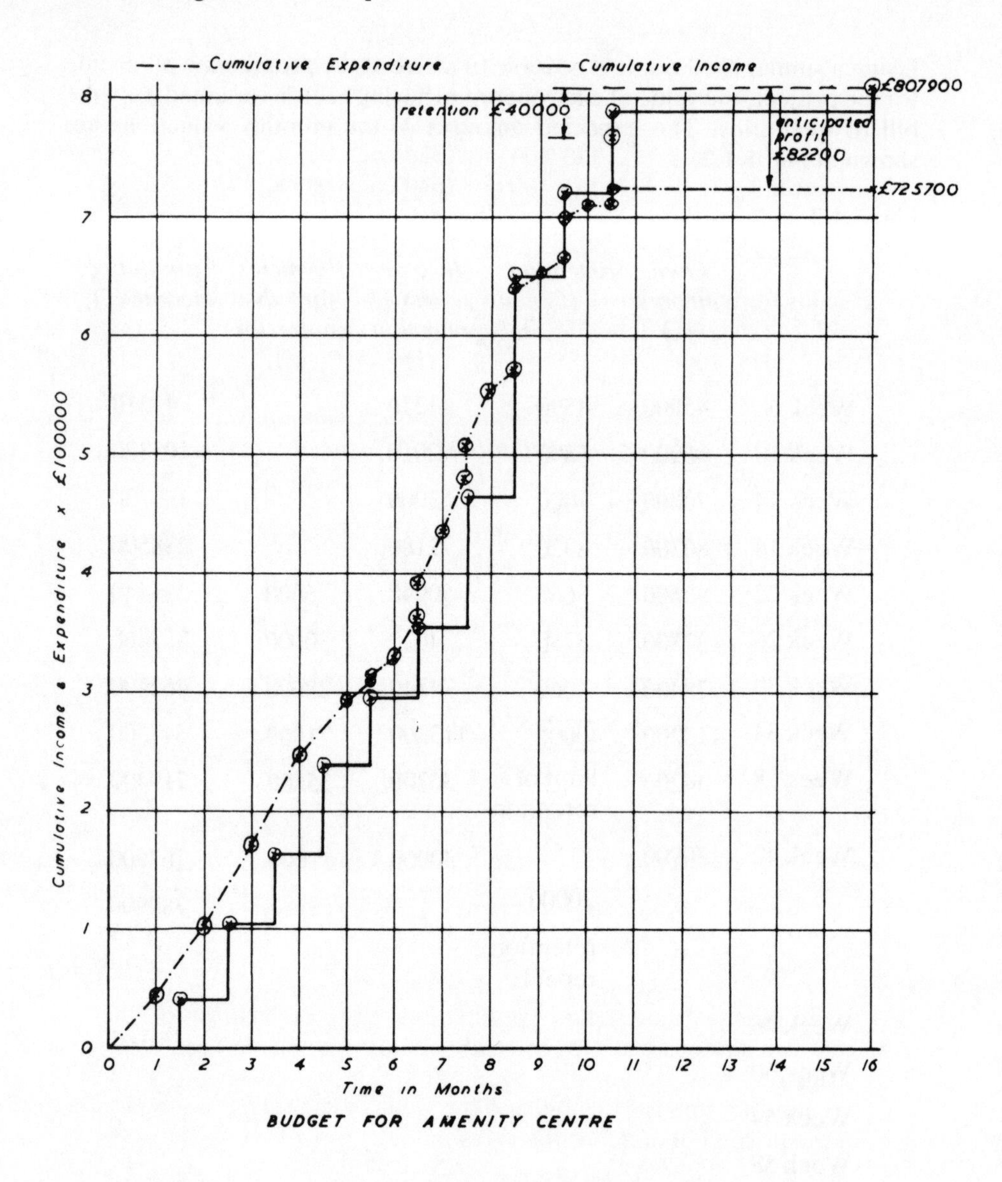

Fig. 5.1.3

5.1.5.2 Repetitive construction

It is assumed that a contract has been entered into with the client and that payment will be made in stages for each unit, e.g. excavation, concrete foundations, brickwork, roof carcass, and roof finishes. Payment will be received one week after completion of a stage, and the contract makes no provision for retention money.

It has been ascertained that the total cost of completing these shells is £136 550, which is built up in the following manner:

Excavation	£1650	= £165 each block
Concrete foundations	£31 900	= £3190 each block
Brickwork	£63 400	= £6340 each block
Roof construction	£13 200	= £1320 each block
Roof finish	£26 400	= £2640 each block
	£136 550	

By allocating these costs to the programme of work it can be seen that the distribution of cost will be that shown in Fig. 5.2.1.

Fig. 5.2.1

	Weekly expenditure (£)	*Cumulative expenditure (£)*
Week 1	2805.0	2805.0
Week 2	10740.0	13545.0
Week 3	14635.0	28180.0
Week 4	17275.0	45455.0
Week 5	16862.50	62317.50
Week 6	16862.50	79180.0
Week 7	16862.5	96042.5
Week 8	16862.5	112905.0
Week 9	14470.0	127375.0
Week 10	6535.0	133910.0
Week 11	52640.0	136550.0

(N.B. In practice these figures would be taken to the nearest £10 but here precise calculations have been made so that the reader can follow the working.)

The contract sum is £157 100 and the value of each stage (as ascertained from the priced bill of quantities) is as follows:

Excavation	£1 900	= £190 each block
Concrete foundations	£36 700	= £3670 each block
Brickwork	£72 900	= £7290 each block
Roof construction	£15 200	= £1520 each block
Roof finish	£30 400	= £3040 each block
	£157 100	

By allocating these values to the programme of work it can be seen that the distribution of the value will be as shown in Fig. 5.2.2 (only completed stages will be measured at the end of the week).

Fig. 5.2.2

	Weekly expenditure (£)	*Cumulative expenditure (£)*
Week 1	0	0
Week 2	380	380
Week 3	15200	15580
Week 4	12860	28440
Week 5	16090	44530
Week 6	15520	60050
Week 7	31040	91090
Week 8	15520	106610
Week 9	15520	122130
Week 10	15520	137650
Week 11	16410	154060
Week 12	43040	157100
	157100	

Both sets of figures can be plotted on a chart field (Fig. 5.2.3) and by subtracting the income from the expenditure the capital required at that particular time can be found.

5.1.5.3 Project budget and cash flow

Figure 5.3.1 shows the programme for a small office building. All operations are programmed at their earliest times, but the amount of float available in each operation is indicated by a dotted line. The expenditure on and income for each operation has been ascertained and is shown in Fig. 5.3.2.

It is assumed that a contract has been entered into with the client and that payment for the work carried out will be made one month after the end of the month in which the work is carried out. The retention will be 5% throughout. Half the retention will be released one month after the completion. By allocating the planned expenditure and anticipated income to the master programme (see Fig. 5.3.3, operations at earliest times), it is possible to arrive at the expenditure and income for each month. These figures can then be processed further to arrive at the cumulative amounts for expenditure and income, having made the necessary adjustments for retention and payment delays. The figures for income and expenditure have been recorded graphically in Fig. 5.3.4.

The cash flow for the project with all the operations taking place at their earliest times can now be ascertained by plotting the monthly planned expenditure and anticipated income as shown in Fig. 5.3.5.

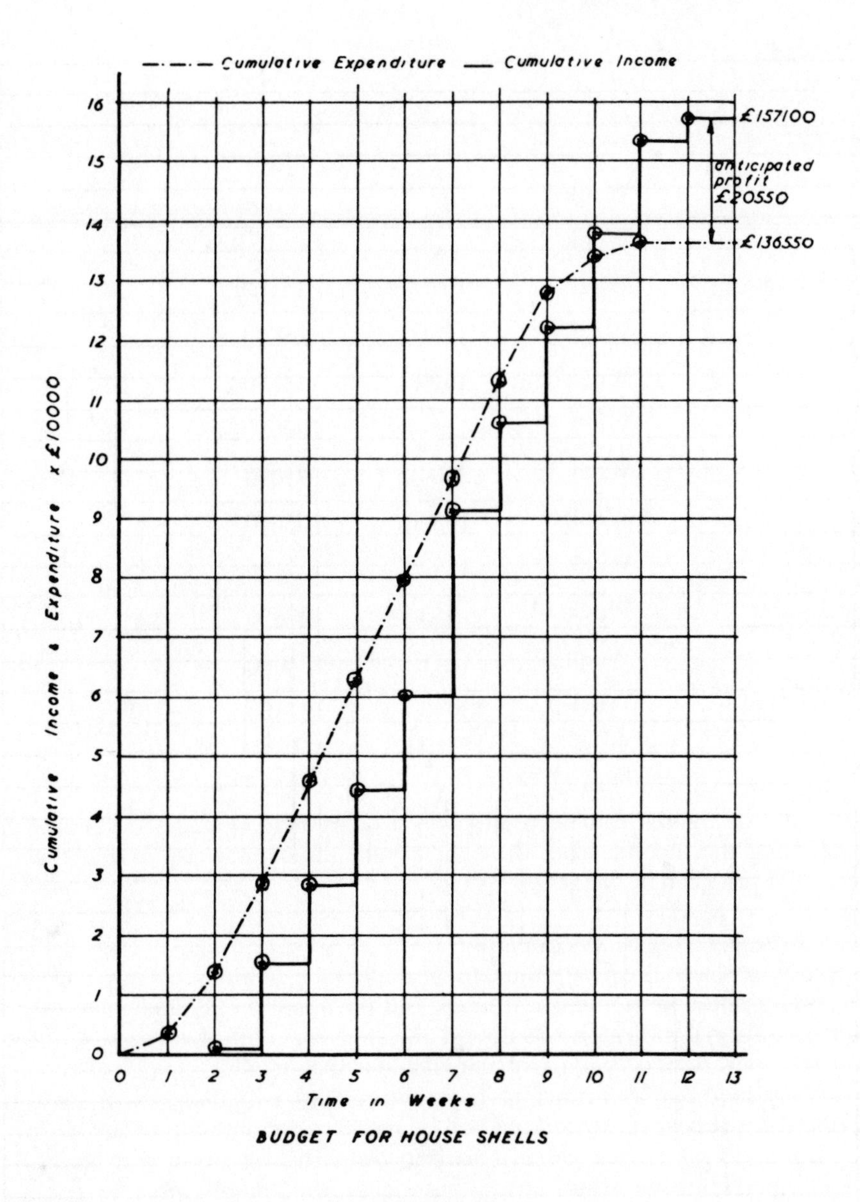

Fig. 5.2.3

With all operations starting at their earliest time, it can be seen that the maximum amount of working capital required is £37 781 at the end of month 4, and that the project becomes self financing at the end of month 6.

For the purpose of comparison the process can be repeated to ascertain the effect on the budget and cash flow of carrying out all operations at their latest times. Fig. 5.3.6 shows the new allocation of planned expenditure and anticipated income with the appropriate processing at the foot of the chart.

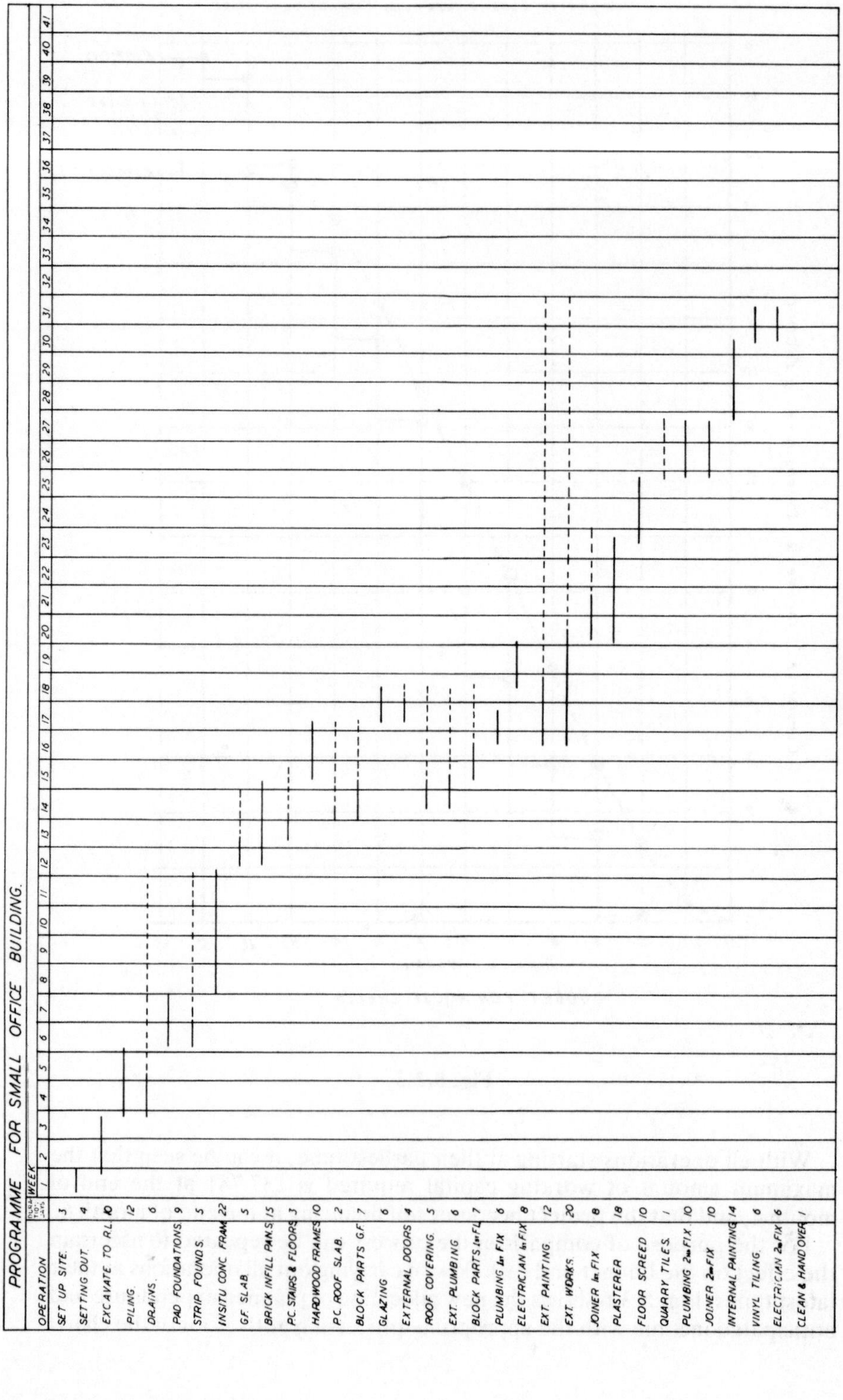
PROGRAMME FOR SMALL OFFICE BUILDING.
OPERATION
DURATION DAYS
WEEK
1 2 3 4 5 6 7 8 9 10 11 12 13 14 15 16 17 18 19 20 21 22 23 24 25 26 27 28 29 30 31 32 33 34 35 36 37 38 39 40 41
SET UP SITE. 3
SETTING OUT. 4
EXCAVATE TO RL 10
PILING. 12
DRAINS. 6
PAD FOUNDATIONS. 9
STRIP FOUNDS. 5
INSITU CONC FRAM 22
G.F. SLAB. 5
BRICK INFILL PANS. 15
P.C. STAIRS & FLOORS 3
HARDWOOD FRAMES 10
P.C. ROOF SLAB. 2
BLOCK PARTS G.F. 7
GLAZING 6
EXTERNAL DOORS 4
ROOF COVERING. 5
EXT. PLUMBING 6
BLOCK PARTS 1st FL 7
PLUMBING 1st FIX 6
ELECTRICIAN 1st FIX 8
EXT. PAINTING 12
EXT. WORKS 20
JOINER 1st FIX 8
PLASTERER 18
FLOOR SCREED 10
QUARRY TILES. 1
PLUMBING 2nd FIX. 10
JOINER 2nd FIX 10
INTERNAL PAINTING 14
VINYL TILING. 6
ELECTRICIAN 2nd FIX 6
CLEAN & HAND OVER 1

Fig. 5.3.1

Operation	*Planned expenditure*	*Anticipated income*
Set up site	210	240
Setting out	60	80
Excavate to R L	1000	1200
Piling	3600	4800
Drains	300	360
Pad foundations	630	720
Strip foundations	500	600
In situ conc frame	22000	33000
G.F. slab	800	900
Brick infill panels	8700	9750
PC stairs and floors	2400	2700
Hardwood frames	1000	1200
PC roof slab	2000	2500
Block parts GF	1750	2100
Glazing	240	300
Ext doors	1600	1800
Roof covering	1000	1200
Ext plumbing	180	210
Block parts 1st floor	2100	2450
Plumbing 1st fix	900	1050
Electrician 1st fix	400	480
Ext painting	240	300
Ext works	7564	9000
Joiner 1st fix	800	880
Plasterer	1800	2160
Floor screed	600	700
Quarry tiles	100	140
Plumbing 2nd fix	1000	1200
Joiner 2nd fix	1000	1200
Int painting	420	504
Vinyl tiling	240	300
Electrician 2nd fix	360	540
Clean and handover	50	60
Prelims	13175	13950

Fig. 5.3.2

PROGRAMME FOR SMALL OFFICE BUILDING - EXPENDITURE & INCOME																	
OPERATION	DURATION DAYS	PLANNED EXPENDITURE	ANTICIPATED INCOME	WEEKS 1	2	3	4	5	6	7	8	9	10	11	12	13	14
SET UP SITE.	3	210	240	210/240													
SETTING OUT	4	60	80	60/80													
EXCAVATE TO R.L.	10	1000	1200	100/120	500/600	400/480											
PILING	12	3600	4800			300/400	1500/2000	1500/2000	300/400								
DRAINS	6	300	360			50/60	250/300										
PAD FOUNDATIONS.	9	630	720						280/320	350/400							
STRIP FOUNDATIONS	5	500	600						400/480	100/120							
INSITU CONCRETE FRAME.	22	22000	33000								5000/7500	5000/7500	5000/7500	5000/7500	2000/3000		
GROUND FLOOR SLAB	5	800	900												480/540	320/360	
BRICK INFILL PANELS.	15	8700	9750												1740/1950	2900/3250	2900/3250
PRECAST STAIRS & FLOORS.	3	2400	2700													2400/2700	
HARDWOOD FRAMES	10	1000	1200														
PRECAST ROOF SLAB.	2	2000	2500														2000/2500
BLOCK PARTITIONS. G.F.	7	1750	2100														1250/1500
GLAZING	6	240	300														
EXTERNAL DOORS	4	1600	1800														
ROOF COVERING	5	1000	1200														600/720
EXTERNAL PLUMBING.	6	180	210														90/105
BLOCK PARTS 1ST FLOOR	7	2100	2450														
PLUMBING 1ST FIX	5	900	1050														
ELECTRICIAN 1ST FIX.	8	400	480														
EXTERNAL PAINTING.	12	240	300														
EXTERNAL WORKS.	20	7564	9000														
JOINER 1ST FIX.	8	800	880														
PLASTERER.	18	1800	2160														
FLOOR SCREED.	10	600	700														
QUARRY TILE.	1	100	140														
PLUMBING 2ND FIX	10	1000	1200														
JOINERY 2ND FIX.	10	1000	1200														
INTERNAL PAINTING.	14	420	504														
VINYL TILING.	6	240	300														
ELECTRICIAN 2ND FIX	6	360	540														
CLEAN & HANDOVER	1	50	60														
PRELIMINARIES		13175	13950	425/450	425/450	425/450	425/450	425/450	425/450	425/450	425/450	425/450	425/450	425/450	425/450	425/450	425/450
PLANNED EXPENDITURE		78719		795	925	1175	2175	1925	1405	875	5425	5425	5425	5425	4645	6045	7265
ANTICIPATED INCOME			98574	890	1050	1390	2750	2450	1650	970	7950	7950	7950	7950	5940	6760	8525
PLANNED EXPEND (MONTHLY)							5070				9630				20920		
ANTICIPATED INCOME (MONTHLY)							6080				13020				29790		
5% RETENTION ON ANTICIPATED INCOME							304				651				1490		
INCOME LESS RETENTION TIME ADJUSTED											5776				12369		
CUMULATIVE EXPENDITURE							5070				14700				35620		
CUMULATIVE INC. TIME ADJ											5776				18145		
DIFFERENCE BETWEEN EXPENDITURE & INCOME							-5070				-14700				-29844		

Fig. 5.3.3

AT EARLIEST TIMES.

15	16	17	18	19	20	21	22	23	24	25	26	27	28	29	30	31	32	33	34	35	36	37	38
1160/1300																							
300/360	500/600	200/240																					
500/600																							
		120/150	120/150																				
		1200/1350	400/450																				
400/480																							
90/105																							
900/1050	1200/1400																						
	300/350	600/700																					
			100/120	250/300	50/60																		
	40/50	100/125	100/125																				
	756/900	1891/2250	1891/2250	1891/2250	1135/1350																		
					400/440	400/440																	
					400/480	500/600	500/600	400/480															
								60/70	300/350	240/280													
										100/140													
										100/120	500/600	400/480											
										100/120	500/600	400/480											
												30/36	150/180	150/180	90/108								
															80/100	160/200							
															120/180	240/360							
																50/60							
425/450	425/450	425/450	425/450	425/450	425/450	425/450	425/450	425/450	425/450	425/450	425/450	425/450	425/450	425/450	425/450	425/450							
3775	3221	4536	3036	2566	2410	1325	925	885	725	965	1425	1255	575	575	715	875							
4345	3750	5265	3545	3000	2780	1490	1050	1000	800	1110	1650	1446	630	630	838	1070							
	20306				12548				3860				4220			2165							
	23380				14590				4340				4836			2538							
	1169				730				217				242			127							
	28300				22211				13860				4123				4594				2411		
	55926				68474				72334				76554			78719							
	46445				68656				82516				86639				91233				93644 2466	2465 RET'N MONTH 15	
	−37781				−22029				−3678				+5962				+7920				+12514 +17390		

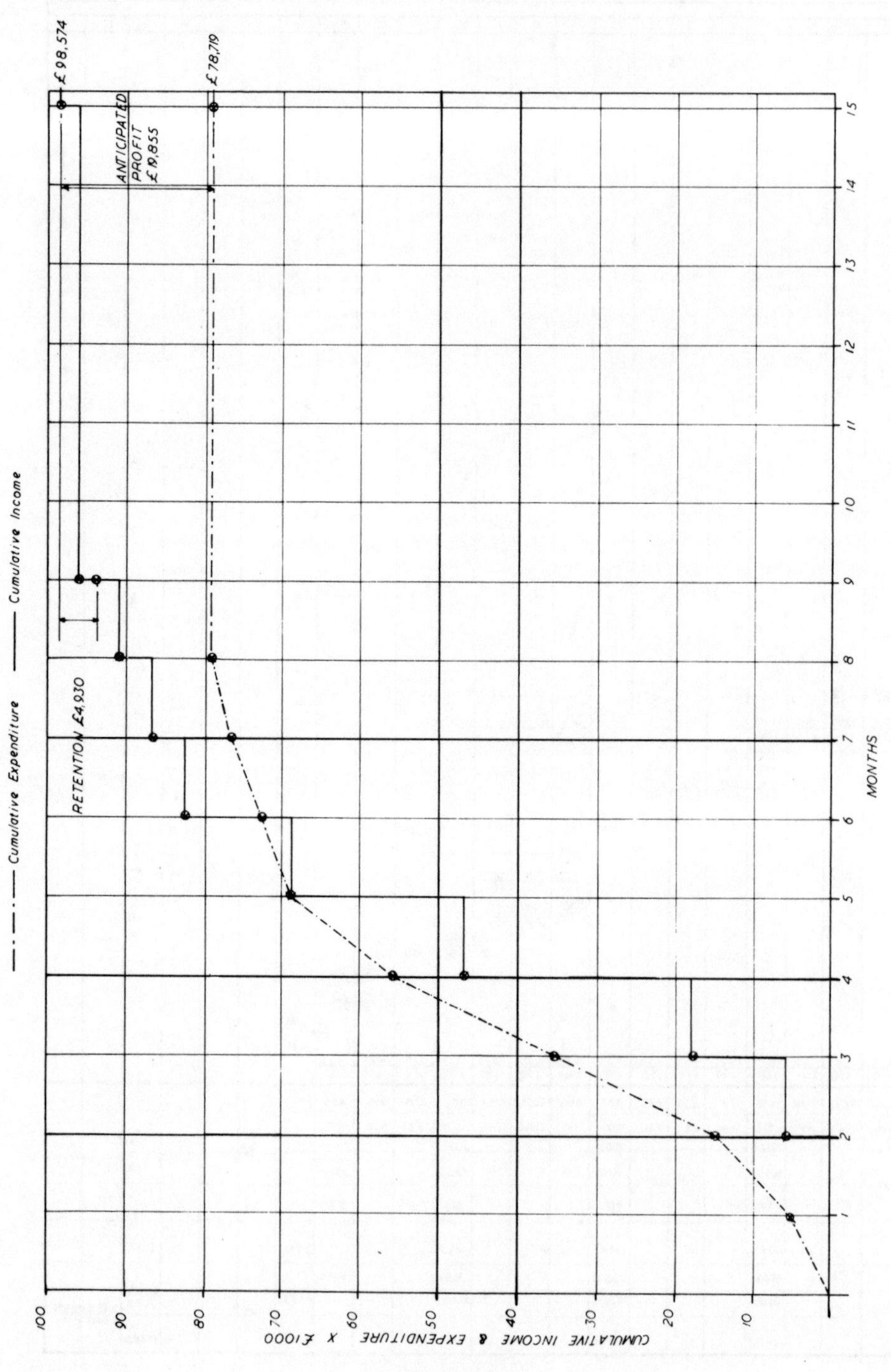

CUMULATIVE INCOME & EXPENDITURE [EARLIEST TIMES.]
Cumulative Expenditure
Cumulative Income
£98,574
£78,719
ANTICIPATED PROFIT £19,855
RETENTION £4,930
MONTHS
1 2 3 4 5 6 7 8 9 10 11 12 13 14 15
CUMULATIVE INCOME & EXPENDITURE x £1000
10 20 30 40 50 60 70 80 90 100

Fig. 5.3.4

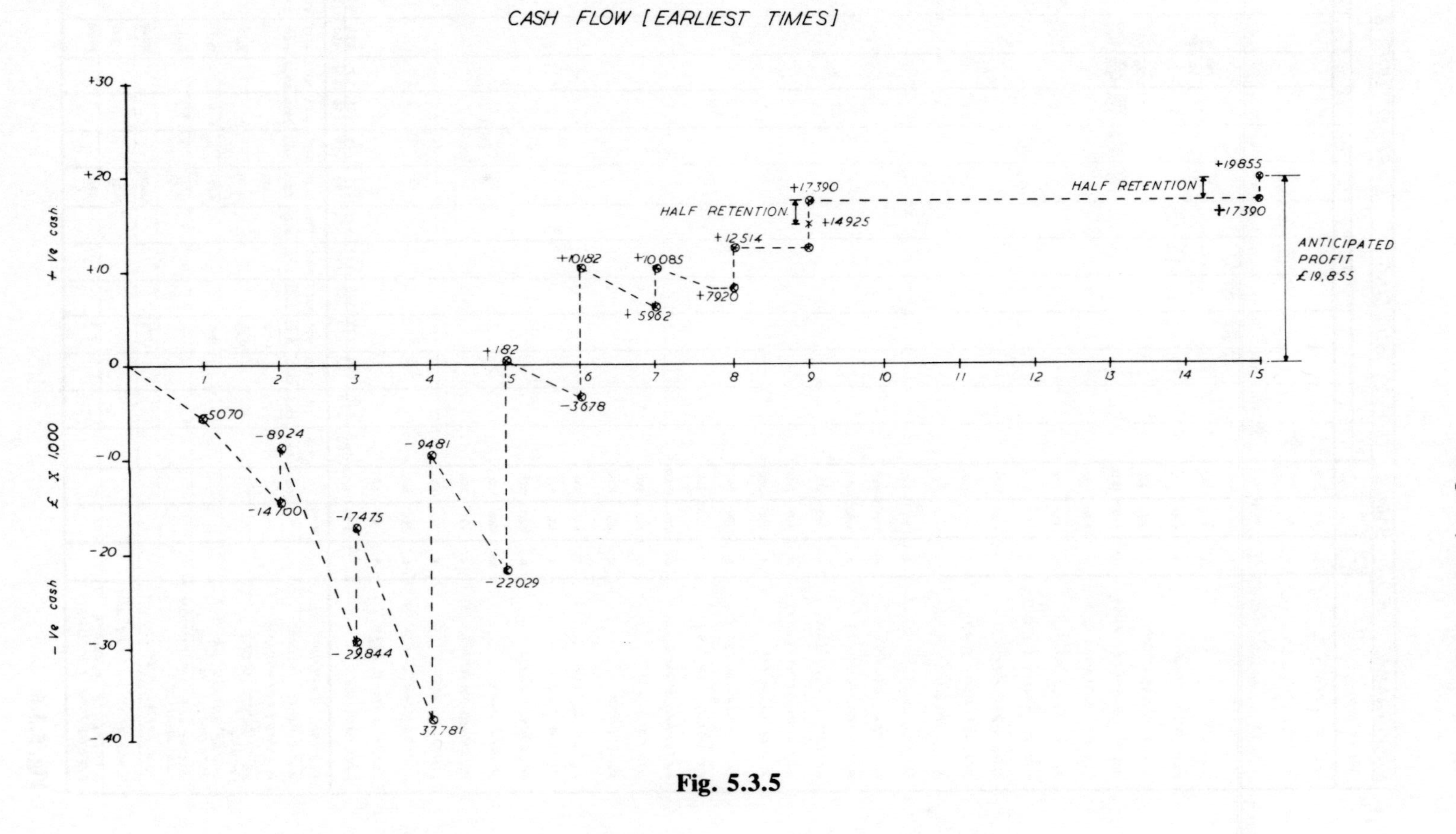
CASH FLOW [EARLIEST TIMES]
+ve cash
£ X 1,000
−ve cash
+30
+20
+10
0
−10
−20
−30
−40
1
2
3
4
5
6
7
8
9
10
11
12
13
14
15
5070
−8924
−14700
−17475
−29844
−9481
37781
+182
−22029
+10182
−3678
+10085
+5962
+7920
+12514
HALF RETENTION
+17390
+14925
HALF RETENTION
+19855
+17390
ANTICIPATED PROFIT £19,855

Fig. 5.3.5

PROGRAMME FOR SMALL OFFICE BUILDING – EXPENDITURE & INCOME																		
OPERATION	DURATION DAYS	PLANNED EXPENDITURE	ANTICIPATED INCOME	WEEKS 1	2	3	4	5	6	7	8	9	10	11	12	13	14	15
SET UP SITE	3	210	240	210 / 240														
SETTING OUT	4	60	80	60 / 80														
EXCAVATE TO REDUCED LEV	10	1000	1200	100 / 120	500 / 600	400 / 480												
PILING	12	3600	4800			300 / 400	1500 / 2000	1500 / 2000	300 / 400									
DRAINS	6	300	360											250 / 300	50 / 60			
PAD FOUNDATIONS	9	630	720						280 / 320	350 / 400								
STRIP FOUNDATIONS	5	500	600											400 / 480	100 / 120			
INSITU CONCRETE FRAMES	22	22000	33000								5000 / 7500	5000 / 7500	5000 / 7500	5000 / 7500	2000 / 3000			
GROUND FLOOR SLAB	5	800	900														800 / 900	
BRICK INFILL PANELS	15	8700	9750												1740 / 1950	2900 / 3250	2900 / 3250	1160 / 1300
PRECAST STAIRS & FLOORS	3	2400	2700															2400 / 2700
HARDWOOD FRAMES	10	1000	1200															300 / 360
PRECAST ROOF SLAB	2	2000	2500															
BLOCK PARTITIONS G.F.	7	1750	2100															
GLAZING	6	240	300															
EXTERNAL DOORS	4	1600	1800															
ROOF COVERING	5	1000	1200															
EXTERNAL PLUMBING	6	180	210															
BLOCK PARTITIONS 1ST FL.	7	2100	2450															
PLUMBING 1ST FIX	6	900	1050															
ELECTRICIAN 1ST FIX	8	400	480															
EXTERNAL PAINTING	12	240	300															
EXTERNAL WORKS	20	7564	9000															
JOINER 1ST FIX	8	800	880															
PLASTERER	18	1800	2160															
FLOOR SCREED	10	600	700															
QUARRY TILES	1	100	140															
PLUMBING 2ND FIX	10	1000	1200															
JOINER 2ND FIX	10	1000	1200															
INTERNAL PAINTING	14	420	504															
VINYL TILING	6	240	300															
ELECTRICIAN 2ND FIX	6	360	540															
CLEAN & HAND OVER	1	50	60															
PRELIMINARIES		13175	13950	425 / 450	425 / 450	425 / 450	425 / 450	425 / 450	425 / 450	425 / 450	425 / 450	425 / 450	425 / 450	425 / 450	425 / 450	425 / 450	425 / 450	425 / 450
PLANNED EXPENDITURE		78719		795	925	1125	1925	1925	1005	775	5425	5425	5425	6075	4315	3325	4125	4285
ANTICIPATED INCOME			98574	890	1050	1330	2450	2450	1170	850	7950	7950	7950	8730	5580	3700	4600	4810
PLANNED EXPENDITURE (MONTHLY)							4770				9130				21240			
ANTICIPATED INCOME (MONTHLY)							5720				12420				30210			
5% RETENTION ON ANTICIPATED INCOME							286				621				1511			
INCOME LESS RETENTION TIME ADJUSTED											5434				11789			
CUMULATIVE EXPENDITURE							4770				13900				35140			
CUMULATIVE INC. TIME ADJ.											5434				17233			
DIFFERENCE BETWEEN EXPENDITURE & INCOME							-4770				-13900				-29706			

Fig. 5.3.6

AT LATEST TIMES.

16	17	18	19	20	21	22	23	24	25	26	27	28	29	30	31	32	33	34	35	36	37	38	39
500 / 600	200 / 240																						
1000 / 1250	1000 / 1250																						
1250 / 1500	500 / 600																						
	120 / 150	120 / 150																					
	1200 / 1350	400 / 450																					
	600 / 720	400 / 480																					
	90 / 105	90 / 105																					
	900 / 1050	1200 / 1400																					
300 / 350	600 / 700																						
		100 / 120	250 / 300	50 / 60																			
													40 / 50	100 / 125	100 / 125								
												1891 / 2250	1891 / 2250	1891 / 2250	1891 / 2250								
						300 / 330	500 / 550																
				400 / 480	500 / 600	500 / 600	400 / 480																
							60 / 70	300 / 350	240 / 280														
											100 / 140												
									100 / 120	500 / 600	400 / 480												
									100 / 120	500 / 600	400 / 480												
											30 / 36	150 / 180	150 / 180	90 / 108									
														80 / 100	160 / 200								
														120 / 180	240 / 360								
															50 / 60								
425 / 450	425 / 450	425 / 450	425 / 450	425 / 450	425 / 450	425 / 450	425 / 450	425 / 450	425 / 450	425 / 450	425 / 450	425 / 450	425 / 450	425 / 450	425 / 450								
3475	5635	2735	675	875	925	1225	1385	725	865	1425	1355	2466	2506	2706	2866								
4150	6615	3155	750	990	1050	1380	1550	800	970	1650	1586	2880	2930	3213	3445								
15210				9920				4260				6111			8078								
17260				11510				4780				7086			9588								
863				576				239				354			479								
28699				16397				10934				4541				6732				9108			
50350				60270				64530				70641			78719								
45932				62329				73263				77804				84536				93645 2465	2465 RETN		
−33117				−14338				−2201				+2622				−915				+5817 +17390	MONTH 15		

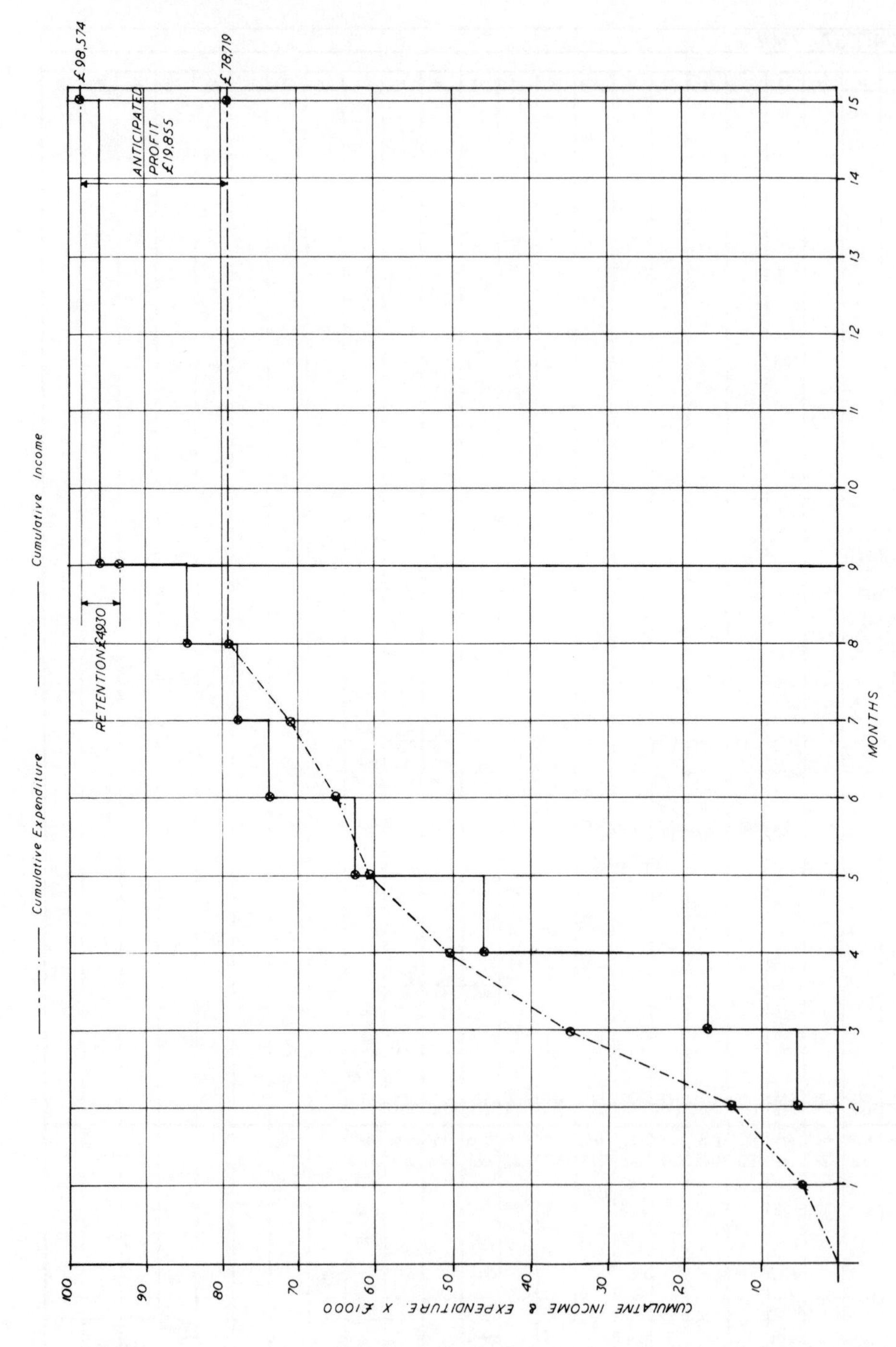
CUMULATIVE INCOME & EXPENDITURE [LATEST TIMES]
Cumulative Expenditure
Cumulative Income
£98,574
£78,719
ANTICIPATED PROFIT £19,855
RETENTION £4930
MONTHS
1
2
3
4
5
6
7
8
9
10
11
12
13
14
15
CUMULATIVE INCOME & EXPENDITURE. X £1000
10
20
30
40
50
60
70
80
90
100

Fig. 5.3.7

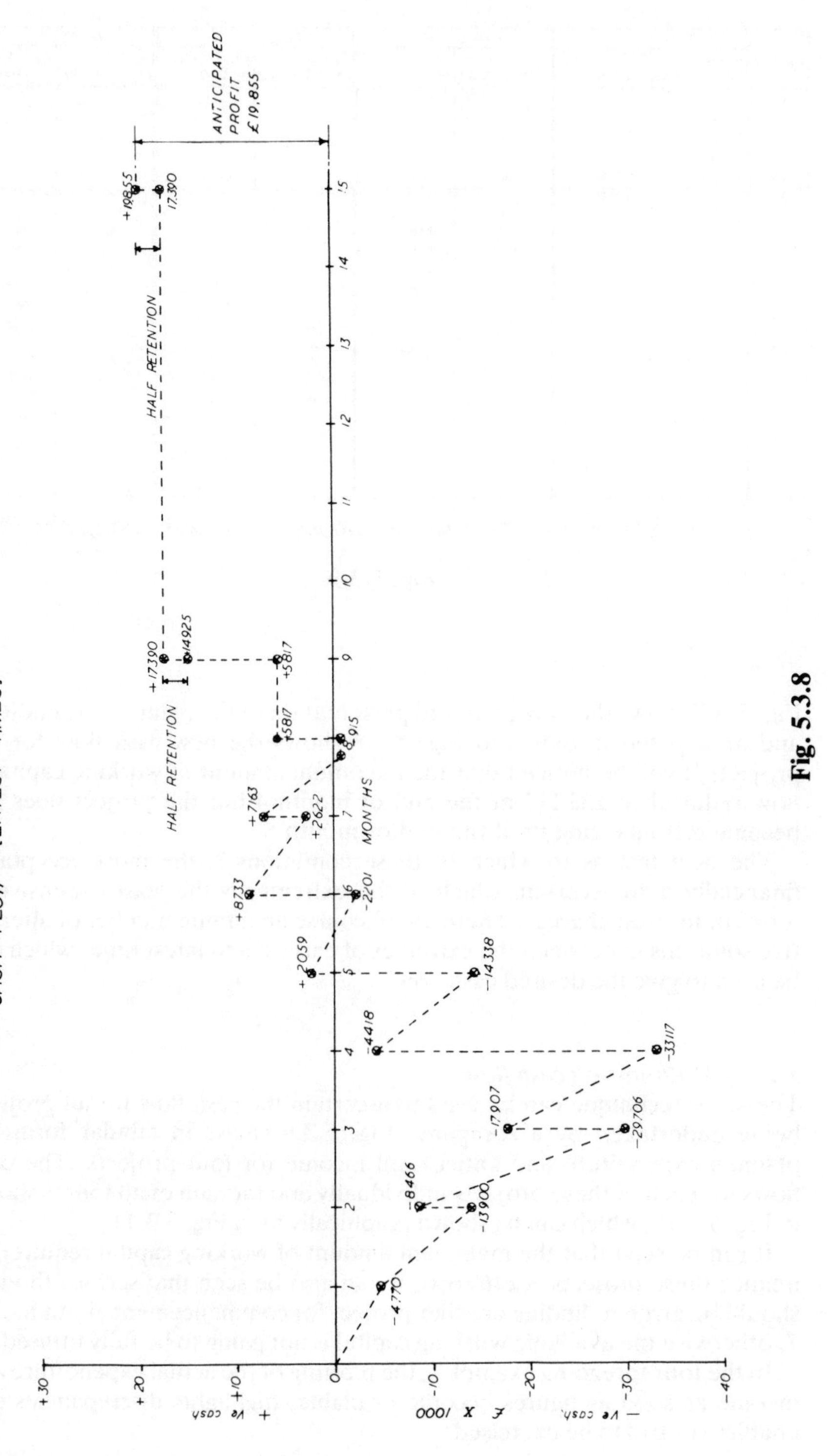
CASH FLOW [LATEST TIMES]
HALF RETENTION
HALF RETENTION
ANTICIPATED PROFIT £19,855
MONTHS
£ X 1000
+ ve cash
− ve cash
+30
+20
+10
0
-10
-20
-30
-40
1
2
3
4
5
6
7
8
9
10
11
12
13
14
15
-4770
-8466
-13900
-17907
-29706
-4418
-33117
+2059
-14338
+8733
-2201
+7163
+2622
-915
5817
+5817
+17390
14925
+19855
17390

Fig. 5.3.8

MONTH	PROJECT A		PROJECT B		PROJECT C		PROJECT D	
	PLANNED EXPENDITURE	ANTICIPATED INCOME	PLANNED EXPENDITURE	ANTICIPATED INCOME	PLANNED EXPENDITURE	ANTICIPATED INCOME	PLANNED EXPENDITURE	ANTICIPATED INCOME
1	6500							
2	17500							
3	34500	7500	13000					
4	36500	20100	35000					
5	27500	39700	69000	14600				
6	17500	42000	73000	39200			2750	
7	5500	31600	55000	77300			8000	
8	4500	20100	36000	81800	5500		17000	3300
9		6300	11000	61600	16000		18750	9500
10		5200	9000	40300	34000	6400	16500	20200
11				12300	37500	18700	13500	22300
12				10100	33000	39800	9250	19600
13					27000	43900	7250	16100
14					18500	38600	3250	11000
15					14500	31600	1500	8600
16					6500	21600	500	3900
17					3000	17000		1800
18					1000	7600		600
19						3500		
20						1200		

MONTHLY EXPENDITURE & INCOME FOR ALL PROJECTS

Fig. 5.3.9

Fig. 5.3.7 shows the new graphical presentation of the planned expenditure and anticipated income and Fig. 5.3.8 shows the new cash flow for the project. It will be noticed that the maximum amount of working capital is now reduced to £33 117 at the end of month 4 but the project does not become self financing until the end of month 8.

The acid test as to which of these conditions is the most acceptable financially is to ascertain which of the extremes is the least expensive in terms of interest charges. There are of course an infinite number of alternative solutions in between the extremes of earliest and latest times which can be used to give the desired cash flow.

5.1.5.4 Multi-project cash flow

The same technique can be used to ascertain the cash flow for all projects being undertaken by a company. Fig. 5.3.9 shows in tabular form the planned expenditure and anticipated income for four projects. The cash flows for each of these projects individually and the sum of all four is shown in Fig. 5.3.10, which can be shown graphically as in Fig. 5.3.11.

It can be seen that the maximum amount of working capital required to finance these projects is £169 650. It can also be seen that serious thought should be given to finding another project for commencement about month 7, otherwise the available working capital is not going to be fully utilised.

In the four foregoing examples, the plotting of the actual expenditure and income as soon as figures become available, highlights discrepancies and enables control to be exercised.

MONTH	PROJECT A					PROJECT B					PROJECT C					PROJECT D					TOTAL CASH FLOW FOR ALL PROJECTS
	PLANNED EXPENDITURE	ANTICIPATED INCOME	CUMULATIVE PLANNED EXPENDITURE	CUMULATIVE ANTICIPATED INCOME	CASH FLOW	PLANNED EXPENDITURE	ANTICIPATED INCOME	CUMULATIVE PLANNED EXPENDITURE	CUMULATIVE ANTICIPATED INCOME	CASH FLOW	PLANNED EXPENDITURE	ANTICIPATED INCOME	CUMULATIVE PLANNED EXPENDITURE	CUMULATIVE ANTICIPATED INCOME	CASH FLOW	PLANNED EXPENDITURE	ANTICIPATED INCOME	CUMULATIVE PLANNED EXPENDITURE	CUMULATIVE ANTICIPATED INCOME	CASH FLOW	
1	6500		6500		-6500																- 6500
2	17500		24000		-24000																- 24000
3	34500	7500	58500	7500	-51000	13000		13000		-13000											- 64000
4	36500	20100	95000	27600	-67400	35000		48000		-48000											- 115400
5	27500	39700	122500	67300	-55200	69000	14600	117000	14600	-102400											- 157600
6	17500	42000	140000	109300	-30700	73000	39200	190000	53800	-136200						2750		2750		-2750	- 169650
7	5500	31600	145500	140900	-4600	55000	77300	245000	131100	-113900						8000		10750		-10750	- 129250
8	4500	20100	150000	161000	+11000	36000	81800	281000	212900	-68100	5500		5500		-5500	17000	3300	27750	3300	-24450	- 87050
9		6300		167300	+17300	11000	61600	292000	274500	-17500	16000		21500		-21500	18750	9500	46500	12800	-33700	- 55400
10		5200		172500	+22500	9000	40300	301000	314800	+13800	34000	6400	55500	6400	-49100	16500	20200	63000	33000	-30000	- 42800
11							12300		327100	+26100	37500	18700	93000	25100	-67900	13500	22300	76500	53300	-23200	- 42500
12							10100		337200	+36200	33000	39800	126000	64900	-61100	9250	19600	85750	74900	-10850	- 13250
13											27000	43900	153000	108800	-44200	7250	16100	93000	91000	-2000	+ 12500
14											18500	38600	171500	147400	-24100	3250	11000	96250	102000	+5750	+ 40350
15											14500	31600	186000	179000	-7000	1500	8600	97750	110600	+12850	+ 64550
16											6500	21600	192500	200600	+8100	500	3900	98250	114500	+16250	+ 83050
17											3000	17000	195500	217600	+22100		1800		116300	+18050	+ 98850
18											1000	7600	196500	225200	+28700		600		116900	+18650	+ 106050
19												3500		228700	+32200						+ 109550
20												1200		229900	+33400						+ 110750

CUMULATIVE MONTHLY EXPENDITURE AND INCOME AND TOTAL CASH FLOW FOR ALL PROJECTS

Fig. 5.3.10

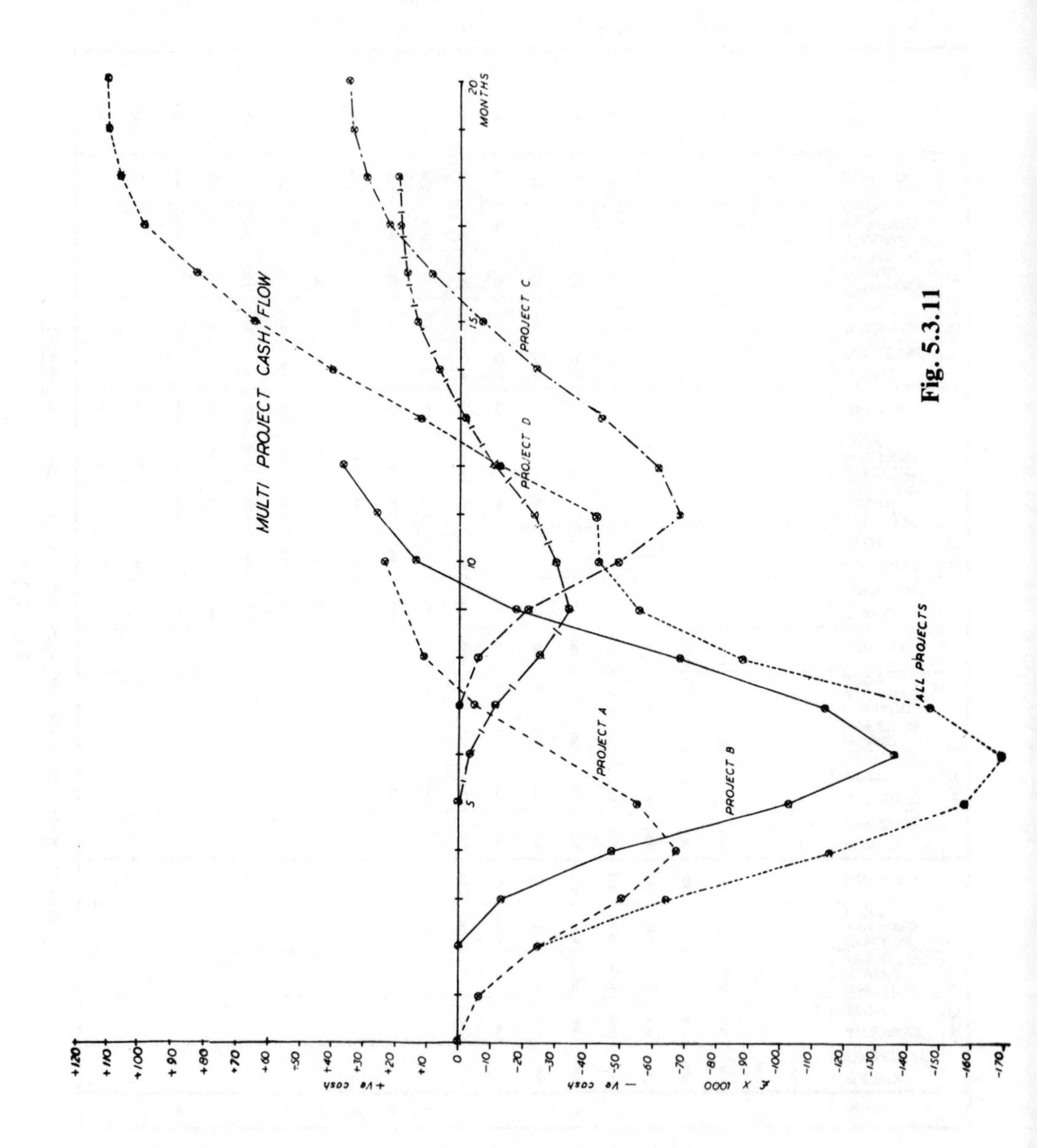
MULTI PROJECT CASH FLOW
PROJECT A
PROJECT B
PROJECT C
PROJECT D
ALL PROJECTS
20 MONTHS
+ve cash
— ve cash
£ X 1000

Fig. 5.3.11

5.1.5.5 Cost of financing projects

Once the cash flow has been established for the project, it is a simple step to determine the cost of providing the working capital. Whether the cash resources are available within the company or have to be borrowed from external sources (e.g. bank overdraft), the cost of interest payments should be taken into account. If the company's own funds are used, it is equally important to allow for the cost of interest payments, as the money could have been available for investment elsewhere.

Taking the project shown in Fig. 5.3.1 as an example, it is possible to ascertain which of the solutions, earliest or latest start times, is the most effective when considering the cost of the interest payments.

Take first the graph showing the cash flow for the earliest start times in Fig. 5.3.5. The area between the horizontal time axis and the graph below this axis represents the working capital requirement over the duration of the contract in £-months. The area above the line represents the surplus on the contract in £-months. We are at present only interested in the area below the line as we are attempting to ascertain the cost of borrowing this working capital.

If the interest rate is 12% per annum, then the cost of providing the working capital for this project under these conditions is as follows:

$$76939 \times \frac{12\%}{12} = £769.39$$

Fig. 5.4.1 Cost of working capital – earliest start times (based on Fig. 5.3.5)

Month	*Calculation for each month (£-months)*	*Area (£-months)*
1	$\frac{(0 + 5070)}{2}$	2535
2	$\frac{(5070 + 14700)}{2}$	9885
3	$\frac{(8924 + 29844)}{2}$	19384
4	$\frac{(17475 + 37781)}{2}$	27628
5	$\frac{(9481 + 22029)}{2}$	15755
6	$\frac{(0 + 3678)}{2} \times \frac{3678}{3860}$	1752
		76939

Fig. 5.4.2 Cost of working capital – latest start times (based on Fig. 5.3.8)

Month	*Calculation for each month (£-months)*	*Area (£-months)*
1	$\frac{(0 + 4710)}{2}$	2355
2	$\frac{(4710 + 13900)}{2}$	9305
3	$\frac{(8466 + 29706)}{2}$	19086
4	$\frac{(17907 + 33117)}{2}$	25512
5	$\frac{(4418 + 14338)}{2}$	9378
6	$\frac{(0 + 2201)}{2} \times \frac{2201}{4260}$	568.5
7	nil	nil
8	$\frac{(915 + 915)}{2} \times \frac{1}{4}$	228.75
		66433.25

If the interest rate is 12% per annum, then the cost of providing the working capital for this project under these conditions is as follows:

$$66433.25 \times \frac{12\%}{12} = £664.33$$

It can be seen that the latest start times solution, of the two extremes considered, gives the cheapest solution. Once the project becomes self financing, the surpluses which result can either be invested on the money market on, say, 7 days recall or they can be used to finance other projects.

5.1.5.6 Tabular presentation of contract budgets

Contract budgets can also be presented and controlled in tabular form as opposed to the graphical methods shown previously.

The elements of work selected for inclusion are those which need very careful control, i.e. labour, material and plant. Cost data must be collected in such a way that it is readily identifiable with each of the elements. Overheads and profit can also be controlled using this method but usually only site overheads are included if the control is carried out on site. If Head Office overheads and profit are included, then this work will invariably be carried out by very senior levels of management. The chart is divided into two main parts, the right hand side gives the current month's figures whilst the left hand side gives the cumulative figures to date, i.e. including the current month. These charts can be a little more elaborate and show the current month and the cumulative figures before and after the current month. The technique enables a check to be kept on the current performance, total performance and trends. Any significant variances should be investigated and appropriate action taken to bring the work back into line with the budget.

Fig. 5.5.1 shows the early stages of a project presented in this manner. As this project is being controlled at site level, all the figures are costs, target costs being set by the production control department and the actual cost figures being collected on site as work proceeds.

The total cost column records the total cost in the whole project for the particular element. The target cost of work done records what the actual amount of work done should have cost, which can be compared with the actual cost shown in the next column. The variance column records the difference between the figures shown in the previous two columns and the variance % column is found by dividing the variance by the target figures.

The totals are arrived at by totalling the columns vertically. In the case of the variance column, the figures in brackets are unfavourable variances (negative values) and the ones without brackets are favourable variances (positive values). The variance % total is not arrived at by totalling this column but by dividing the total variance by the total target cost.

A more comprehensive version of the use of this technique is shown in the following example.

The budget for the shell of a two-storey in situ concrete framed building is shown in Fig. 5.5.2. The progress achieved on the project and the cumulative costs after the first four months are as shown in Fig. 5.5.3.

At the end of month 5 the progress achieved and the associated costs are shown in Fig. 5.5.4. With this information to hand, the cost being incurred on the contract can be analysed, highlighting what variances are occurring and indicating where action is needed. This analysis is shown in Fig. 5.5.5.

5.1.5.7 Cash flow forecasting

The cash flow calculations presented previously have all been based upon a detailed programme for the project, and detailed figures for the income and expenditure on the project. It is obvious therefore that this type of calculation would be carried out after the contract had been awarded to the contractor, this being the earliest time that such detailed information was available. It would, however, be very helpful if the contractor could produce an approximate forecast of the cash flow requirements for a potential contract without having to go to such elaborate lengths. If a forecast of the cash flow can be produced for a project, it will enable the contractor to decide whether he has sufficient financial resources to undertake the contract.

One method of forecasting cash flow for the project is based on establishing an ideal reference curve derived from cash flow curves for a reasonable-sized sample of similar projects carried out in the past, and using this reference curve to predict the likely cash flow for future projects.

						MONTH 4			
	£	£	£	£	%	£	£	£	%
ELEMENT	TOTAL COST OF WORK	TARGET COST OF WORK DONE	ACTUAL COST OF WORK DONE.	VARIANCE.	VARIANCE	TARGET COST OF WORK DONE	ACTUAL COST OF WORK DONE.	VARIANCE.	VARIANCE.
DIRECT LABOUR.									
EXCAVATION OF BASEMENT AND FOUNDATIONS.	1800	1800	1973	[173]	[9·61]	—	—	—	—
CONCRETE BLINDING.	1600	1600	1437	163	10·19	—	—	—	—
CB & FIX REINF. IN FOUNDS.	2500	2050	1790	260	12·68	450	395	55	12·22
CONCRETE FOUNDATIONS.	3500	2800	2560	240	8·57	940	900	40	4·25
CB & FIX REINF. IN SLAB.	3100	1320	1560	[240]	[18·18]	540	530	10	1·85
CONCRETE SLAB.	5000	1820	1780	40	2·19	1,630	1800	[170]	[10·42]
STOP END & EDGE OF FORMWORK	900	310	275	35	11·29	405	95	310	765
DIRECT MATERIALS.									
CONCRETE MATERIALS.									
[Blinding]	4800	4800	5150	[350]	[7·29]	—	—	—	—
[Foundation]	10 500	8400	7680	720	8·57	2800	2700	100	3·57
[Slab].	15 000	5460	5300	160	2·93	4900	5400	[500]	[10·20]
REINF. IN FOUNDATIONS.	11 250	10 250	8950	1300	12·68	2250	1975	275	12·22
REINF. IN SLAB.	13 950	5940	7020	[1080]	[18·18]	2400	2390	10	0·41
TIMBER IN FORMWORK.	700	245	300	[55]	[22·44]	450	600	[150]	[33·33]
PLANT.									
EXCAVATION OF BASEMENT AND FOUNDATIONS.	12000	13500	12400	1100	·8·14	—	—	—	—
PUMPING WATER.	NIL	NIL	340	[340]	—	—	70	[70]	—
SITE ON COSTS.	8 660	6030	6450	[420]	[6·96]	1685	1720	[35]	[2·07]
	95260	66 325	64 965	1360	2·05	18 450	18 575	[125]	[·68]

Fig. 5.5.1

Fig. 5.5.2 Budget

Element	*Labour*	*Plant*	*Material*	*Site O/H*	*Contribution*	*Total*
Strip site	1650	8250	—	1200	2160	13260
Excavate for foundations and slab	3300	16500	—	2500	4500	26800
Foundations	14400	3200	19800	4700	8460	50560
G.F. slab	7400	3300	9900	2600	4680	27880
G.F. columns, stairs and first floor slab	15400	5700	14900	4500	8100	48600
First floor columns and roof slabs	16900	5900	15800	4800	8640	52040
Roof cladding	2400	510	5100	1000	1800	10810
Brick exterior walls	19000	9400	25000	6700	1200	61300
	80450	52760	90500	28000	39540	291250

Fig. 5.5.3 Progress at month 4

Element	% *Complete*	*Labour*	*Plant*	*Material*	*Site O/H*	*Contribution*	*Total*
Strip site	100	1800	8300	—	1200	1960	13260
Excavate for foundations and slab	100	2700	17000	—	2500	4600	26800
Foundations	100	14500	2900	19600	4700	8860	50560
G.F. slab	80	6620	2440	7800	2080	3364	22304
G.F. columns, stairs and first floor slab	60	9940	3120	9000	2700	4400	29160
First floor columns and roof slab	40	6560	2260	6420	1920	3656	20816
Roof cladding	25	550	148	1275	250	479.5	2702.5
Brick exterior walls	10	2000	840	2600	670	20	6130
		44670	37008	46695	16020	27339.5	171732.5

Fig. 5.5.4 Progress during month 5

Element	*% completed during the month*	*Labour*	*Plant*	*Material*	*Site O/H*	*Contribution*	*Total*
Strip site	—	—	—	—	—	—	—
Excavate for foundations and slab	—	—	—	—	—	—	—
Foundations	—	—	—	—	—	—	—
G.F. slab	20	1680	760	1880	520	736	5576
G.F. columns, stairs and first floor slab	15	2110	1000	2135	675	1370	7290
First floor columns and roof slab	10	1890	540	1600	480	694	5204
Roof cladding	10	270	81	500	100	130	1081
Brick exterior walls	15	2650	1210	3750	1005	580	9195
		8600	3591	9865	2780	3510	28346

ELEMENT	TOTAL COST OF WORK.	CUMULATIVE TO END OF MONTH 4				MONTH 5				CUMULATIVE TO END OF MONTH 5.			
		TARGET COST OF WORK DONE	ACTUAL COST OF WORK DONE	VARIANCE £	VAR. %	TARGET COST OF WORK DONE	ACTUAL COST OF WORK DONE	VARIANCE £	VAR. %	TARGET COST OF WORK DONE	ACTUAL COST OF WORK DONE	VARIANCE £	VAR. %
LABOUR													
Strip site.	1650	1650	1800	(150)	(9·09)	–	–	–	–	1650	1800	(150)	(9·09)
Excavate for founds & slab.	3300	3300	2700	600	18·18	–	–	–	–	3300	2700	600	18·18
Foundations.	14400	14400	14500	(100)	(0·69)	–	–	–	–	14400	14500	(100)	(0·69)
G.F. slab.	7400	5920	6620	(700)	(11·82)	1480	1680	(200)	(13·51)	7400	8300	(900)	(12·16)
G.F. columns, stairs & F.F. slab	15400	9240	9940	(700)	(7·57)	2310	2110	200	8·66	11550	12050	(500)	(4·33)
F.F. columns & roof slab.	16900	6760	6560	200	2·96	1690	1890	(200)	(11·83)	8450	8450	0	0
Roof cladding	2400	600	550	50	8·33	240	270	(30)	(12·50)	840	820	20	2·38
Brick exterior walls.	19000	1900	2000	(100)	(5·26)	2850	2650	200	7·02	4750	4650	100	2·11
PLANT													
Strip site	8250	8250	8300	(50)	(0·61)	–	–	–	–	8250	8300	(50)	(0·61)
Excavate for foundations & slab.	16500	16500	17000	(500)	(3·03)	–	–	–	–	16500	17000	(500)	(3·03)
Foundations.	3200	3200	2900	300	9·37	–	–	–	–	3200	2900	300	9·37
G.F. slab.	3300	2640	2440	200	7·57	660	760	(100)	(15·15)	3300	3200	100	3·03
G.F. columns, stairs & F.F. slab.	5700	3420	3120	300	8·77	855	1000	(145)	(16·96)	4275	4120	155	3·63
F.F columns & roof slab.	5900	2360	2260	100	4·23	590	540	50	8·47	2950	2800	150	5·08
Roof cladding.	510	127·50	148	(20·5)	(16·08)	51	81	(30)	(58·82)	178 50	229	(50·50)	(28·29)
Brick exterior walls.	9400	940	840	100	10·63	1410	1210	200	14·18	2350	2050	300	12·76
MATERIALS													
Foundations	19800	19800	19600	200	1·01	–	–	–	–	19800	19600	200	1·01
G.F. slab.	9900	7920	7800	120	1·52	1980	1880	100	5·05	9900	9680	220	2·22
G.F columns, stairs & F.F. slab.	14900	8940	9000	(60)	(0·67)	2235	2135	100	4·47	11175	11135	40	0·36
F.F. columns & roof slab.	15800	6320	6420	(100)	(1·58)	1580	1600	(20)	(1·26)	7900	8020	(120)	(1·52)
Roof cladding.	5100	1275	1275	0	0	510	500	10	1·96	1785	1775	10	0·56
Brick exterior walls.	25000	2500	2600	(100)	(4·00)	3750	3750	0	0	6250	6350	(100)	(1·60)
SITE OVERHEADS													
Strip site.	1200	1200	1200	0	0	–	–	–	–	1200	1200	0	0
Excavate for foundations & slab.	2500	2500	2500	0	0	–	–	–	–	2500	2500	0	0
Foundations	4700	4700	4700	0	0	–	–	–	–	4700	4700	0	0
G.F. slab.	2600	2080	2080	0	0	520	520	0	0	2600	2600	0	0
G.F. columns, stairs & F.F. slab.	4500	2700	2700	0	0	675	675	0	0	3375	3375	0	0
F.E columns & roof slab.	4800	1920	1920	0	0	480	480	0	0	2400	2400	0	0
Roof cladding	1000	250	250	0	0	100	100	0	0	350	350	0	0
Brick exterior walls.	6700	670	670	0	0	1005	1005	0	0	1675	1675	0	0
		143982·5	144393	(410·5)	(0·29)	24971	24836	13 5	0·54	168953·5	169229	(275·5)	(0·16)
	Total contribution	Target contribution.	Actual contribution	Variance	%	Target contribution.	Actual contribution.	Variance	%	Target contribution	Actual contribution	Variance	%
CONTRIBUTION													
Strip site.	2160	2160	1960	(200)	(9·26)	–	–	–	–	2160	1960	(200)	(9·26)
Excavate for foundations & slab.	4500	4500	4600	100	2·22	–	–	–	–	4500	4600	100	2·22
Foundations.	8460	8460	8860	400	4·73	–	–	–	–	8460	8860	400	4·73
G.F. slab.	4680	3744	3364	(380)	(10·15)	936	736	(200)	(21·36)	4680	4100	(580)	(12·39)
G.F. columns, stairs & F.F. slab.	8100	4860	4400	(460)	(9·47)	1215	1370	155	12·75	6075	5770	(305)	(5·02)
F.F. columns & roof slab.	8640	3456	3656	200	5·79	864	694	(170)	(19·67)	4320	4350	30	0·69
Roof cladding	1800	450	479·5	29·5	6·55	180	130	(50)	(27·77)	630	609·5	(20·5)	(3·25)
Brick exterior walls	1200	120	20	(100)	(83·33)	180	580	400	222·00	300	600	300	100·00
	39540	27750	27339·5	(410·5)	(1·48)	3375	3510	135	4·00	31125	30849·5	(275·5)	(0·89)

Fig. 5.5.5

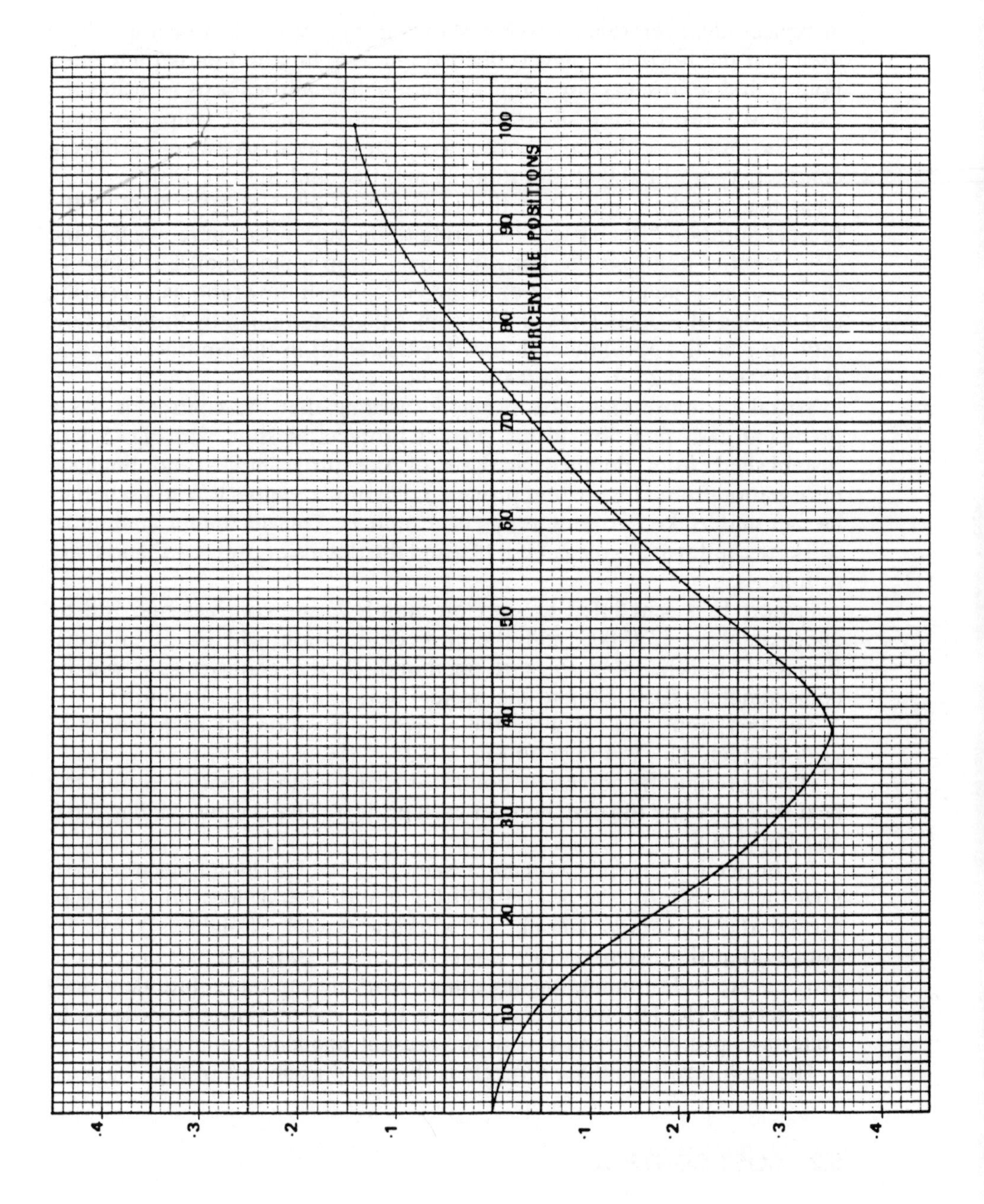

Fig. 5.6.1 Reference cash flow curve

A typical ideal reference curve is shown in Fig. 5.6.1, in which it will be seen that the horizontal axis is divided into one hundred parts (percentile points) and represents any project duration. The vertical axis indicates the cash flow ratio, i.e. a proportion of the contract sum. A simple example will illustrate how this reference curve can be used. Assume we have a contract, value £100 000, duration 24 months, and that we wish to know what the cash flow will be at the 12 month point, i.e. after 50% of the project duration has elapsed. Referring to the reference curve, we find the 50 percentile point on the horizontal axis, drop a perpendicular down to the curve and then read back onto the verical cash flow ratio axis, arriving at a reading of −0.24. The contract sum is then multiplied by this cash flow ratio to give the absolute cash flow for this point in the contract duration (£100 000 × −0.24 = −£24 000).

To take a more elaborate example, assume that the contractor has a predicted cash flow for all his current projects as shown in Fig. 5.3.10 (established from detailed analysis of the programmes and financial information for all his current projects). He now requires to know whether he can undertake two new projects, E and F. The maximum amount of working capital available is £170 000 and the details of projects E and F are as follows:

Project E is scheduled to start on month 7, the contract sum will be approximately £270 000 and the project duration will be 18 months. Project F is scheduled to start on month 9, the contract sum will be approximately £330 000 and the project duration will be 21 months.

The first step is to calculate the percentile position for each month of each project. This is shown in Fig. 5.6.2.

Having found the percentile points for each month on both projects, the next step is to find the cash flow for each project. This is shown in Fig. 5.6.3 and Fig. 5.6.4.

By inspecting Fig. 5.6.5 it can be seen that the maximum negative cash flow for all six projects is £169 650 for month 6 and does not therefore exceed the maximum amount of working capital available. It will be noticed that this occurs before projects E and F start and that once these projects are both running the maximum working capital requirement does not exceed £155 440, which is on month 13. It will therefore be possible to undertake projects E and F with the present financial resources.

The cash flow for all six projects is therefore as shown in Fig. 5.6.5.

5.2 COST CONTROL

5.2.1 Introduction

Costing has been carried out in the building industry for a number of years, but the type of costing used by some firms only gives the total cost of the project when it has been completed. This is then compared with the amount received, but if the project has lost money there is little that can be done about it. It is obviously a distinct advantage if a contractor can ascertain

Fig. 5.6.2

No. of months	*Percentile points corresponding to end of each month for projects of the following durations*	
	18 months	*21 months*
1	6	5
2	11	10
3	17	14
4	22	19
5	28	24
6	33	29
7	39	33
8	44	38
9	50	43
10	56	48
11	61	52
12	67	57
13	72	62
14	78	67
15	83	71
16	89	76
17	94	81
18	100	86
19		90
20		95
21		100

which section of a project is in deficit and know when it starts losing money. Cost control techniques have been developed for this purpose.

5.2.2 Objectives

The main objectives of cost control are:

1. To see that the company's policy with regard to production is carried out, which in turn will ensure that planned profit margins are maintained.
2. To arrive at the cost of each stage, operation (in the case of repetitive

Fig. 5.6.3 Cash flow for Project E (Contract sum £270 000 — duration 18 months)

Month relative to Fig. 5.3.10	*Number of months from start of project*	*Percentile position with respect to last column*	*Cash flow ratio from Fig. 5.6.1*	*Absolute cash flow in (£)*
7	1	6	−0.018	−0.018 × 270 000 = −4860
8	2	11	−0.05	−0.05 × 270 000 = −13500
9	3	17	−0.115	−0.115 × 270 000 = −31050
10	4	22	−0.195	−0.195 × 270 000 = −52650
11	5	28	−0.273	−0.273 × 270 000 = −73710
12	6	33	−0.32	−0.32 × 270 000 = −86400
13	7	39	−0.347	−0.347 × 270 000 = −93690
14	8	44	−0.31	−0.31 × 270 000 = −83700
15	9	50	−0.237	−0.237 × 270 000 = −63990
16	10	56	−0.17	−0.17 × 270 000 = −45900
17	11	61	−0.12	−0.12 × 270 000 = −32400
18	12	67	−0.065	−0.065 × 270 000 = −17550
19	13	72	−0.025	−0.025 × 270 000 = −6750
20	14	78	+0.025	+0.025 × 270 000 = +6750
21	15	83	+0.065	+0.065 × 270 000 = +17550
22	16	89	+0.105	+0.105 × 270 000 = +28350
23	17	94	+0.125	+0.125 × 270 000 = +33750
24	18	100	+0.14	+0.14 × 270 000 = +37800

construction), or unit, and to carry out a continuous comparison with the target to ascertain the gain or loss on each. This information must be available early enough for corrective action to be taken.

3. To provide information on cost for use in future estimating.

5.2.3 Cost control systems

Before deciding upon the degree of sophistication of a system, consideration

Fig. 5.6.4 Cash flow for Project F. (Contract sum £330 000 — duration 21 months)

Month relative to Fig. 5.3.10	*Number of months from start of project*	*Percentile position with respect to last column*	*Cash flow ratio from Fig. 5.6.1*	*Absolute cash flow in (£)*
9	1	5	−0.015	−0.015 × 330 000 = −4950
10	2	10	−0.043	−0.043 × 330 000 = −14190
11	3	14	−0.08	−0.08 × 330 000 = −26400
12	4	19	−0.15	−0.15 × 330 000 = −49500
13	5	24	−0.225	−0.225 × 330 000 = −74250
14	6	29	−0.285	−0.285 × 330 000 = −94050
15	7	33	−0.32	−0.32 × 330 000 = −105600
16	8	38	−0.347	−0.347 × 330 000 = −114510
17	9	43	−0.32	−0.32 × 330 000 = −105600
18	10	48	−0.265	−0.265 × 330 000 = −87450
19	11	52	−0.215	−0.215 × 330 000 = −70950
20	12	57	−0.16	−0.16 × 330 000 = −52800
21	13	62	−0.11	−0.11 × 330 000 = −36300
22	14	67	−0.065	−0.065 × 330 000 = −21450
23	15	71	−0.032	−0.032 × 330 000 = −10560
24	16	76	+0.01	+0.01 × 330 000 = +3300
25	17	81	+0.05	+0.05 × 330 000 = +16500
26	18	86	+0.082	+0.082 × 330 000 = +27060
27	19	90	+0.108	+0.108 × 330 000 = +35640
28	20	95	+0.128	+0.128 × 330 000 = +42240
29	21	100	+0.14	+0.14 × 330 000 = +46200

Fig. 5.6.5

Month	*Projects A,B,C & D*	*Project E*	*Project F*	*Projects A,B,C,D,E & F*
1	−6500			−6500
2	−24000			−24000
3	−64000			−64000
4	−115400			−115400
5	−157600			−157600
6	−169650			−169650
7	−120250	−4860		−125110
8	−87050	−13500		−100550
9	−55400	−31050	−4950	−91400
10	−42800	−52650	−14190	−109640
11	−42500	−73710	−26400	−142610
12	−13250	−86400	−49500	−149150
13	+12500	−93690	−74250	−155440
14	+40350	−83700	−94050	−137400
15	+64550	−63990	−105600	−105040
16	+83050	−45900	−114510	−77360
17	+98850	−32400	−105600	−39150
18	+106050	−17550	−87450	+1050
19	+109550	−6750	−70950	+31850
20	+110750	+6750	−52800	+64700
21	+110750	+17550	−36300	+92000
22	+110750	+28350	−21450	+117650
23	+110750	+33750	−10560	+133940
24	+110750	+37800	+3300	+151850
25	+110750	+37800	+16500	+165050
26	+110750	+37800	+27060	+175610
27	+110750	+37800	+35640	+184190
28	+110750	+37800	+42240	+190790
29	+110750	+37800	+46200	+194750
30	+110750	+37800	+46200	+194750

must be given to the cost of operating it and the benefits it provides. Clearly a balance should be reached between the two.

Systems can vary from those which control the work on a section or stage basis to those which control it on a unit basis, and in some systems only certain sections of the project are selected for control. It is usual to limit site cost control to the control of labour and plant as these are the areas where there is likely to be the greatest amount of variability.

5.2.4 Timing of presentation of cost information

When cost control techniques are used the system must be administered in such a manner that the site manager is made aware of any deviations from planned cost as soon as they are recognized. Ideally the costs should be checked daily, but on most building operations this would be extremely expensive and the cost of the work done is in consequence usually checked weekly.

5.2.5 Uses of cost information

Cost control information is invaluable for future estimating. Feedback must be accompanied by a full description of the conditions under which the work was carried out, as conditions can vary a great deal from one contract to another.

Another very important use of cost control information is in the pricing of variations. If the variations can be identified before the work commences and a separate record is kept of work costs on them, the contractor will have factual information to assist him in settling a rate for the work done.

The weekly measure carried out for the cost control system can also be used for the monthly valuation and thus eliminates the need for the work to be re-measured at the end of each month, although materials on site would have to be ascertained separately for the purpose of the monthly valuation.

5.2.6 Unit cost control

In this chapter it is proposed to deal with only two systems of cost control. The first system is one which could be applied to non-repetitive construction and is usually referred to as a unit cost control system. In the example given, this has been applied to the amenity centre (chapter 1, section 1.3.3).

It has already been seen that for unit costs to be of value as a control tool they must be issued as soon as possible after their compilation, which is usually the week following that to which they refer. Any unit costs in excess of or below the target by more than a predetermined margin require further investigation and should be brought to the attention of management, thus applying the principle of management by exception. The information is collected by means of the allocation sheet and the weekly summary sheet, together with actual measurement of work done.

5.2.6.1 Allocation sheet (Fig. 5.7)

This is the daily record of the hours worked by an operative or gang from which the hours spent on an operation can be obtained.

The sheets are the basis of the cost control system and must therefore be

DAILY ALLOCATION SHEET

CONTRACT	FOREMAN/GANGER
CONTRACT NO.	TRADE
DATE	SHEET NO. OF

*BLOCK NO. & OPERATION / NAME & CLOCK NO.											DAILY TOTAL FOR OPERATIVE	STOPPAGES [HOURS]
	T	L	T	L	T	L	T	L	T	L		
TOTALS												

PLANT	PLANT NO.	O T	S T	O T	S T	O T	S T	O T	S T	O T	S T	DAILY TOTALS FOR PLANT OPERATING	STANDING
TOTALS													

REMARKS	
WEATHER CONDITIONS	SIGNED FOREMAN/GANGER

*BLOCK NO. ONLY APPLIES WHEN THIS FORM IS BEING USED FOR HOUSING

Fig. 5.7

accurately compiled by the man in charge of the gang. To achieve this accuracy, they should be completed daily and handed in to the site office where the operation descriptions, time per operation and total time per operation will be checked.

When plant is involved in operations, this should be recorded by the plant operator if the machine has a permanent operator and does not work with a gang, e.g. the excavator. If the item does not have a permanent operator (for example the vibrating roller) or always works in conjunction with a gang (the concrete mixer), then it should be booked by the ganger.

Remarks should be made about working conditions, any special methods used, etc. Note should also be made of how non-productive time is made up (i.e. how much is standing time and how much breakdown time).

5.2.6.2 Weekly summary sheet (Fig. 5.8)

This sheet is used to collect all the information on operational hours from the daily allocation sheets, and will be compiled by the cost surveyor at the end of each week. The hours recorded on the allocation sheets are ordinary hours and net overtime spent on operations, so that the weekly summary sheet presents the total basic hours spent on an operation in that week. This information is required for calculating the amount of bonus payable as well as the insertion in the weekly cost sheet.

5.2.6.3 Measurement of work done

Measurement of work done should be carried out by a person who fully understands the demarcation between operations so that the measure can be directly related to the costs arrived at through the allocation sheets and summary sheets.

In the type of work involved in the amenity centre many of the operations will be non-repetitive and these will have to be measured physically by tape and/or from working drawings. A record of the weekly measure should be kept in the form of record drawings which will be marked up weekly with the amount of work done. This record will be particularly useful when bonus is paid, ensures that gangs are not paid twice for the same piece of work, and constitutes a detailed record of progress. Items such as columns can be measured by simply observing their location and the number completed, the quantities being worked out from the drawings.

5.2.6.4 The weekly cost control sheet (Fig. 5.9)

This sheet is used for the collection of all the information arrived at via the allocation sheets, summary sheets, and records of measure. The unit cost for each operation can then be calculated and this facilitates control by comparison with set unit targets.

Columns A and B

These columns list the operation numbers and operation description for the work carried out during the week in question.

WEEKLY SUMMARY SHEET

CONTRACT — WEEK NO. — WEEK ENDING

CONTRACT NO. — SHEET NO. — OF

*BLOCK NO. — COMPILED BY

OPERATION :-	DAY	M	TU	W	TH	F	SAT	SUN	TOTAL
	LAB. HRS.								
PLANT USED :-	TRADE. HRS.								
	PLANT HRS.								

OPERATION :-	DAY	M	TU	W	TH	F	SAT	SUN	TOTAL
	LAB. HRS.								
PLANT USED:-	TRADE.HRS.								
	PLANT.HRS.								

OPERATION:-	DAY	M	TU	W	TH	F	SAT	SUN	TOTAL
	LAB. HRS.								
PLANT USED:-	TRADE.HRS.								
	PLANT.HRS.								

OPERATION :-	DAY	M	TU	W	TH	F	SAT	SUN	TOTAL
	LAB. HRS.								
PLANT USED :-	TRADE.HRS.								
	PLANT.HRS.								

OPERATION :-	DAY	M	TU	W	TH	F	SAT	SUN	TOTAL
	LAB. HRS.								
PLANT USED:-	TRADE.HRS.								
	PLANT.HRS.								

*BLOCK NO. ONLY APPLIES WHEN THIS FORM IS BEING USED FOR HOUSING

Fig. 5.8

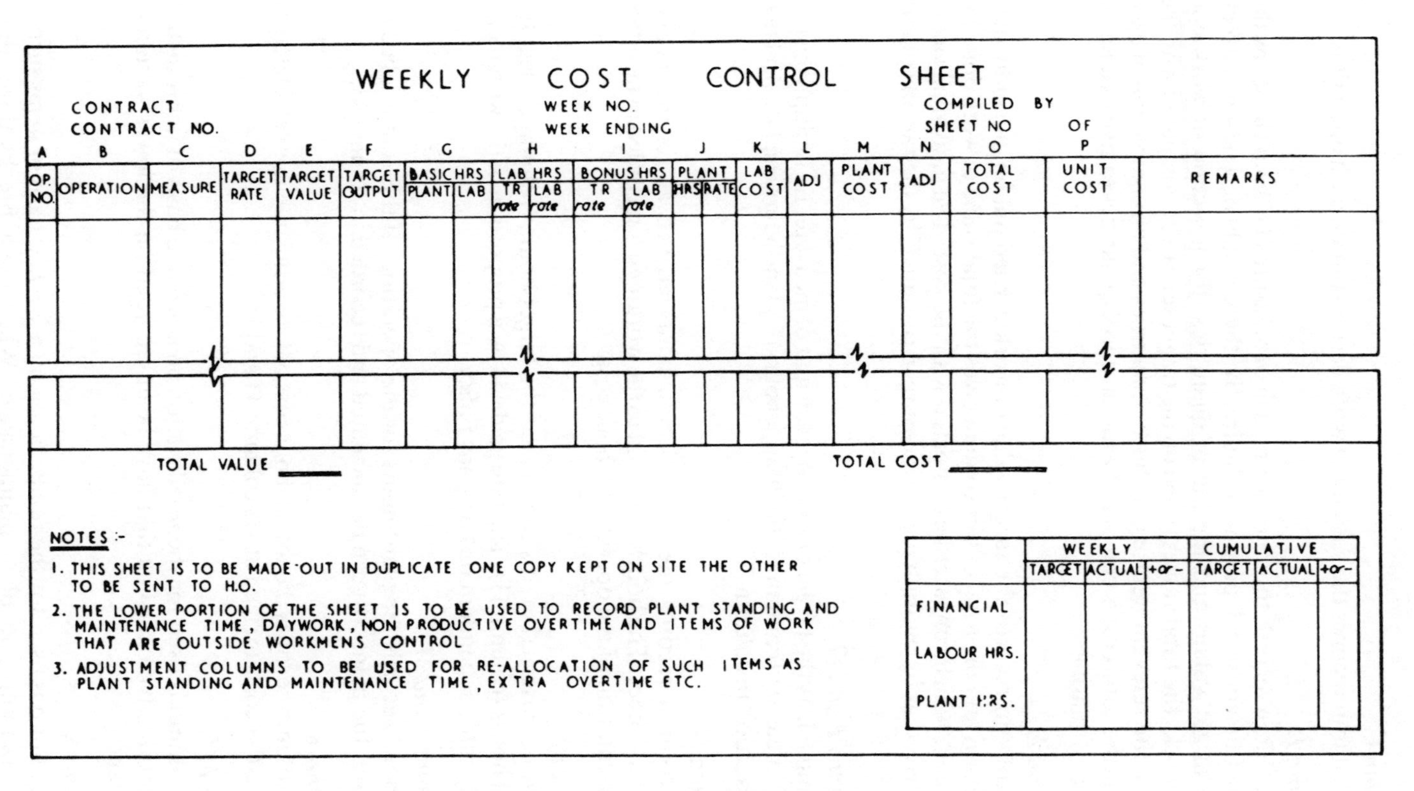

WEEKLY COST CONTROL SHEET

CONTRACT
CONTRACT NO.
WEEK NO.
WEEK ENDING
COMPILED BY
SHEET NO OF

A	B	C	D	E	F	G		H		I		J		K	L	M	N	O	P	
OP. NO.	OPERATION	MEASURE	TARGET RATE	TARGET VALUE	TARGET OUTPUT	BASIC HRS		LAB HRS		BONUS HRS		PLANT		LAB COST	ADJ	PLANT COST	ADJ	TOTAL COST	UNIT COST	REMARKS
						PLANT	LAB	TR rate	LAB rate	TR rate	LAB rate	HRS	RATE							

TOTAL VALUE ____ TOTAL COST ____

NOTES :-

1. THIS SHEET IS TO BE MADE OUT IN DUPLICATE ONE COPY KEPT ON SITE THE OTHER TO BE SENT TO H.O.
2. THE LOWER PORTION OF THE SHEET IS TO BE USED TO RECORD PLANT STANDING AND MAINTENANCE TIME, DAYWORK, NON PRODUCTIVE OVERTIME AND ITEMS OF WORK THAT ARE OUTSIDE WORKMENS CONTROL
3. ADJUSTMENT COLUMNS TO BE USED FOR RE-ALLOCATION OF SUCH ITEMS AS PLANT STANDING AND MAINTENANCE TIME, EXTRA OVERTIME ETC.

	WEEKLY			CUMULATIVE		
	TARGET	ACTUAL	+or–	TARGET	ACTUAL	+or–
FINANCIAL						
LABOUR HRS.						
PLANT HRS.						

Fig. 5.9

Column C
This column records the amount of work carried out on each operation.

Column D
This column records the value of one unit of measure and would be set by the production control department. Indirectly, there would be a tie-up between these target values and the bill of quantities: for a section of work e.g. brickwork, the total values based on the target set can be compared with the total of the relevant items in the bill of quantities and on completion of each section the total cost is compared with the bill of quantities to give a guide for future estimating.

Column E
This column records the target value of work carried out on one individual operation and the total of this column gives the total value of work done for the week, which can later be compared with the cost. The individual values are arrived at by multiplying the measure in column C by the rate of column D.

Columns F and G
Column F lists the target output for each unit of measure. By multiplying this by the measure (column C) the basic labour- or plant-hours can be calculated for insertion in column G.

Column H
This column records the actual labour-hours arrived at via the weekly summary sheet. The hourly rate paid to the workmen is recorded to facilitate the calculation of the operation labour cost.

Column I
The hours inserted in this column are the target hours (column G) less the actual hours (column H). The rate paid will be a proportion of the workmen's hourly rate, dependent on company policy.

Columns J and M
Column J records the plant-hours for the operation. If this is multiplied by the rate, the plant cost can be calculated and recorded in column M.

Column K
This column records the total labour cost before adjustment is calculated by adding the costs arrived at via columns H and I.

Column L
This column is used for the re-allocation on a *pro rata* basis of items such as non-productive overtime that have been recorded on the lower section of the sheet.

Column N
This column is used for the re-allocation on a *pro rata* basis of items such as plant standing and plant maintenance. When plant breaks down there should be no hire charge to the site, as the cost of the breakdown is borne by either the plant department or the plant hire firm.

Column O
This is the total weekly cost for an operation and is arrived at by adding columns K, L, M and N. By adding up the values given in this column, the total cost for the week is arrived at.

Column P
This records the actual cost of producing one unit of measure on an operation, e.g. the cost of one cubic yard of concrete. It is arrived at by dividing the total cost for an operation (column O) by the total measure for that operation (column C); this unit rate can then be compared with the target rate in column D.

5.2.7 Operational cost control

The second system of cost control can be applied to repetitive construction and is usually classed as an operational cost control system. The example has been applied to the housing site at Tenbury (chapter 1, section 1.10.4).

The object of operational cost control is to arrive at the cost of completing each operation on each block and then to compare this with a target for each. An operational cost in excess of or below the target by more than a predetermined margin requires further investigation and should be brought to the attention of management. As with unit cost control, information is collected by means of daily allocation sheets, weekly summary sheets, and actual measurement of the work done.

5.2.7.1 Daily allocation and weekly summary sheets
The daily allocation sheets and weekly summary sheets for this system will be similar to those for the unit control system (Figs. 5.7 and 5.8).

5.2.7.2 Measurement of work done
On repetitive construction it is not necessary to measure the work done by taking physical dimensions on site. Instead, the project is divided into repetitive operations of known work content (ascertained from drawings) and the measurement of work done is carried out by observing the number of operations completed and the block number where these observations were made. The exception to this rule is with work below the damp-proof course, where variations in the work content may occur from one block to another, necessitating physical measurement of the work.

5.2.7.3 Weekly operational cost sheet (Fig. 5.10)
This sheet is used to arrive at the cost of completed operations, and is a means of converting the information presented in the weekly summary sheet into monetary terms. It is very similar to the unit cost sheet with the exception of columns A and C: column A simply records the block number, whilst column C lists the target hours for bonus purposes.

Only completed operations are entered on the operational cost sheet, the hours spent on uncompleted operations being retained on the weekly summary sheet and carried forward to the next weekly summary sheet. When

WEEKLY OPERATIONAL COST SHEET

CONTRACT

CONTRACT NO.

WEEK NO. WEEK ENDING

SHEET NO. OF.

COMPILED BY

A	B		C	D		E		F		G	H	I	J	K	
BLOCK NO.	OPERATION		BONUS TARGET HRS.	ACTUAL LAB. HRS.		BONUS HRS.		PLANT		LAB COST	ADJ	PLANT COST	ADJ	TOTAL COST	REMARKS
	OP. NO	OP.		TR. rate	LAB. rate	TR. rate	LAB. rate	HRS	RATE						

NOTES

1 THE LOWER PORTION OF THE SHEET IS TO BE USED TO RECORD PLANT STANDING AND MAINTENANCE TIME, DAYWORK, NON PRODUCTIVE OVERTIME AND ITEMS OF WORK THAT ARE OUTSIDE WORKMENS CONTROL

2 ADJUSTMENT COLUMNS TO BE USED FOR RE-ALLOCATION OF SUCH ITEMS AS PLANT STANDING AND MAINTENANCE TIME, EXTRA OVERTIME ETC.

3 ONLY COMPLETED OPERATIONS ARE TO BE ENTERED ON THIS SHEET

Fig. 5.10

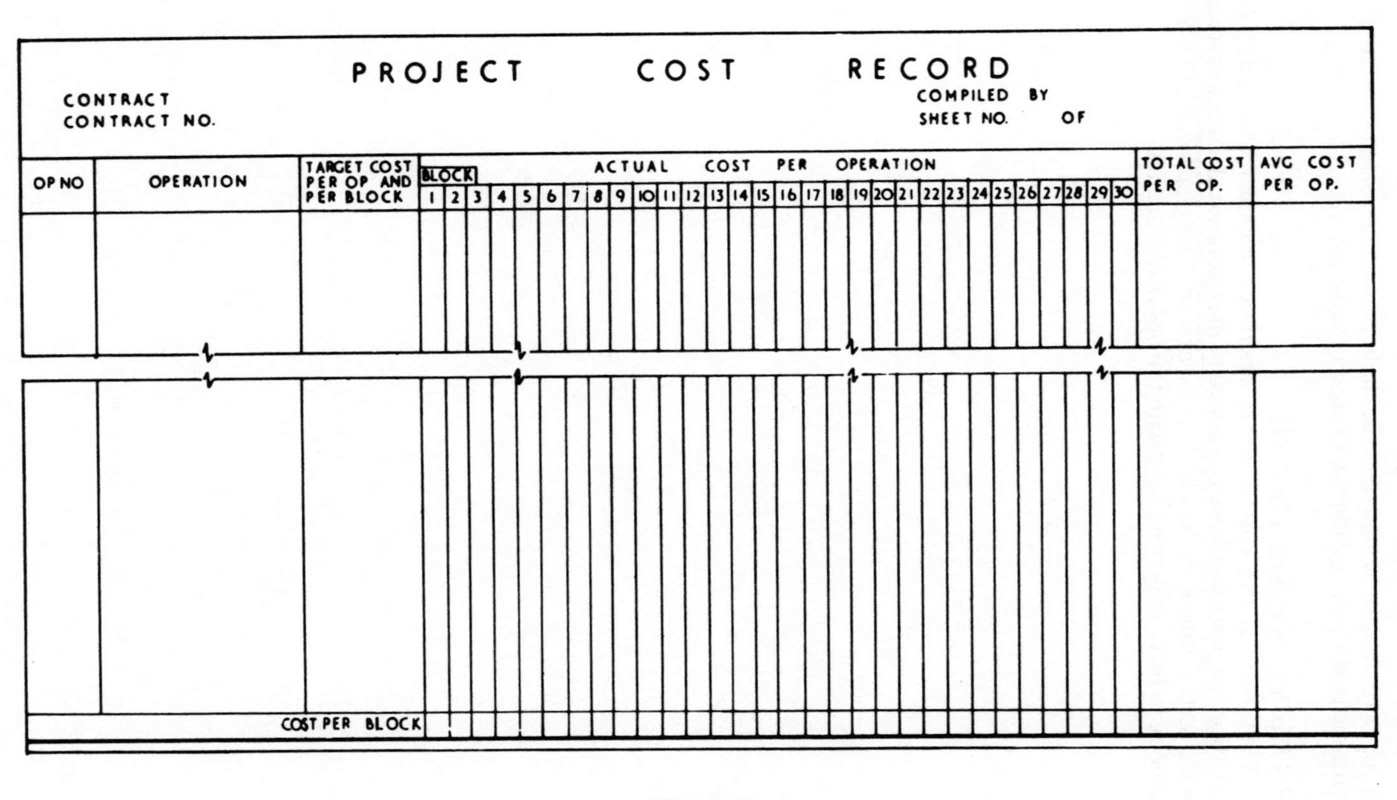
PROJECT COST RECORD
CONTRACT
CONTRACT NO.
COMPILED BY
SHEET NO. OF
OP NO
OPERATION
TARGET COST PER OP AND PER BLOCK
BLOCK
ACTUAL COST PER OPERATION
1 2 3 4 5 6 7 8 9 10 11 12 13 14 15 16 17 18 19 20 21 22 23 24 25 26 27 28 29 30
TOTAL COST PER OP.
AVG COST PER OP.
COST PER BLOCK

Fig. 5.11

these operations are completed the hours are transferred first to the operational cost sheet and then to the project cost record.

5.2.7.4 Project cost record (Fig. 5.11)
This sheet records the cost of every operation completed to date. As information becomes available it is added to the operational cost sheet week by week, so that this sheet enables the cost of completing each operation and each block to be readily compared with the original estimates.

6 Operational Research

6.1 INTRODUCTION

Like the other management techniques mentioned in this book, Operational Research is another example of the application of scientific method in the solution of management problems.

Decisions ultimately made by management are taken after consideration of relative advantages and disadvantages of various courses of action.

Small problems can be solved manually as shown in the examples which follow, but for more complex problems computers are far more economical and are of course faster and more accurate. Many standard computer packages are available in which case, once the problem has been formulated, the provision of the solution is a relatively simple process.

When using Operational Research; a scientific model, i.e. a set of equations, or a computer model, etc., is developed, and by applying data, decisions can be made more realistically. Operational Research is therefore dynamic.

6.1.1 Steps taken in applying operational research

The steps taken in applying Operational Research are:

1. Define the problem.
2. Collect and record all relevant data.
3. Examine and analyse data and develop a mathematical or systematic model of the real life situation.
4. Check that model is valid.
5. Test model under differing conditions.
6. Select the optimum solution using the model.
7. Implement the results.
8. Keep a check on the model and see it is still valid under changing circumstances.

It can be seen that these steps are very similar to those used in Work Study.

It very often happens that Operational Research will not give a single optimum solution and the manager's judgement will be necessary to decide what action should be taken.

6.2 LINEAR PROGRAMMING

6.2.1 Introduction

Linear programming is probably the most used of the mathematical methods. It can be used for solving many problems, examples being:

1. Optimising use of resources.
2. Determining most satisfactory product mix.
3. Division of work between different units.
4. Transportation problems.
5. Personnel assignment.
6. Establishing best processing schedules.
7. Determining optimum size of bid.
8. Location of new production plants, offices and warehouses.

The problem is to minimise or maximise some particular feature, this may be maximising profit or minimising loss, etc.

Problems containing only two variables can be solved graphically. Problems with three variables can also be solved graphically, if a three dimension graph can be drawn, but this can be very complicated. With more than three variables, a technique known as 'Simplex' is used and this is the most used method of solution.

6.2.2 Example of use of linear programming

A builder has a hydraulic press for producing paving slabs. The press is capable of producing 125 slabs per day. He has just bought some land on which to manufacture unpressed precast slabs but due to limitation in space, he can only produce 200 precast slabs per day.

The builder employs 18 men on this type of operation and does not wish to increase his labour force. 8 man days are required to produce 100 precast slabs and 4 man days for 100 hydraulically pressed slabs.

It is calculated that the profit will be £8 per 100 for precast concrete and £6 per 100 for hydraulically pressed slabs.

The problem is to maximise total profit.

6.2.3 The solution

Let S_1 be daily production of precast slabs (in hundreds).

Let S_2 be daily production of hydraulically pressed slabs (in hundreds).

This is a Linear Programming problem with two decision variables, S_1 and S_2. The builder can produce as many of each as he likes. If he makes S_1 precast slabs and S_2 hydraulically pressed slabs he will make a daily profit of

$$8S_1 + 6S_2.$$

This is known as the objective function and will give the total profit which he wishes to maximise.

The maximum number of each type which can be produced is 200 and 125 respectively.

$$\therefore \quad S_1 \leq 2$$

$$S_2 \leq 1\tfrac{1}{4}.$$

There is also a side constraint because of the limit in the labour available.

The maximum labour available is 18 men, and it takes 8 man days to produce 100 precast slabs and 4 man days to produce 100 hydraulically pressed slabs. Therefore the total programme requirements are

$$8S_1 + 4S_2 \text{ man days}$$

and $$8S_1 + 4S_2 \leq 18.$$

NOTE: if he wished to produce 200 precast and 125 hydraulically pressed slabs, this would require

$$8 \times 2 + 4 \times 1\tfrac{1}{4} = 21 \text{ man days}$$

and this is not feasible due to labour force constraint. It is of course possible to produce 175 precast and 100 hydraulically pressed slabs. This would keep all labour occupied and would leave excess capacity on presses and in the yard.

$$(8 \times 1\tfrac{3}{4} + 4 \times 1 = 18).$$

The profit in this case would be

$$8 \times 1\tfrac{3}{4} + 6 \times 1 = £20.$$

6.2.3.1 Graphical solution

As there are only two decision variables, the problem can be represented graphically. The area of possible production programme is limited by the capacity of the presses (125 slabs per day) and the space available for precast slabs (200 slabs per day). See Fig. 6.1.

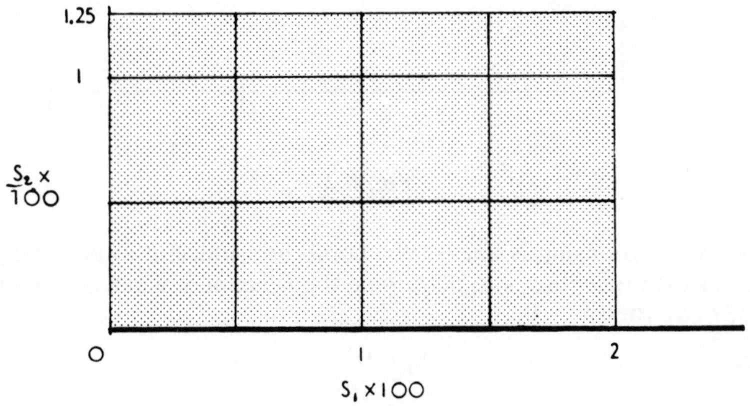

Fig. 6.1

Commonsense dictates that $S_1 \geq 0$ and $S_2 \geq 0$.

Fig. 6.2 shows the effect of the constraint

$$8S_1 + 4S_2 \leq 18.$$

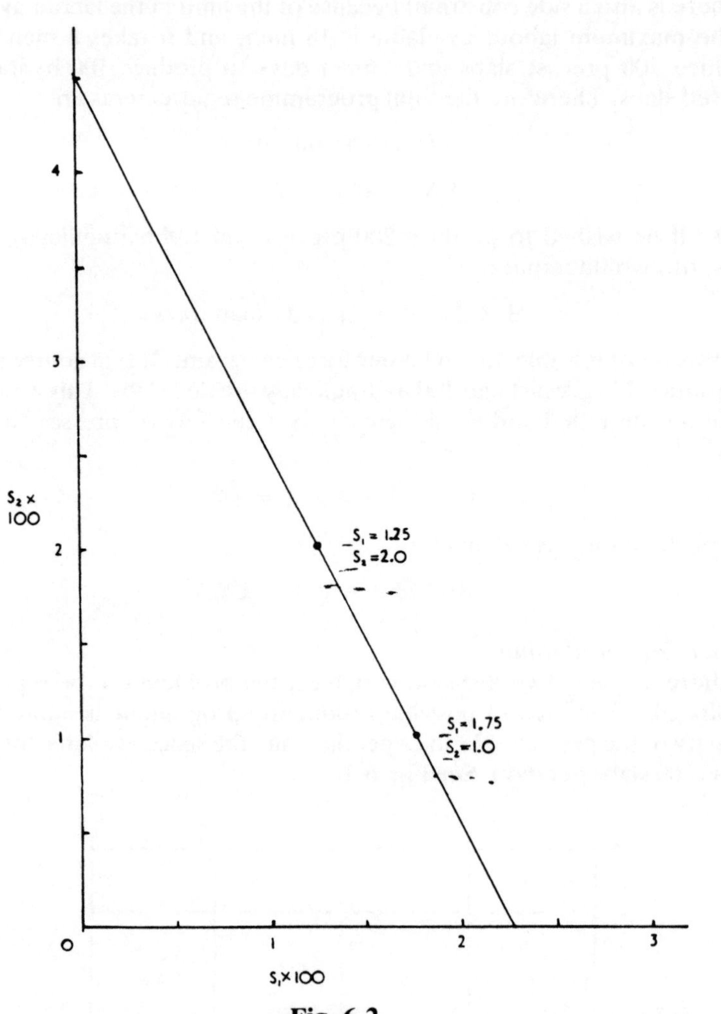

Fig. 6.2

As the relationship is linear, two points are sufficient to plot the graph. The easiest procedure is to determine the value of S_1 when $S_2 = 0$ and the value of S_2 when $S_1 = 0$.

Let $$S_2 = 0$$
$$8S_1 + 0 = 18$$
$$S_1 = \frac{18}{8} = 2\tfrac{1}{4}$$

Let $$S_1 = 0$$
$$0 + 4S_2 = 18$$
$$S_2 = 4\tfrac{1}{2}.$$

These values are plotted on Fig. 6.2.

To verify that this is a straight line, other points can be plotted, e.g.

Let $S_2 = 1$

$$8S_1 + 4 = 18$$

$$S_1 = \frac{18 - 4}{8} = 1\tfrac{3}{4}$$

Let $S_2 = 2$

$$8S_1 + 8 = 18$$

$$S_1 = \frac{18 - 8}{8} = 1\tfrac{1}{4}.$$

The manpower requirements are now restricted to the area between this line and the origin since any point within the triangle or on the edge satisfies the constraint $8S_1 + S_2 \leq 18$ whilst any point outside fails to satisfy it.

If Figs 6.1 and 6.2 are combined (see Fig. 6.3) it can be seen clearly that the area of choice is now limited to the area shown shaded. Any point outside this area will require either excessive manpower or greater production of slabs.

The objective function $8S_1 + 6S_2$ will now be considered. Take any particular profit margin and calculate what combination of S_1 and S_2 will give this profit; assume a profit of £8,

i.e. $8S_1 + 6S_2 = £8$

If $S_2 = 0$

$$8S_1 + 0 = £8$$

$$S_1 = 1$$

If $S_1 = 0$

$$0 + 6S_2 = £8$$

$$S_2 = 1\tfrac{1}{3}.$$

This can be plotted on the graph (see Fig. 6.4).

For each point on the line the profit is £8 per day

e.g. $S_2 = \tfrac{1}{2}$

$$8S_1 + 3 = £8$$

$$S_1 = \tfrac{5}{8}.$$

Now take a profit of £12 per day,

i.e. $8S_1 + 6S_2 = £12$

If $S_2 = 0$

$$S_1 = 1\tfrac{1}{2}$$

If $$S_1 = 0$$
$$S_2 = 2.$$

This is also plotted on Fig. 6.4.

It can be seen that the lines representing profits of £8 and £12 respectively are parallel and that the £12 profit line is further from the origin. These lines of variable profit may be likened to contour lines and the further these are from the origin the greater the profit margin will be. In this case it can be seen that the greatest profit margin allowed by the constraints lies on the line EF at the point B, i.e. $S_1 = 1\frac{5}{8}$, $S_2 = 1\frac{1}{4}$ this represents a profit of

$$8S_1 + 6S_2$$
$$= 8 \times 1\tfrac{5}{8} + 6 \times 1\tfrac{1}{4}$$
$$= 13 + 7\tfrac{1}{2}$$
$$= £20.50 \text{ per day}$$

by making 162½ precast slabs and 125 hydraulically pressed slabs per day.

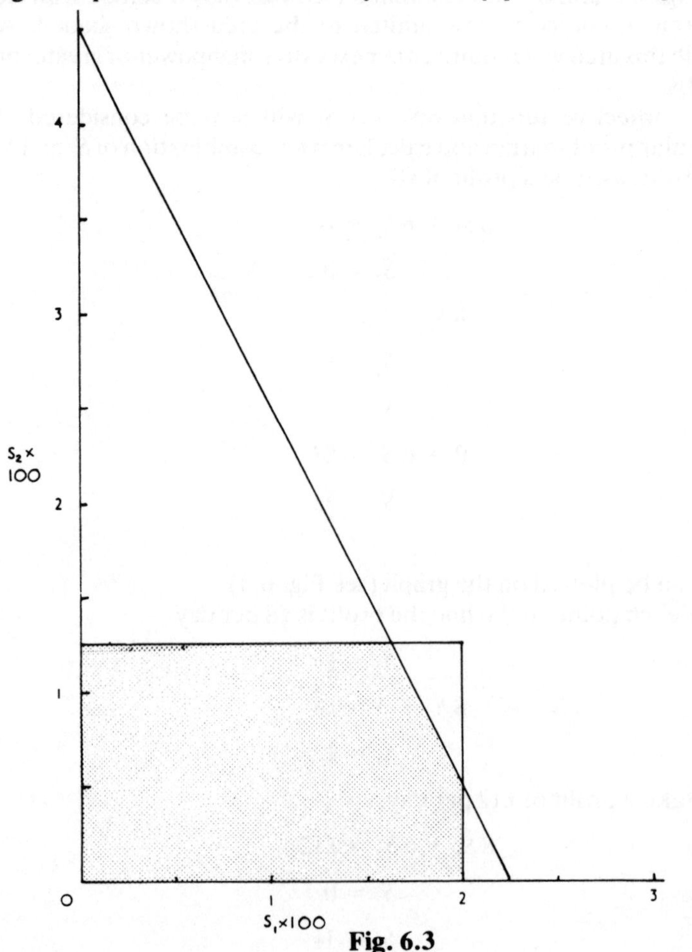

Fig. 6.3

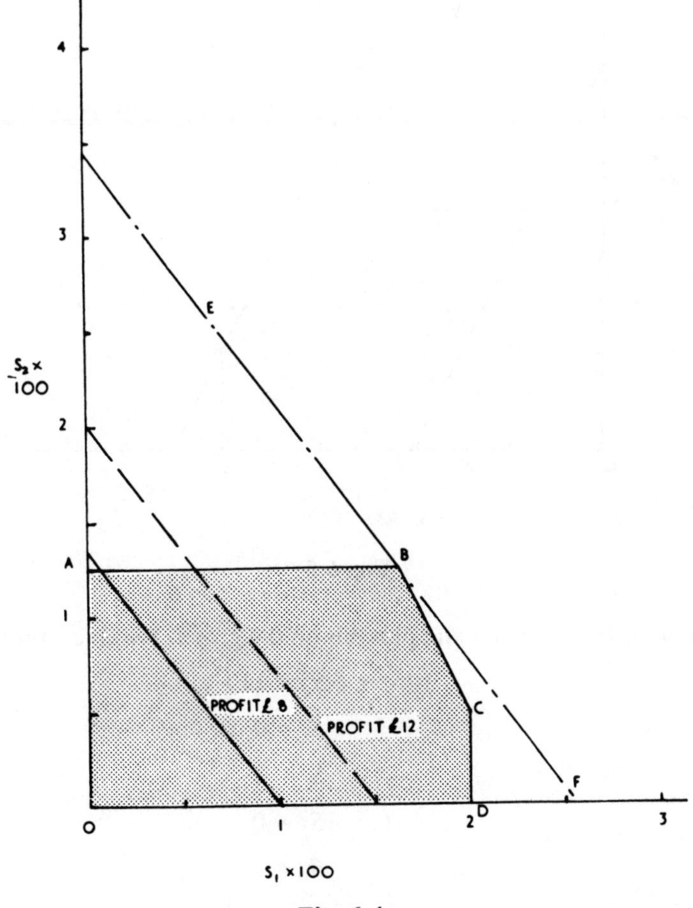

Fig. 6.4

Obviously it is not possible to make 162½ slabs but this would represent 325 every two days.

This is the solution to this problem.

6.2.3.2 Varying the profit margin

Once the solution has been found for one profit margin it is very easy to provide a solution showing the effect of other profit margins.

Example 1

If the profit on precast slabs is to be £10 per 100, and on hydraulically pressed slabs is to be £5 per 100, the graphical solution will be as shown in Fig. 6.5.

The objective function would become

$$10\,S_1 + 5\,S_2.$$

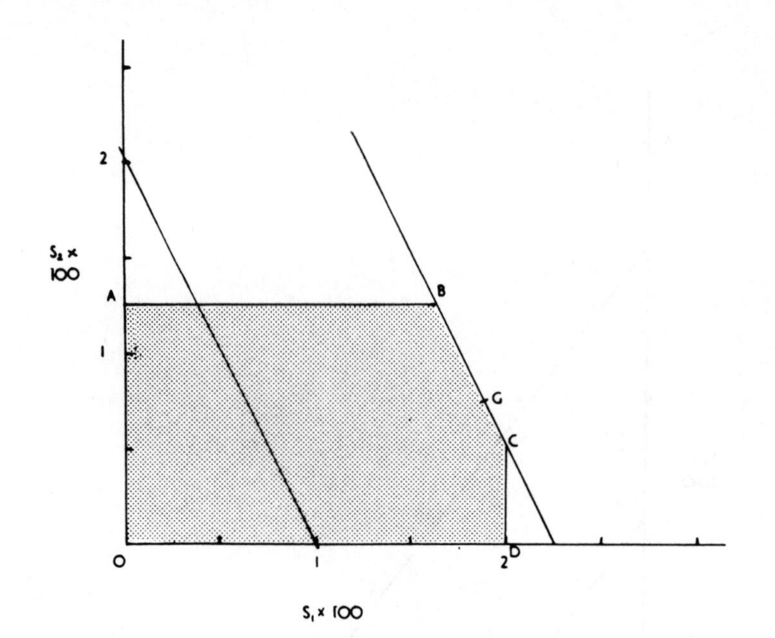

Fig. 6.5

Take a particular profit and plot this on the graph. Assume a profit of £10

i.e. $$10S_1 + 5S_2 = £10$$

If $$S_1 = 0$$
$$S_2 = 2$$

If $$S_2 = 0$$
$$S_1 = 1.$$

In this case the greatest profit margin can be achieved in a number of different ways, each one falling on the line BC

e.g. at B $S_1 = 1\frac{5}{8}$ $S_2 = 1\frac{1}{4}$

$\therefore$ profit $= 10 \times 1\frac{5}{8} + 5 \times 1\frac{1}{4} = 22\frac{1}{2}$

at C $S_1 = 2$ $S_2 = \frac{1}{2}$

$\therefore$ profit $= 10 \times 2 + 5 \times \frac{1}{2} = 22\frac{1}{2}$

at G $S_1 = 1\frac{7}{8}$ $S_2 = \frac{3}{4}$

$\therefore$ profit $= 10 \times 1\frac{7}{8} + 5 \times \frac{3}{4} = 22\frac{1}{2}$.

Example 2

If the profit on the precast slabs is to be £12.50 and on hydraulically pressed slabs £5 the graphical solution will be as shown in Fig. 6.6.

The objective function would be $12.5\,S_1 + 5\,S_2$.

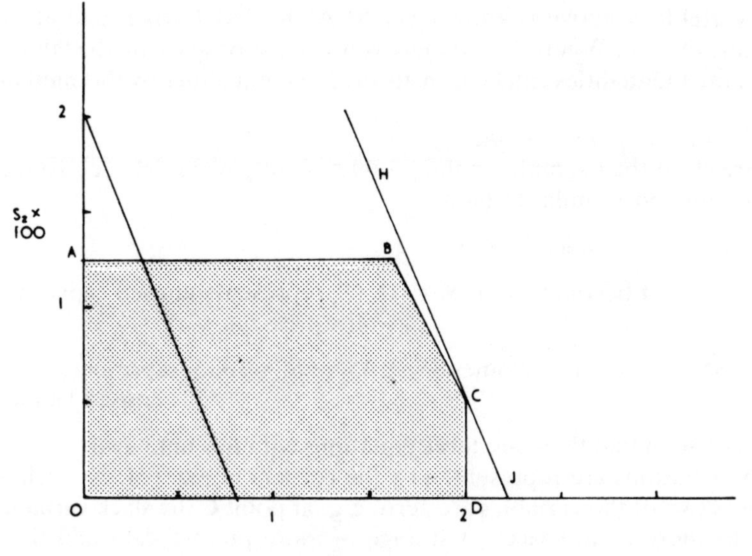

Fig. 6.6

Take a profit of £10.

i.e. $\quad 12.50\,S_1 + 5\,S_2 = £10$

If $\quad S_1 = 0 \qquad S_2 = 2$

If $\quad S_2 = 0 \qquad S_1 = 0.8.$

The greatest profit margin is now at point C and this represents a profit of

$$12.50\,S_1 + 5\,S_2$$

$$= 12.50 \times 2 + 5 \times \tfrac{1}{2} = £27.50.$$

It can be seen that no matter what profit margin is used the solution always lies on the boundary of the feasable region and usually on a corner point. This is a fundamental property of linear programming formulations.

6.2.4 The simplex technique

6.2.4.1 Introduction

As stated earlier, problems with more than two variables are not normally solved graphically and an algebraic solution is necessary. In order to solve the problem algebraically it is necessary to transform all the inequalities into equalities, e.g. $S_1 \leq 2$ is an inequality but by introducing S_3 to represent the excess of space on land bought by the builder this can be turned into an equality thus: $S_1 + S_3 = 2$.

6.2.4.2 Slack variables

The variable S_3 above is known as a SLACK VARIABLE and like other variables $S_3 \geq 0$. Where $S_3 = 0$ there is no excess of space on the land.

All the inequalities can be transformed into equalities by this method.

6.2.4.3 Paving slabs example

Referring to the example on the production of paving slabs, equations can be obtained in a similar fashion:

e.g. $S_1 \leq 2$ becomes $S_1 + S_3 = 2$

$S_2 \leq 1\frac{1}{4}$ becomes $S_2 + S_4 = 1\frac{1}{4}$ (S_4 represents unused capacity of the presses)

$8S_1 + 4S_2 \leq 18$ becomes $8S_1 + 4S_2 - S_5 = 18$ (S_5 represents unused labour)

It can be seen that there are three equations with five unknowns.

The solutions are represented by the corners in Fig. 6.4. At each of the corners two of the variables are zero, e.g. at point C the slack variable S_3 is zero as there is no space for storage of more precast slabs and the slack variable S_5 is zero because the labour force is fully utilised.

The solution normally lies on a corner point and at the corners two variables are zero and three are positive (in this case). The number of positive variables is normally the same as the number of side constraints. N.B. In some cases the slack variables may be positive and the decision variables zero.

As the solution lies on a corner point the aim is to find the corner points. Generally speaking, when there are *n* linear equations and *n* unknowns there is only one solution. There are exceptions to this rule but it is not necessary to consider the exceptions in this text.

The solution to this type of problem can be found by elimination but this method is very laborious and the *Simplex technique* is preferable in this situation.

6.2.4.4 Application of the simplex technique

Start at the origin when no precast or hydraulically pressed slabs are being produced (see Fig. 6.7).

$$S_1 = 0,\ S_2 = 0,\ S_3 = 2,\ S_4 = 1\tfrac{1}{4},\ S_5 = 18 \qquad (1)$$

i.e. at origin 0 profit $= 8\,S_1 + 6\,S_2 = 0$.

Proceed by replacing zero variables (S_1 and S_2) with positive variables (S_3, S_4 and S_5). At each stage exchange one positive for one zero variable so that the objective function increases or remains as it was – if all possible switches decrease the value of the objective function then the solution has been found.

It is usual to increase the variable giving the largest increase in profit first, i.e. S_1.

$$S_1 = 2,\ S_2 = 0,\ S_3 = 0,\ S_4 = 1\tfrac{1}{4},\ S_5 = 2 \qquad (2)$$

i.e. at point D profit $= 8S_1 + 6S_2 = £16$.

$$S_1 = 2,\ S_2 = \tfrac{1}{2},\ S_3 = 0,\ S_4 = \tfrac{3}{4},\ S_5 = 0 \qquad (3)$$

i.e. at point C profit $= 8S_1 + 6S_2 = £19$.

$$S_1 = 1\tfrac{5}{8},\ S_2 = 1\tfrac{1}{4},\ S_3 = \tfrac{3}{8},\ S_4 = 0,\ S_5 = 0 \qquad (4)$$

i.e. at point B profit $= 8S_1 + 6S_2 = £20.50$.

$$S_1 = 0,\ S_2 = 1\tfrac{1}{4},\ S_3 = 2,\ S_4 = 0,\ S_5 = 13 \qquad (5)$$

i.e. at point A profit $= 8S_1 + 6S_2 = £7.50$.

6.2.4.5 Note

Figure 6.7 is used simply to illustrate the movement from corner to corner. As stated earlier, in problems having a number of variables it would not be possible to illustrate the method using a diagram but the procedure used to calculate the optimum solution is as shown above.

It is simple to calculate the value of having one extra man in the gang by calculating the profit with one extra man and thus calculate the cost of employing him. This will show whether it is economical to do so.

Using the simplex technique it is possible to find the cost of restrictions as well as the solution to the problem. A linear programming problem is characterised by three requirements:

1. There must be a function, linear in a number of decision variables which has to be maximised or minimised.
2. Decision variables must *not* assume negative values.
3. They must satisfy a number of linear side conditions, usually in the form of inequalities. Often the most difficult task is to recognise a problem as a linear programming problem.

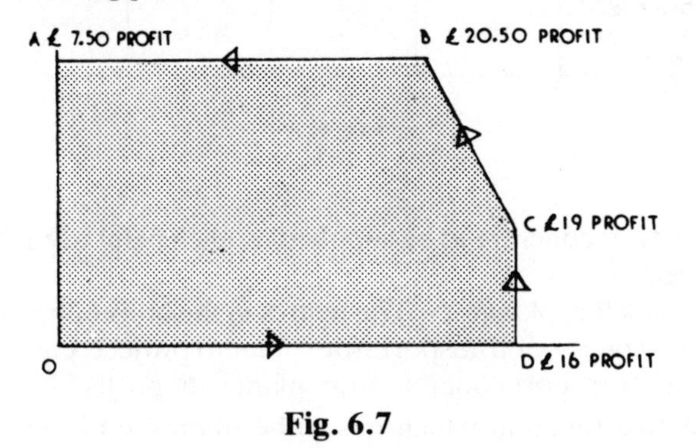

Fig. 6.7

6.3 THE TRANSPORTATION PROBLEM

6.3.1 Introduction

Transportation problems can be solved using the Simplex Technique but a better method has been developed. This method is sometimes called the 'Transportation Technique of Linear Programming'. It can be applied in situations where products or materials are produced in a number of different places for distribution to other places. The object is to optimise the effectiveness of the operation.

Basically the optimum solution is arrived at by progressively comparing the cost of routes initially selected with those not selected and systematically adjusting the schedule where costs are lower.

In general terms the objective is to allocate jobs to facilities in such a way as to optimise the effectiveness in terms of cost.

6.3.2 Example of transportation problem

A pre-mixed concrete firm has to supply concrete to three different projects, 1, 2 and 3. The projects require 200, 350 and 400 cubic metres of concrete respectively in a particular week. The firm has three plants, A, B and C producing concrete and these can provide 250, 400 and 350 cubic metres respectively.

The problem is one of minimising costs, and the costs of transporting concrete from each plant to each project is given in Fig. 6.8.

TRANSPORT FROM MIXING PLANT	TRANSPORT TO PROJECT AT			CAPACITY OF MIXING PLANT
	1	2	3	
A	40	20	30	250
B	40	50	100	400
C	20	80	30	350
DAILY REQTS BY PROJECTS IN M^3	200	350	400	

Fig.6.8

The quantity of concrete to be supplied by each plant to each project must be determined.

Figure 6.8 is self explanatory. The figures in the body of the table represent the cost in new pence of transport from plant to project, e.g. it costs 80p per cubic metre to transport concrete from plant C to project 2.

The objective function which has to be minimised is $40x_{A_1} + 20x_{A_2} + 30x_{A_3} + 40x_{B_1} + 50x_{B_2} + 100x_{B_3} + 20x_{C_1} + 80x_{C_2} + 30x_{C_3}$. x represents the quantity of concrete be sent from plant to project, i.e. x_{A_1} represents the quantity of concrete sent from plant A to project 1.

When all the costs are added, this will give the total cost of transportation.

The side conditions which must be imposed due to the limited capacity of each plant are

$$x_{A_1} + x_{A_2} + x_{A_3} \leq 250 \qquad (1)$$

$$x_{B_1} + x_{B_2} + x_{B_3} \leq 400 \quad (2)$$

$$x_{C_1} + x_{C_2} + x_{C_3} \leq 350. \quad (3)$$

The side conditions which must be imposed by the requirements of each project are

$$x_{A_1} + x_{B_1} + x_{C_1} = 200 \quad (4)$$

$$x_{A_2} + x_{B_2} + x_{C_2} = 350 \quad (5)$$

$$x_{A_3} + x_{B_3} + x_{C_3} = 400. \quad (6)$$

And a side condition imposed by the excess capacity of the premixed concrete plant is

$$x_{A_4} + x_{B_4} + x_{C_4} = 50. \quad (7)$$

6.3.2.1 Slack variables

As the side conditions imposed by the limited capacity of each plant are inequalities, slack variables must be introduced. The side conditions can then be expressed as equations, thus:

$$x_{A_1} + x_{A_2} + x_{A_3} + x_{A_4} = 250 \quad (8)$$

$$x_{B_1} + x_{B_2} + x_{B_3} + x_{B_4} = 400 \quad (9)$$

$$x_{C_1} + x_{C_2} + x_{C_3} + x_{C_4} = 350 \quad (10)$$

i.e. x_{A_4} represents the excess capacity of plant A
x_{B_4} represents the excess capacity of plant B
x_{C_4} represents the excess capacity of plant C.

The decision variables must of course be non-negative.

The number of positive variables in a transportation problem is normally $m + n - 1$, where m is the number of rows and n is the number of columns – in this case $3 + 4 - 1 = 6$. This is because whilst there are 7 side conditions, one can easily be found if we know the remaining six, so there are in fact only six independent side conditions and therefore six decision variables, i.e. there are as many positive variables as there are side conditions.

6.3.2.2 Finding an initial solution using the Northwest Corner Rule

An initial solution must be found and this can be done using the table shown in Fig. 6.9 which has an additional column in which the excess capacity available at the mixing plants can be shown.

The selection of the decision variables in the body of the table can now be made using the Northwest Corner Rule, thus:

1. Start at the northwest corner (x_{A_1} representing the quantity of concrete to go from plant A to project 1). Plant A produces 250 cubic metres, project 1 requires 200 cubic metres. Send the maximum amount to project 1, i.e. 200 cubic metres. This satisfies the requirements of project 1 and thus completes the first column (Fig. 6.10). Now move progressively right and down until x_{C_4} is reached thus.

TRANSPORT FROM MIXING PLANT	TRANSPORT TO PROJECT AT			SURPLUS	CAPACITY OF MIXING PLANT
	1	2	3		
A					250
B					400
C					350
DAILY REQTS. BY PROJECTS IN M^3	200	350	400	50	

Fig. 6.9

TRANSPORT FROM MIXING PLANT	TRANSPORT TO PROJECT AT			SURPLUS	CAPACITY OF MIXING PLANT
	1	2	3		
A	200				250
B	0				400
C	0				350
DAILY REQTS BY PROJECTS IN M^3	200	350	400	50	

Fig. 6.10

2. Plant A still has 50 cubic metres excess capacity. This is sent to project 2. This exhausts the supply from plant A and the first row is complete (Fig. 6.11).

TRANSPORT FROM MIXING PLANT	TRANSPORT TO PROJECT AT			SURPLUS	CAPACITY OF MIXING PLANT
	1	2	3		
A	200	50	0	0	250
B	0				400
C	0				350
DAILY REQTS BY PROJECTS IN M^3	200	350	400	50	

Fig. 6.11

3. The new northwest corner is x_{B_2}. Project 2 needs 300 cubic metres more and this quantity is supplied from plant B. The second column is now complete (Fig. 6.12).

TRANSPORT FROM MIXING PLANT	TRANSPORT TO PROJECT AT			SURPLUS	CAPACITY OF MIXING PLANT
	1	2	3		
A	200	50	0	0	250
B	0	300			400
C	0	0			350
DAILY REQTS BY PROJECTS IN M³	200	350	400	50	

Fig. 6.12

4. Plant B still has 100 cubic metres excess capacity. This is now sent to project 3 and second row is now complete (Fig. 6.13).

TRANSPORT FROM MIXING PLANT	TRANSPORT TO PROJECT AT			SURPLUS	CAPACITY OF MIXING PLANT
	1	2	3		
A	200	50	0	0	250
B	0	300	100	0	400
C	0	0			350
DAILY REQTS BY PROJECTS IN M³	200	350	400	50	

Fig. 6.13

5. The new northwest corner is x_{C_3}. Project 3 needs 300 cubic metres more and this quantity is supplied from plant C (Fig. 6.14).

6. Plant C still has 50 cubic metres excess capacity and this is shown in the surplus column (Fig. 6.15).

This gives a solution with exactly the right number of variables, which satisfies all the side conditions.

TRANSPORT FROM MIXING PLANT	TRANSPORT TO PROJECT AT			SURPLUS	CAPACITY OF MIXING PLANT
	1	2	3		
A	200	50	0	0	250
B	0	300	100	0	400
C	0	0	300		350
DAILY REQTS BY PROJECTS IN M³	200	350	400	50	

Fig. 6.14

TRANSPORT FROM MIXING PLANT	TRANSPORT TO PROJECT AT			SURPLUS	CAPACITY OF MIXING PLANT
	1	2	3		
A	200	50	0	0	250
B	0	300	100	0	400
C	0	0	300	50	350
DAILY REQTS BY PPOJECTS IN M³	200	350	400	50	

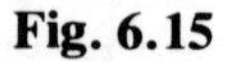

Fig. 6.15

NOTES

Whilst the Northwest Corner Rule gives an initial solution, it takes no account of cost. The total cost of transport in the above solution can be found thus:

$$x_{A_1} = 200,\ x_{A_2} = 50,\ x_{B_2} = 300,\ x_{B_3} = 100,$$
$$x_{C_3} = 300, \text{ and } x_{C_4} = 50.$$

Therefore the total cost will be:

$(200 \times 40) + (50 \times 20) + (300 \times 50) + (100 \times 100) + (300 \times 30) + (50 \times 0)$

$= 8000 + 1000 + 15\,000 + 10\,000 + 9000 + 0$

$= 43\,000$p

$=$ £430.

6.3.2.3 Alternative method of finding an initial solution

It is in fact possible to develop an initial solution which is cheaper. To do this the table of costs (Fig. 6.8) is used and the procedure is as follows:

1. Begin with the least cost. In this case two costs are the same, A_2 and C_1. Use A_2 and make this 250 which completes the first row (Fig. 6.16). Then continue by distributing concrete

TRANSPORT FROM MIXING PLANT	TRANSPORT TO PROJECT AT			SURPLUS	CAPACITY OF MIXING PLANT
	1	2	3		
A		250			250
B	200	100	100		400
C			300	50	350
DAILY REQTS BY PROJECTS IN M³	200	350	400	50	

Fig. 6.16

(a) to the site which received some but not all, concrete required at the end of the previous step or from the plant at which some remained at the end of the previous step.

(b) At least cost thus:

2. Complete second column by filling in x_{B_2} or x_{C_2}, choose the one which costs least, i.e. x_{B_2} and make this 100 which completes the second column.

3. Complete the second row by filling in x_{B_1} or x_{B_3}, choose the one with the least cost, i.e. x_{B_1} and make this 200 which is the total requirement for site 1. Now complete the second row by making x_{B_3} 100 and this exhausts the supply from plant B.

4. Complete the third column by filling in x_{C_1} or x_{C_3}. In this case the least cost is the same and we can therefore choose either. Choose x_{C_3} and make this 300 which is the total requirement of site 3.

5. Complete the third row by making x_{C_4} 50 which is the excess capacity of plant C.

The total cost in this case will be:

$$(250 \times 20) + (200 \times 40) + (100 \times 50) + (100 \times 100) + (300 \times 30) + (50 \times 0)$$

$$= 5000 + 8000 + 5000 + 10\,000 + 9000 = 0$$

$$= 37\,000\text{p}$$

$$= £370.$$

This is a saving of £60.

6.3.2.4 The optimum solution to transportation problems

It is necessary to know if this solution is the optimum solution; and the following check will determine this. In a transportation problem the optimum solution remains so when all the results in a row or column are increased or decreased by the same amount, e.g. using the table of costs (Fig. 6.8). Deduct 50p from all the costs in the second row and add 50p to all entries in the third column the result will be as shown in Fig. 6.17. The optimum solution for Fig. 6.8 and Fig. 6.17 is the same, i.e. in both cases it is necessary to send the same amount of concrete from each plant to each site. It is obviously impossible to have negative costs but it is convienient to use negative costs to determine whether a solution is optimum or not.

TRANSPORT FROM MIXING PLANT	TRANSPORT TO PROJECT AT			SURPLUS	CAPACITY OF MIXING PLANT
	1	2	3		
A	40	20	80	0	250
B	−10	0	100	0	400
C	20	80	80	0	350
DAILY REQTS BY PROJECTS IN M^3	200	350	400	50	

Fig. 6.17

6.3.2.5 Checking for optimum solutions

To find whether the solution found is the optimum one the table of costs is adjusted so that the allocations used have a cost of zero. The signs of the costs not used are then considered. If the signs of costs not used are all positive, the solution is optimum. If any signs are negative the solution is not optimum. To do this calculation the table of costs Fig. 6.8 and the initial solution Fig. 6.16 are used and the procedure is as follows:

1. Set out a table of costs for the distribution used, with an extra column and row (Fig. 6.18).

	SITE 1 REQTS. 200	SITE 2 REQTS 350	SITE 3 REQTS 400	SURPLUS 50	
PLANT A CAPACITY 250		20			
PLANT B CAPACITY 400	40	50	100		40
PLANT C CAPACITY 350			30	0	
	0				

Fig. 6.18

2. Make these costs zero by adding or deducting constant numbers. (This will not alter the optimum solution.)

(i) Start by inserting zero in the first column of the table of costs, Fig. 6.18.

(ii) Cost of 40p (B_1) must become zero, therefore deduct 40 p from the second row, shown in last column.

(iii) Cost of 50p (B_2) must also become zero, 40p is already shown in the second row, therefore deduct a further 10p, shown in second column. (40p + 10p) = 50p (Fig. 6.19).

(iv) Similarly given 40p in the second row we need to deduct a further 60p shown in the third column (40p + 60p) = 100p (Fig. 6.19).

(v) Proceed as above to give – 30p in the third row (60p – 30p) = 30p, 30p in the fourth column (30p – 30p) = 0, 10p in the first row (10p + 10p) = 20p (Fig. 6.19).

	SITE 1 REQTS. 200	SITE 2 REQTS. 350	SITE 3 REQTS. 400	SURPLUS 50	
PLANT A CAPACITY 250		20			10
PLANT B CAPACITY 400	40	50	100		40
PLANT C CAPACITY 350			30	0	-30
	0	10	60	30	

Fig. 6.19

3. Return to the original table of costs and insert the numbers just calculated in the bottom row and last column (Fig. 6.20).

	SITE 1 REQTS. 200	SITE 2 REQTS. 350	SITE 3 REQTS. 400	SURPLUS 50	
PLANT A CAPACITY 250	40	20	30	0	10
PLANT B CAPACITY 400	40	50	100	0	40
PLANT C CAPACITY 350	20	80	30	0	-30
	0	10	60	30	

Fig. 6.20

Deduct from all costs the cost in the bottom row and last column, e.g. cost 40p (A_1) becomes 40p − 10p − 0 = 30p. The complete table is shown in Fig. 6.21.

When all unused allocations are positive or zero the solution will be optimum. In this case there are negative costs and the costs can be further reduced.

The objective function is

$$30x_{A_1} + 0x_{A_2} - 40x_{A_3} - 40x_{A_4}$$
$$0x_{B_1} + 0x_{B_2} + 0x_{B_3} - 70x_{B_4}$$
$$50x_{C_1} + 110x_{C_2} + 0x_{C_3} + 0x_{C_4}$$

	SITE 1 REQTS. 200	SITE 2 REQTS. 350	SITE 3 REQTS. 400	SURPLUS 50	
PLANT A CAPACITY 250	40−10−0 = 30	20−10−10 = 0	30−10−60 = −40	0−10−30 = −40	10
PLANT B CAPACITY 400	40−40−0 = 0	50−40−10 = 0	100−40−60 = 0	0−40−30 = −70	40
PLANT C CAPACITY 350	20−30−0 = 50	80−−30−10 = 110	30−−30−60 = 0	0−−30−30 = 0	−30
	0	10	60	30	

Fig. 6.21

6.3.2.6 Finding the optimum solution

To find the optimum solution the value of the objective function must be decreased by making x_{A_3}, x_{A_4} or x_{B_4} positive. Choose x_{B_4} as it is multiplied by the largest negative number. This must be increased as much as possible taking account of the side constraints. Every increase decreases the value of the objective function. The procedure is as follows:

(i) Using the last solution (Fig. 6.16) add a + sign to the zero of x_{B_4} (Fig. 6.22) as this will be increased to make it positive.

TRANSPORT FROM MIXING PLANT	TRANSPORT TO PROJECT AT 1	2	3	SURPLUS	CAPACITY OF MIXING PLANT
A	0	250	0	0	250
B	200	100	100$^-$	0$^+$	400
C	0	0	300$^+$	50$^-$	350
DAILY REQTS. BY PROJECTS IN M^3	200	350	400	50	

Fig. 6.22

(ii) If concrete is retained at plant B (excess capacity) it will be necessary to send less to site 1, 2 or 3 (B_1, B_2 or B_3) and it will be necessary to retain less concrete at plant C. Attach a minus sign to C_4. If less concrete is retained at plant C more must be sent to site 3 (C_3). If more concrete is sent to site 3 from plant C (C_3) less will be required from plant B. This determines that it will be site 3 and not site 1 or 2 which receives less concrete from B.

(iii) x_{B_4} can be increased by a maximum of 50 cubic metres; if it was increased by more than this x_{C_4} would become negative. All allocations with a plus sign are then increased by 50 and those with a minus sign decreased by 50 (Fig. 6.23). The total cost is now:

$(250 \times 20) + (200 \times 40) + (100 \times 50) + (50 \times 100) + (50 \times 0) + (350 \times 30)$

$= 5000 + 8000 + 5000 + 5000 + 0 + 10\,500$

$= 33\,500$p

$= £335.$

A further saving of £35.

TRANSPORT FROM MIXING PLANT	TRANSPORT TO PROJECT AT			SURPLUS	CAPACITY OF MIXING PLANT
	1	2	3		
A		250		0	250
B	200	100	50	50	400
C			350	0	350
DAILY REQTS. BY PROJECTS IN M³	200	350	400	50	

Fig. 6.23

To check whether this is the optimum solution the latest cost schedule is used (Fig. 6.24) and the same procedure is repeated (Figs. 6.24, 6.25 and 6.26).

This still leaves a negative cost and the solution is not therefore optimum, x_{A_3} must be increased and the procedure above (i) to (iii) is followed (Figs 6.27 and 6.28).

The total costs is now

$(200 \times 20) + (50 \times 30) + (200 \times 40) + (150 \times 50) + (50 \times 0) + (350 \times 30)$

$= 4000 + 1500 + 8000 + 7500 + 0 + 10\,500$

$= 31\,500$p

$= £315.$

A further saving of £20.

	SITE 1 REQTS. 200	SITE 2 REQTS. 350	SITE 3 REQTS. 400	SURPLUS 50	
PLANT A CAPACITY 250		20			10
PLANT B CAPACITY 400	40	50	100	0	40
PLANT C CAPACITY 350			30		−30
	0	10	60	−40	

Fig. 6.24

	SITE 1 REQTS. 200	SITE 2 REQTS. 350	SITE 3 REQTS. 400	SURPLUS 50	
PLANT A CAPACITY 250	40	20	30	0	10
PLANT B CAPACITY 400	40	50	100	0	40
PLANT C CAPACITY 350	20	80	30	0	−30
	0	10	60	−40	

Fig. 6.25

	SITE 1 REQTS. 200	SITE 2 REQTS. 350	SITE 3 REQTS. 400	SURPLUS 50	
PLANT A CAPACITY 250	30	0	−40	30	10
PLANT B CAPACITY 400	0	0	0	0	40
PLANT C CAPACITY 350	50	100	0	70	−30
	0	10	60	−40	

Fig. 6.26

TRANSPORT FROM MIXING PLANT	TRANSPORT TO PROJECT AT			SURPLUS	CAPACITY OF MIXING PLANT
	1	2	3		
A	0	250⁻	0⁺	0	250
B	200	100⁺	50⁻	50	400
C	0	0	350	0	350
DAILY REQTS. BY PROJECTS IN M^3	200	350	400	50	

Fig. 6.27

TRANSPORT FROM MIXING PLANT	TRANSPORT TO PROJECT AT			SURPLUS	CAPACITY OF MIXING PLANT
	1	2	3		
A	0	200	50	0	250
B	200	150	0	50	400
C	0	0	350	0	350
DAILY REQTS BY PROJECTS IN M^3	200	350	400	50	

Fig. 6.28

A check is again made to determine whether this is the optimum solution (Figs. 6.29, 6.30 and 6.31). This leaves no negative number and is therefore the optimum solution.

	SITE 1 REQTS. 200	SITE 2 REQTS. 350	SITE 3 REQTS. 400	SURPLUS 50	
PLANT A CAPACITY 250	—	20	30	—	10
PLANT B CAPACITY 400	40	50	—	0	40
PLANT C CAPACITY 350	—	—	30	—	10
	0	10	20	−40	

Fig. 6.29

	SITE 1 REQTS. 200	SITE 2 REQTS. 350	SITE 3 REQTS. 400	SURPLUS 50	
PLANT A CAPACITY 250	40	20	30	0	10
PLANT B CAPACITY 400	40	50	100	0	40
PLANT C CAPACITY 350	20	80	30	0	10
	0	10	20	−40	

Fig. 6.30

	SITE 1 REQTS. 200	SITE 2 REQTS. 350	SITE 3 REQTS. 400	SURPLUS 50	
PLANT A CAPACITY 250	30	0	0	30	10
PLANT B CAPACITY 400	0	0	40	0	40
PLANT C CAPACITY 350	10	60	0	30	10
	0	10	20	−40	•

Fig. 6.31

6.4 THE ASSIGNMENT PROBLEM

6.4.1 Introduction

Like transportation problems, assignment problems can be solved using the Simplex technique, but again a better method has been developed.

This technique can be applied in situations where a fixed number of operatives are available to do a fixed number of jobs or operations. The object is to allocate jobs to facilities in such a way as to optimise the overall effectiveness.

6.4.2 Example in the construction industry

Assume the majority of a contractor's work is local authority housing and he wishes to expand and carry out larger housing projects. He has decided to use the Line of Balance method of planning and his carpenters will therefore be specialising in a limited area of work. At present his carpenter gangs each work on the whole range of operations, and on one project, the contractor has analysed the feed back data relating to these gangs as shown in Fig. 6.32.

The problem is to decide which gangs should specialise in the various operations.

6.4.3 Procedure for solving problem

1. Select the operation which takes the minimum time to perform in each column and fill in the time required for this operation beneath the column (see Fig. 6.32), e.g. in column 4 first fix takes the least time to perform at 33 hours. Fill in 33 beneath column 4.

Operation	*Gang No.*	*1*	*2*	*3*	*4*	*5*	*6*
Carcassing (floors and roofs)		110	100	95	105	115	120
First fix		36	30	40	33	42	32
Second fix		112	100	105	115	120	85
Finishings		30	25	40	35	35	28
Floorboards and staircase		40	37	45	48	50	35
First floor partitions		90	105	90	78	85	100
		30	25	40	33	35	28

Fig. 6.32

2. Subtract the minimum time in each column from all the other operation times (see Fig. 6.33), e.g. in column 1 carcassing takes 110 hours. Subtract 30 hours and this leaves 80 hours.

3. Select the minimum time in each row and fill in the time at the end of the row (see Fig. 6.33), e.g. in the first row 55 is the minimum operation time, fill in 55 at end of row.

Operation	*Gang No.*	*1*	*2*	*3*	*4*	*5*	*6*	
Carcassing (floors and roof)		80	75	55	72	80	92	55
First fix		6	5	0	0	7	4	0
Second fix		82	75	65	82	85	57	57
Finishings		0	0	0	2	0	0	0
Floorboards and staircase		10	12	5	15	15	7	5
First floor partitions		60	80	50	45	50	72	45

Fig. 6.33

4. Subtract the minimum time in each row from all the other times (see Fig. 6.34).

5. Draw vertical and horizontal lines through the rows and columns containing zeros, starting with those with most zeros. A solution has been found when a total of six lines are required to cover all the zeros. In Fig. 6.34 there are only four lines and a solution therefore has not been found.

6. Select the smallest number not covered by a line, i.e. five and subtract this number from all numbers not covered. Where numbers are covered twice, i.e. at intersections, add 5 (see Fig. 6.35).

Operation	*Gang No.*	*1*	*2*	*3*	*4*	*5*	*6*
Carcassing (floors and roof)		25	20	0	17	25	37
First fix		6	5	0	0	7	4
Second fix		25	18	8	25	28	0
Finishings		0	0	0	2	0	0
Floorboards and staircase		5	7	0	10	10	2
First floor partitions		15	35	5	0	5	20

Fig. 6.34

Operation	*Gang No.*	*1*	*2*	*3*	*4*	*5*	*6*
Carcassing (floors and roof)		20	15	0	17	20	37
First fix		1	0	0	0	2	4
Second fix		20	13	8	25	23	0
Finishings		0	0	5	7	0	5
Floorboards and staircase		0	2	0	10	5	2
First floor partitions		10	30	5	0	0	27

Fig. 6.35

7. Repeat stage 5 as shown in Fig. 6.35. This gives six lines which will result in a solution. To find a solution, proceed as follows:

(a) Find a row with a single zero and assign that operation to the gang in that column. E.g. first row has one zero under gang 3 therefore carcassing is done by gang 3.
(b) Cancel all zeros in gang 3 column.
(c) Find next row with a single zero and assign as in (a) above, e.g. third row has one zero under gang 6, therefore second fix is done by gang 6.
(d) Cancel all zeros under gang 6.
(e) Proceed as (c), thus assigning floor boards and staircase to gang 1.
(f) No other rows have a single zero. Inspect columns for a single zero. In this case all remaining rows and columns have more than one zero. There will therefore be more than one solution. Allocate finishings to gang 2.
(g) Cancel all zeros in gang 2 column.
(h) Proceed as above allocating first fix to gang 4 and first floor partitions to gang 5. This gives assignments as follows:

Gang 1 will specialise in Floor Boards and Staircase
Gang 2 will specialise in Finishings
Gang 3 will specialise in Carcassing
Gang 4 will specialise in First Fix
Gang 5 will specialise in First Floor Partitions
Gang 6 will specialise in Second Fix

The total work content of this solution is

$$40 + 25 + 95 + 33 + 85 + 85 = 363 \text{ hours.}$$

An alternative solution could be

Gang 1 will specialise in Floor Boards and Staircase
Gang 2 will specialise in First Fix
Gang 3 will specialise in Carcassing
Gang 4 will specialise in First Floor Partitions
Gang 5 will specialise in Finishings
Gang 6 will specialise in Second Fix

This also gives a work content of

$$40 + 30 + 95 + 78 + 35 + 85 = 363 \text{ hours.}$$

It may not always be possible to use the gangs in accordance with these solutions the whole of the time, but wherever possible the allocations shown should be used.

6.5 QUEUING THEORY

6.5.1 Introduction

Queuing situations occur every day in one form or another. There are queues for buses, queues in the supermarket and other less obvious queues such as callers queuing for a vacant line on telephone switchboards and components queuing up waiting to be assembled in a factory. In fact in a factory, queues of partly assembled products are there by design to act as a buffer stock and thereby ensure continuity of work for operatives. This type of queue also occurs in the production of repetitive units in the construction industry and can be seen in the form of activity buffers, and stage buffers in the Line of Balance Method (see Fig. 1.38).

In some cases however queues of work waiting to be done can be very expensive. An example here could be repairs to plant, as this could cause holdups to other operatives.

In repetitive work such as housing, it is important to realise the effects of queues and their influence on production planning and control.

All queuing problems have one thing in common, a service and a customer, i.e. someone or something waiting to be served, e.g. houses waiting to be tiled by a roofer, excavators waiting to be repaired by mechanics, etc. The problem is usually one of economics, to compare the cost of providing an immediate service with the cost of keeping the 'customer' waiting.

Queuing Theory is based on probability theory because customers usually arrive at random intervals and usually take differing lengths of time to serve.

6.5.2 Queuing models

Mathematical models can be used to represent the real life situations and calculations made to find the likely waiting time in a queue, and the effect on this waiting time and costs if the service is improved, or reduced. It is important to consider other possible effects such as loss of customers due to excessive waiting, etc., where this is relevant.

6.5.3 Traffic intensity

Traffic Intensity or *Utilisation factor* is the ratio of mean overall service time to mean interarrival time (mean interval between arrivals).

i.e. $$\text{Traffic Intensity } (\rho) = \frac{\text{mean service time}}{\text{mean interarrival time}} \text{ or } \frac{S}{A}.$$

6.5.4 Mean waiting time

In queuing problems the mean waiting time is often required and this is found as follows:

Mean waiting time including time being served (total time in the system) $$= \frac{1}{1-\rho} \times S = \frac{1}{1-\frac{S}{A}} \times S = \frac{1}{\frac{A-S}{A}} = \frac{AS}{A-S}.$$

If time being served is to be excluded, the formula is

$$\frac{AS}{A-S} - S = \frac{AS - S(A-S)}{A-S} = \frac{AS - AS - S^2}{A-S}$$

$$= \frac{S^2}{A-S}.$$

These formulae apply to situations where there is one service point and only apply after the system has become stabilised. This depends on the traffic intensity which must be less than 1. The nearer traffic intensity is to 1, then the longer it takes for the system to become stabilised. The formulae also apply strictly for random arrivals and service times which have a negative exponential distribution.

As the traffic intensity approaches 1 the mean waiting time increases rapidly and hold-up results. Obviously with traffic intensities greater than one, the queue will grow indefinitely. If however low traffic intensities are used in order to meet demand immediately it occurs, high non-productive time results for the server. A traffic intensity of 0.6 means that the server is busy for only 60% of the time on average.

It is possible to have higher traffic intensities for normal periods wherever extra help can be brought in or overtime can be worked at busy periods. If this is not possible, it is better to have traffic intensities of say 0.7 to 0.8.

Waiting time can be reduced if arrivals can be made more regular and/or service times can be made approximately equal, i.e. a small standard deviation.

If arrivals come at exactly equal intervals and service times were constant, there would of course be no queue even with traffic intensities of 1.

Conversely, the more irregular the arrivals and service times the longer the waiting times will be with a given traffic intensity.

6.5.5 Example in the construction industry

A planning engineer in a construction firm is specialising in pre-tender planning. He has been working excessive overtime and still cannot completely meet the demand. The problem is to decide the best course of action in these circumstances. Information has been collected to obtain the average time necessary for a planning engineer to complete his part of a pre-tender plan and this was found to be 21 hours or 3 normal working days. The time normally available for him to complete his part of the plan is 9 working days from the time he receives information from the estimator. Information was also collected of the period of time elapsing between receipts of successive requests for pre-tender planning. The results are shown in Fig. 6.36.

6.5.6 Solution

6.5.6.1 Traffic intensity

Average service time $= 3$ days

Average inter-arrival time $= 4$ days

$\therefore$ Traffic intensity $= \frac{3}{4} = 0.75$.

$\therefore$ On average the planner is busy doing planning only 75% of his normal time.

6.5.6.2 Mean waiting time

The time available from receipt to completion of the work is 9 days. This includes time when the work is actually being done. The formula to use is therefore the one which includes time in the system.

i.e.

$$\text{Mean waiting time} = \frac{AS}{A - S} = \frac{4 \times 3}{1} = 12 \text{ days.}$$

This indicates why the planning engineer has been working so much overtime.

If two planners are used to assisting each other and tackling one project at a time, the traffic intensity would be reduced to $\frac{1\frac{1}{2}}{4} = 0.375$ and the waiting time would be reduced to $\frac{4 \times 1\frac{1}{2}}{2\frac{1}{2}} = 2.4$ days.

This would of course leave both planning engineers unoccupied for a large proportion of their time.

Fig. 6.36

Arrival Number	*Inter Arrival Time (days)*	*Arrival Times (days)*	*Cumulative Average Inter Arrival Time (days)*
1	5	5	5.0
2	6	11	5.5
3	6	17	6.3
4	4	21	5.3
5	3	24	4.8
6	3	27	4.5
7	4	31	4.4
8	2	33	4.1
9	4	37	4.1
10	5	42	4.2
11	6	48	4.4
12	3	51	4.3
13	2	53	4.1
14	1	54	3.9
15	5	59	4.0
16	6	65	4.1
17	4	69	4.1
18	2	71	4.0
19	3	74	3.9
20	5	79	4.0
21	4	83	4.0
22	6	89	4.0
23	3	92	4.0
24	4	96	4.0
25	3	99	4.0
26	7	106	4.1
27	5	111	4.1
28	2	113	4.0
29	2	115	4.0
30	4	119	4.0
31	3	122	3.9
32	6	128	4.0
33	5	133	4.0

6.5.6.3 A compromise solution

A planning engineer who normally specialises in planning after award of the contract could be trained in pre-tender planning and could then assist when necessary.

Figures 6.37 and 6.38 show the effect of traffic intensity on waiting time, assuming inter-arrival time will be 4 days.

6.5.6.4 Assumptions made

It is assumed that the pre-tender planning is completed on a first come, first served basis and that when assistance is required both planners work on the same project.

Fig. 6.37

Inter Arrival Time (days)	*Traffic Intensity*	*Service Time (days)*	*Waiting Time* $\frac{AS}{A - S}$
4	0.1	0.4	0.44
	0.2	0.8	1.0
	0.3	1.2	1.71
	0.4	1.6	2.67
	0.5	2.0	4.0
	0.6	2.4	6.0
	0.7	2.8	9.33
	0.8	3.2	16.00
	0.9	3.6	36

6.5.7 Other examples in the construction industry

Many examples of queuing occur in the construction industry, e.g.

(i) Estimates queuing to be prepared in overworked estimating departments or estimaters idle because few estimates are arriving.
(ii) Lorries queuing up to be filled by excavating machines or excavators standing idle awaiting lorries for disposal of excavated material.
(iii) Concrete gangs idle awaiting completion of formwork and reinforcement.
(iv) Precast concrete panels waiting to be moved by cranes.
(v) The various stages of repetitive housing waiting for operatives to proceed with further operations.

This last example deserves particular attention because a considerable proportion of the total building effort is in housing of various types.

A considerable amount of work has been done on repetitive housing by the Building Research Station and the following is based on their work.

The mean service time is the average time taken to perform a particular operation on a unit. The mean arrival time is the average of the intervals between the completions of the previous unit. Obviously if mean service time is longer than mean arrival time then a queue of houses will result between the two operations and this queue will grow continuously.

If the mean arrival time is longer than the mean service time, operatives will be idle at times. If traffic intensity is 1, i.e. equivalent to balanced gangs and no buffers are provided, variations in productivity, inaccuracies in targets set, poor materials, etc., will result in times when a queue of partially completed houses builds up and times when no houses are ready to be 'served'.

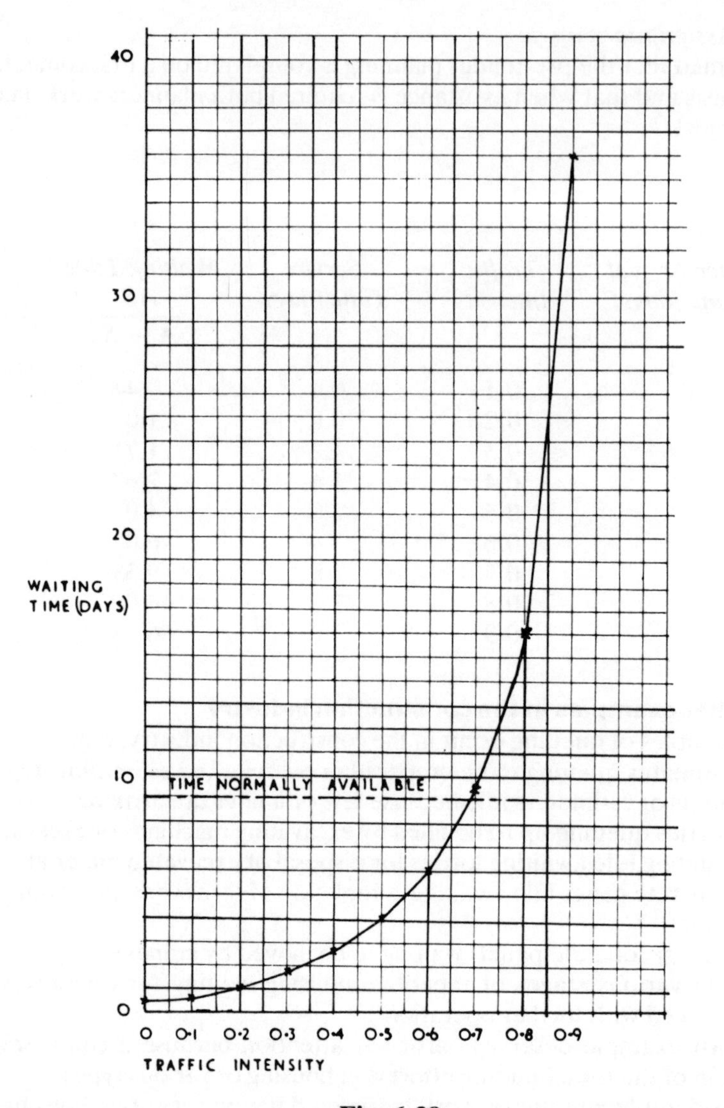

Fig. 6.38

This can be overcome to some degree by allowing a buffer between operations or stages as shown in Fig. 1.29 and Fig. 1.38, i.e. deliberately form a queue. This will of course increase the time of production of a house, resulting in a greater capital investment in many cases.

In practice, the queuing problem becomes very complex, as there may be two gangs engaged on the same operation (see Fig. 6.39), operatives may do more than one job on a particular house (see Fig. 6.40) or two or more operations can be done on a house at one time, which gives a result similar to Fig. 6.39. In these circumstances, a mathematical model becomes very difficult and sometimes impossible to develop and simulation is more useful.

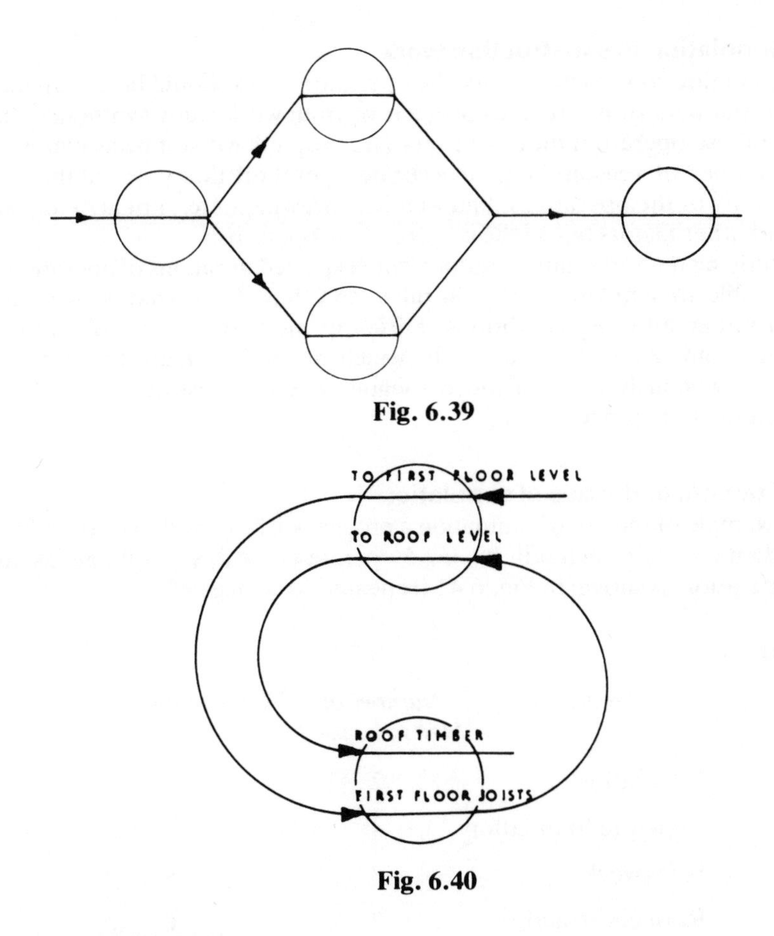

Fig. 6.39

Fig. 6.40

6.6 SIMULATION

6.6.1 Introduction

As stated earlier some problems are too complex to solve mathematically and simulation can often be used in these cases. It can be used to find bottlenecks, labour and plant utilisation, throughput time, etc.

To help solve problems by simulation a model is used which represents the real system as nearly as possible.

A mechanism is necessary to simulate what is likely to happen in practice, i.e. to enable operation under working conditions. This may be discs numbered in accordance with the expected distribution, tables of random numbers, a die, or a computer programme.

Simulation is also used to verify the results of the application of analytical techniques.

6.6.2 Advantages of simulation

One advantage of simulation is that it does not become much more difficult as the problem becomes more complex. It is these complexities which often make analytical techniques impossible.

6.6.3 Simulation in construction work

When planning construction work the operation times should be calculated from output data from previous projects or from work study synthetics. In practice these operation times will vary and progress will not be as planned for a number of reasons, e.g. inaccurate operation times, variability in productivity of the operatives, bad weather, drawings, etc., not arriving on time and other factors out of the control of the operatives.

By studying the variability of output from expected durations of operations it is possible to simulate operation times and thus study what is likely to happen under differing conditions. Different methods of control can be compared and decisions made as to which method is likely to be most effective. Obviously this is a much cheaper way of assessing results than trying them out in practice.

6.6.4 Example of the use of simulation

As an example of the use of simulation a project will be considered consisting of 10 identical units each unit having 5 operations, with separate gangs on each operation as shown in Fig. 6.41 (repeated from page 48).

Fig. 6.41

Operation	*Number of Men to be used*	*Time required*
Excavation	2	2
Concrete foundations	4	4
Brickwork	8	4
Roof construction	3	4
Roof finishing	2	4

The project is planned using the Line of Balance Method with buffers, i.e. the operations will be planned to form a series of queues (see Fig. 6.42).

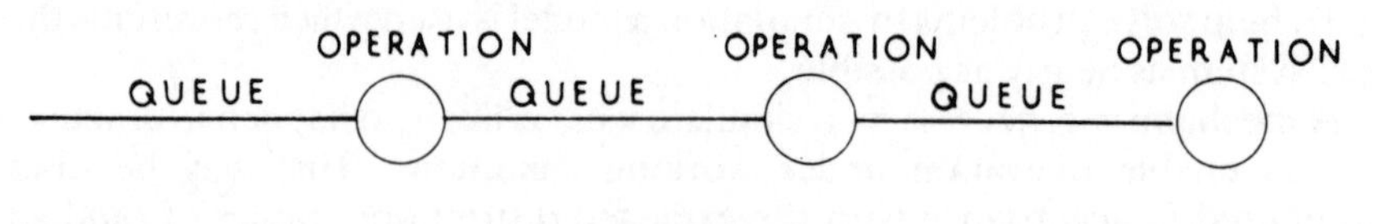

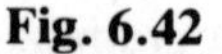

Fig. 6.42

From the Line of Balance Schedule in Fig. 6.43 it can be seen that the project should be completed in 66 days.

Progress can now be simulated using cards numbered to represent the time necessary to complete the respective operations, the numbers having a statistical distribution similar to that found in practice. The Line of Balance Schedule represents the model and the numbered cards are the mechanism.

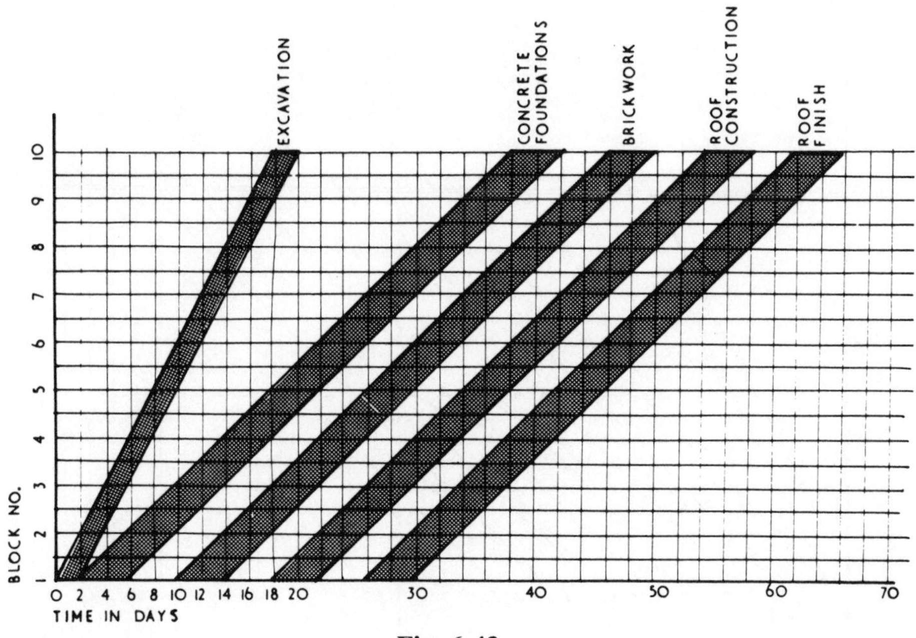

Fig. 6.43

6.6.4.1 Simulating progress

Progress is simulated for each operation on each block and the results are shown on Fig. 6.44. The results are entered onto the Line of Balance Schedule (Fig. 6.45).

Fig. 6.44

Operation *Block No.*	*1*	*2*	*3*	*4*	*5*	*6*	*7*	*8*	*9*	*10*
EXCAVATION										
start	0	2.0	3.6	5.4	7.8	10.2	12.4	14.6	16.4	18.2
simulated operation time	2.0	1.6	1.8	2.4	2.4	2.2	2.2	1.8	1.8	1.6
finish	2.0	3.6	5.4	7.8	10.2	12.4	14.6	16.4	18.2	19.8
CONCRETE FOUNDATIONS										
start	2.0	6.4	10.8	15.4	18.6	22.6	27.4	32.6	36.2	
simulated operation time	4.4	4.4	4.6	3.2	4.0	4.8	5.2	3.6		
finish	6.4	10.8	15.4	18.6	22.6	27.4	32.6	36.2		
BRICKWORK										
start	10.0	14.6	19.8	24.8	30.8	36.6				
simulated operation time	4.6	5.2	5.0	6.0	5.8					
finish	14.6	19.8	24.8	30.8	36.6					
ROOF CARCASS										
start	18.0	21.4	25.2	30.8	36.6					
simulated operation time	3.4	3.8	4.0	4.2						
finish	21.4	25.2	29.2	35.0						
ROOF FINISH										
start	26.0	30.2	35.0							
simulated operation time	4.2	4.8								
finish	30.2	35.0								

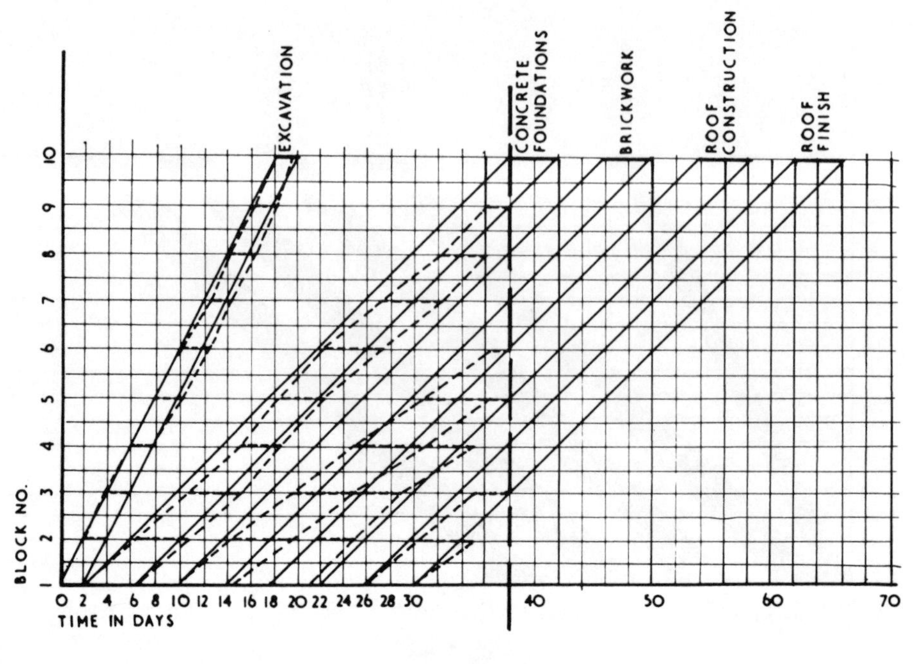

Fig. 6.45

In this particular simulation the results show that progress was as planned until the roof carcass gang want to move into block 4 and they were delayed by 1.6 days. From that time until day 38 (the time up to which progress is shown on the chart), the roof carcass gang is being delayed. This would result in non-productive time and would therefore be costing extra money, unless of course the gangs are adjusted.

6.6.4.2 Comparisons with other schedules

The total cost of the project including overheads and all other factors can be ascertained on completion and a comparison can be made with the planned cost. Comparisons can also be made with a schedule using different buffer times between activities or by using different methods of control, e.g. progress charts in bar chart form or decision rules.

6.6.4.3 The time factor

The simulation of each method would have to be carried out a number of times to get a range of likely results and this would be very time consuming. Use of a computer would cut the time factor quite considerably and allow comparisons to be made more quickly.

6.6.4.4. Effect of buffer on direct costs

If no buffer had been provided between activities, the direct costs would have been much higher than the original schedule because the variability of actual and scheduled times would cause queues of blocks to build up in some cases and operatives to be non-productive in other cases. This is to be expected from what has been discussed in the section on Queuing Theory.

6.6.5 Conclusion

As stated earlier, simulation is extremely useful for helping to solve management problems. It has a very wide range of application and whilst it has already been used in the Construction Industry, there is scope for extension of the use of this technique.

6.7 MANAGEMENT GAMES

6.7.1 Introduction

In management games the participants assume the various management roles and have to make decisions. An analysis of the decisions taken can be made on completion to assess the performance of those taking part. The games simulate real life situations as nearly as possible and rules are imposed to keep the games within acceptable limits.

6.7.2 Example of a management game

The simulation example described in section 6.5.4 could be used as a basis for a management game.

6.7.2.1 The estimate

An estimate could be prepared for the project to include the usual costs met up with in practice, e.g. labour, plant, materials, and overheads.

6.7.2.2 Planning and controlling

The participants can then plan the project and by feeding information to them to represent weekly or daily progress, they up-date their own schedule.

6.7.2.3 Varying operation times

Provision can be made to vary output of operatives by allowing use of overtime or by increasing or decreasing gang sizes as would be possible in practice. This would of course add to the cost of the project.

6.7.2.4 Assessment of results

The results achieved by the participants could be compared with the estimate at intervals and on completion. Comparisons can also be made with other teams playing the game.

6.8 DECISION RULES

6.8.1 Introduction

Decision rules are based on queuing theory. In order to cut down the average waiting time in practice priority can be given to some 'customers', this is known as varying the queue discipline. The remainder of this section on decision rules is taken from 'The Control of Repetitive Construction' by J. F. Nuttall of the Building Research Station. The work done by BRS included simulation exercises but the rules were not tested in practice.

6.8.2 Dynamic control

For a method of control to be dynamic, decisions need to be based on the latest available information. This information may be obtained from a programme adapted by the site manager as necessary. This adaptation of the programme can be formalised which can be very useful to the less experienced site manager. A simple example of this formalisation is to treat the sequence of operations in the programme as the order of priority for the operations.

6.8.3 Decisions based on experience

There are many contracts where no detailed programme exists and the site manager decides from day to day what work is to be done. Thus the decisions on sequence are based on the situation which exists at the moment of decision, as interpreted by the site manager's intuition and experience. This gives a fully dynamic control method which can give good results provided that the site manager is experienced in the type of work and is able to give due consideration to the various factors which should influence his decision.

6.8.4 Formal decision rules

Formal dynamic decision rules do exist which have advantages in being consistent, logical, objective and simple to apply and thus may be of great assistance to a site manager, particularly if he is not very experienced or if the type of construction is complex or unfamilar.

A number of decision rules were tested by the Building Research Station by simulation for a contract for semi-detached houses to give a preliminary assessment of the success they would be likely to have if applied in practice. These rules were used every time a trade was ready to start a fresh operation. The rules took into consideration all the operations which were available to be started at that time and assigned to each of them a certain priority. The operation with the highest priority was then chosen to be performed next.

6.8.4.1 Simple decision rules

The first decision rule to be tested was very simple: it was to choose the operation which would take least time to perform. This gave a fixed order of priority to the operations, e.g. the choice for the bricklayers would always be to do the topping out of any block which had reached that stage, or failing that, to build the ground floor brickwork of the next block due to be started, and the first floor brickwork would only be constructed when there was no other work. However, as in all the decisions rules tested, operations were always carried through to their completion without interruption. It was necessary, with this decision rule, to regard the blocks as becoming available for the first operation at intervals. The second decision rule was also very simple: in it, the priority of the operations depended entirely on the length of time for which they had been available to be performed. Thus, in the queue of houses waiting to be operated on by any particular trade, the men would go first to the house which had been waiting longest since it was vacated by the previous trade. In applying this 'first come, first served' rule it was necessary to prevent the first operation in the houses from always having priority (since a house on which no work had been done would inevitably have been waiting longer than a partly built house). So the houses were released for the first operation periodically, as for the first decision rule.

The third decision rule was partly based on the first. This rule involved looking ahead to the trade which would follow the operation under consideration and giving priority to the trade which had least work ahead of it. When this did not fully determine the operation to be performed, the shortest operation rule was used. Thus in choosing an operation for the bricklayers, if there was less work ready and waiting for the tilers to do than there was for the carpenters, priority was given to topping out so as to release work for the tilers. If there was less work waiting for the carpenters than for the tilers, however, priority was given to ground floor brickwork or, failing that, to first floor brickwork, either of which would release work for the carpenters, ground floor brickwork being the first choice because it was the shorter operation.

6.8.4.2 Decision rules based on the estimated finishing times of the operations

The remaining decision rules which were tested were based on rules suggested by Rowe and Jackson. Each operation was given a priority based on a simple formula which, in effect, gave an approximate estimate of the day on which the operation concerned should be finished.

Suppose that the blocks are numbered 1, 2, 3, . . . in the order in which they are to be constructed and that the operations are numbered 1, 2, 3, . . . on the order in which they must be performed. Suppose further that

S_b = starting date of block b
d_i = estimated duration of operation i in working days per block
c = cycle time (i.e. maximum number of days' work per block for any one gang).

Consider a certain operation, a, on block b and estimate when it can be expected to be finished.

The successive blocks will reach any specified stage of construction at intervals of approximately c working days: in particular, their starting dates will be about c days apart. If dates are measured in working days from the commencement of the first block, which is therefore on day 0, the starting date of block b can be estimated approximately as:

$$S_b = c(b - 1). \tag{1}$$

It would be possible to perform all the operations on block b up to and including operation a within a period of:

$$\sum_{i=1}^{a} d_i \text{ working days}$$

$$\left(\sum_{i=1}^{a} d_i \text{ denotes the sum of the estimated durations of all the operations up to and including operation } a \text{ in working days per block.} \right)$$

However, this assumes that work proceeds continuously on the block without any breaks, whereas in practice there will be periods when the block is waiting between operations because the trade which is to perform the next operation is occupied elsewhere. A simple way of allowing for this is to increase the time allowed for the operations up to and including operation *a* by a factor *f* which is greater than unity, i.e. to allow a period of:

$$f \sum_{i=1}^{a} d_i \text{ working days}$$

for operations 1, 2, . . ., *a* on any block. An approximate estimate of the day on which operation *a* on block *b* should be finished is therefore:

$$P_{ab} = S_b + f \sum_{i=1}^{a} d_i \qquad (2)$$

where S_b is obtained from formula (1).

Since the numbers P_{ab} approximately represent the dates by which the various operations should be completed, it is reasonable to use them as priority numbers for a decision rule, priority being given to the operation which the lowest priority number, i.e. the earliest estimated finishing date.

It might be thought better to take the estimated *starting* dates of the operations as their priority numbers, but the finishing dates have the advantage of giving a certain amount of priority to operations which can be performed quickly and will thus release work more quickly for the following trades.

Formula (2) gives a series of different decision rules, depending on the value which is taken for the factor *f*. In the simulation studies a number of these rules were tested, with values of *f* between one and three, so as to find the rule which gave the best results for the two-storey housing contracts. The priority numbers for the rule with $f = 2$ are given in Fig. 6.46 for the first four blocks to be built.

Fig. 6.46

Priority numbers for decision rule with f = 2

Operation	*Op. no. (a)*	*Trade initial*	*Estimated duration (d_a days)*	fd_a	*Block no. (b)* 1	2	3	4
Approx. starting date (S_b)					0	5½	11	16½
Ground floor brickwork	1	B	2¼	4½	4½	10	15½	21
First floor joists	2	C	1½	3	7½	13	18½	24
First floor brickwork	3	B	2½	5	12½	18	23½	29
Roof timbers	4	C	2½	5	17½	23	28½	34
Topping out	5	B	¾	1½	19	24½	30	35½
Felt, battens & tiles	6	T	5	10	29	34½	40	45½
Staircase	7	C	1½	3	32	37½	43	48½

The information required for obtaining the priority numbers consisted of the first four columns of the table, i.e. a list of the operations in the order in which they were to be performed, with their estimated durations in working days per block for the groups of men who were to perform them. The trade strengths had been chosen so that they were balanced as nearly as possible; thus the bricklayers had a total of $5\frac{1}{2}$ days' work per block, the carpenters $5\frac{1}{2}$ and the tilers 5 days' work. The estimated cycle was therefore $5\frac{1}{2}$ working days per block.

As an alternative to calculating each priority number individually from formula (2) the following procedure could be adopted. First, fill in the starting dates of the blocks, commencing with block 1, which is to start on day 0. The approximate starting dates of the following blocks are obtained in succession by adding the cycle time of $5\frac{1}{2}$ days to the starting date of the previous block.

Next, multiply each estimated duration by the factor f, in this case 2, to obtain the numbers in the column headed 'fd_a'. Now calculate the priority numbers, working down each column in turn and adding the corresponding value fd_a to the number immediately above the number being calculated. For example, for the column headed 'Block 2' the successive calculations are: $\frac{1}{2} + 4\frac{1}{2} = 10$, $10 + 3 = 13$, $13 + 5 = 18$, $18 + 5 = 23$ and so on.

The table of priorities can be calculated before the contract begins and only needs revising if the estimates of the operation durations change, e.g. if the labour available varies from the planned labour strengths.

To use the table in controlling the site work is very simple. Whenever a trade becomes free to undertake a fresh item of work and there is more than one item available for them to do, the item with the smallest priority number in Fig. 6.46 is chosen to be done first. For example, if the bricklayers have just finished the ground floor brickwork of block 3 and have the choice between commencing the ground floor brickwork of block 4, the first floor brickwork of block 2 or topping out on block 1, all of which operations are ready to be started, the decision rule states that the first floor brickwork of block 2 should be done, because this operation has a priority number of 18 compared with 19 and 21 for the other possibilities.

6.8.5 Random decisions

A small scale simulation was also performed to assess the results of determining the sequence of operations by a purely random choice between the operations available at the time. This simulation gave a datum point with which to compare the various control methods tested, the differences between the results for the control methods and the results for the random decision giving a measure of the savings which can be achieved by good control compared with no control of the sequence of operations for this type of contract. An upper limit to the average savings obtainable was provided by the times given in the original programme, which represented an optimum or near-optimum solution for constant operation durations, but in practice this limit would never be attainable because of the impossibility of predicting the future accurately when operation durations vary from block to block.

6.8.6 Results of simulations

The simulations of dynamic control methods were based on the contract for semi-detached houses which had been used for earlier simulations, under the conditions of variation in operation times from block to block. The results are given in Fig. 6.47 and illustrated in Fig. 6.48.

Fig. 6.47 *Comparison between alternative methods of control*

Methods of control	*No. of contracts simulated*	*Contract duration (days)*		*Unproductive time (man-days)*	
		Mean	*Range*	*Mean*	*Range*
Random decision	2	103.3	93¾–112¾	97	79–115
Programme (strict)	18	82.3	77–89	62	22–93
Programme (adapted)	10	81.6	77½–87½	45	23–69
Decision rules:					
Shortest operation 1st	4	87.3	85–90½	75	50–91
First come, first served	5	84.4	81¾–88½	71	56–110
Look ahead to following trade, otherwise shortest operation 1st	4	82.3	76¼–93	52	32–85
Formula with $f = 2$	14	81.7	75¾–84½	47	22–76
$f = 1.7$	5	80.8	76½–83	43	38–51
$f = 1.6$	4	82.7	80¼–86	46	37–56
Times shown in programme	–	78½		29	

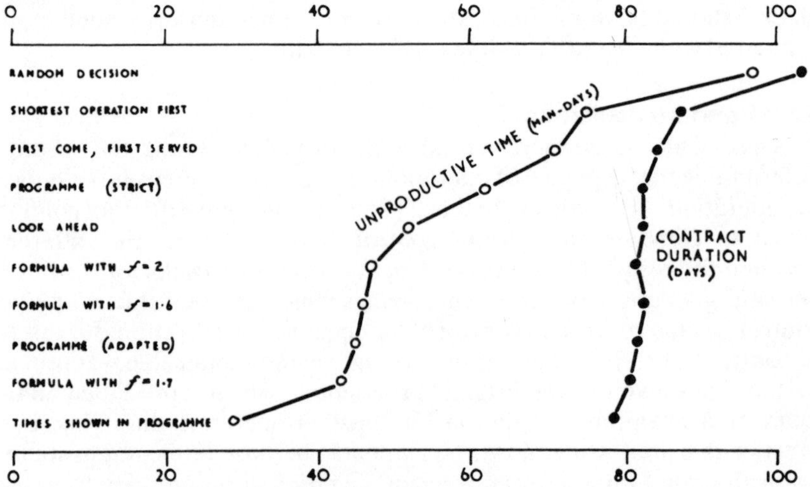

Fig. 6.48

6.9 BIDDING STRATEGY

6.9.1 Introduction

The majority of contracts in this country are obtained via competitive tendering. If a contractor is to be successful he must always be aware of what competitors are doing. If he knew the tender figures of his competitors he would be in a very advantageous position indeed. As he does not know this he must find ways of assessing what their tenders are likely to be to enable him to obtain contracts as profitably as possible, i.e. to enable him to maximise his profit. He must therefore use bidding strategy based on scientific methods. In order to apply the scientific method of statistics to tendering, the contractor needs to know the performance of his competitors in previous competitions. It is the practice of some architects and Local Authorities to inform contractors tendering, of the tenders submitted by all competitors either by opening tenders whilst contractors are present or by informing them after opening the tenders. If the contractor collects this information over a period of time he can learn quite a lot about his competitors.

6.9.2 Factors affecting the approach to the problem

When a contractor is invited to tender for work, he is not normally told who his competitors are. However, it is often possible to find out who is competing for a particular project. Even if the contractor does not know precisely who his competitors are, in certain circumstances he at least knows approximately how many competitors he has particularly if the 'Code of Procedure for Selective Tendering' is being used. In the worst case he does not know who his competitors are nor how many there are. Each of these situations necessitates a slightly different approach.

6.9.3 When names of competing firms can be ascertained

When the contractor can ascertain who the competitors are, he can assess the optimum tender figure to give the maximum expected profit (expected profit being the average profit he would make if he used the same profit margin on a large number of contracts where the estimate of cost was identical and probability of being awarded the contracts remained the same), e.g. if the estimate of cost for a contract is £100 000 and the contractor submits a tender figure of £110 000, if he is awarded the contract he makes a profit of £10 000, if he is not awarded the contract he gets nothing. If the probability of being awarded the contract is 0.4 the expected profit will be £100 000 × 0.4 = £4000. Assume that information has been collected over a period of time with respect to a particular competitor and that these figures have been compared with the contractors estimated cost before profit is added as shown in Fig. 6.49.

6.9.3.1 Frequency of ratios

The information from Fig. 6.49 can be tabulated as shown in Fig. 6.50 which gives the frequency of occurrence of the various ratios.

6.9.3.2 Probability of ratios

From this table the probability of any particular ratio occurring can be found as shown in Fig. 6.51.

Fig. 6.49

Contractors estimate	*Competitors tender*	*Ratio of competitors Tender to contractors Estimate (T/E)*
£120 000	£126 000	1.05
£150 000	£159 000	1.06
£160 000	£166 400	1.04
£140 000	£148 400	1.06
£130 000	£139 100	1.07
£110 000	£111 100	1.01
£160 000	£177 600	1.11
£140 000	£151 200	1.08
£100 000	£107 000	1.07
etc.	etc.	etc.

Fig. 6.50

T/E	*Frequency*
0.98	1
0.99	2
1.00	2
1.01	3
1.02	4
1.03	6
1.04	9
1.05	13
1.06	18
1.07	22
1.08	23
1.09	19
1.10	14
1.11	10
1.12	7
1.13	4
1.14	2
1.15	1
Total	160

Fig. 6.51

T/E	*Probability of occurrence*	*(frequency) / (total)*
0.98	1/160	0.006
0.99	2/160	0.013
1.00	2/160	0.013
1.01	3/160	0.019
1.02	4/160	0.025
1.03	6/160	0.038
1.04	9/160	0.056
1.05	13/160	0.081
1.06	18/160	0.112
1.07	22/160	0.137
1.08	23/160	0.144
1.09	19/160	0.119
1.10	14/160	0.087
1.11	10/160	0.062
1.12	7/160	0.044
1.13	4/160	0.025
1.14	2/160	0.013
1.15	1/160	0.006
		1.000

The contractor has now got a considerable amount of information about this particular competitor. He knows that 0.144 or 14.4% of his competitors tenders were 8% higher than his own estimate of cost, 0.6% were in fact 2% lower than his estimate of cost and 0.6% were 15% higher than his estimate.

6.9.3.3 Cumulative probability distribution

From Fig. 6.51 the cumulative probability distribution for this competitor can be found. The contractor will then know the probability of any tender being lower than this competitor, e.g. the probability of a tender of 0.98 or 98% of the estimate has a probability of being lower than this competitor of $1 - 0.006 = 0.994$. As the next tender/estimate ratio is 0.99 it is assumed that the probability of 0.994 applies for tenders from 0.98 to 0.989, and a tender less than 0.98 of the estimate has a probability of 1.00 of being lower than this competitor.

It is of course possible to carry out the analysis in a more sophisticated way than has been shown with much smaller intervals than the 0.01 or 1% intervals used in this analysis. Fig. 6.52 shows the cumulative probability distribution for this competitor.

Fig. 6.52

T/E	*Probability that tender will be less than competitor*	
0.979		1.000
0.989	1 − 0.006	0.994
0.999	1 − 0.019	0.981
1.009	1 − 0.032	0.968
1.019	1 − 0.051	0.949
1.029	1 − 0.076	0.924
1.039	1 − 0.114	0.886
1.049	1 − 0.170	0.830
1.059	1 − 0.251	0.749
1.069	1 − 0.363	0.637
1.079	1 − 0.500	0.500
1.089	1 − 0.644	0.356
1.099	1 − 0.763	0.237
1.109	1 − 0.850	0.150
1.119	1 − 0.912	0.088
1.129	1 − 0.956	0.044
1.139	1 − 0.981	0.019
1.149	1 − 0.994	0.006
1.159	1 − 1.000	0.000

6.9.3.4 Expected profit

Once the probability of a tender being less than that of the competitors is known, it is simple to calculate the expected profit for each ratio, e.g. if the contractor tenders using a 4.9% profit, i.e. using a ratio of 1.049 then the probability of success is 0.83. His expected profit is therefore $0.049 \times$ estimate $\times\ 0.83$ or $0.049E \times 0.83 = 0.041E$.

Figure 6.53 shows the expected profit for each ratio and from this it can be seen that the maximum expected profit is $0.442E$ or 4.42% using a ratio of 1.059, i.e. a profit margin of 5.9% of the estimated cost.

Fig. 6.53

T/E	*Expected profit*
0.979	$1.000(0.979E - E) = -0.021E$
0.989	$0.994(0.989E - E) = -0.0109E$
0.999	$0.981(0.999E - E) = -0.0010E$
1.009	$0.968(1.009E - E) = 0.0087E$
1.019	$0.949(1.019E - E) = 0.0180E$
1.029	$0.924(1.029E - E) = 0.0268E$
1.039	$0.886(1.039E - E) = 0.0346E$
1.049	$0.830(1.049E - E) = 0.0407E$
1.059	$0.749(1.059E - E) = 0.0442E$
1.069	$0.637(1.069E - E) = 0.0440E$
1.079	$0.500(1.079E - E) = 0.0395E$
1.089	$0.356(1.089E - E) = 0.0317E$
1.099	$0.237(1.099E - E) = 0.0235E$
1.109	$0.150(1.109E - E) = 0.0164E$
1.119	$0.088(1.119E - E) = 0.0105E$
1.129	$0.044(1.129E - E) = 0.0057E$
1.139	$0.019(1.139E - E) = 0.0026E$
1.149	$0.006(1.149E - E) = 0.0009E$
1.159	$0.000(1.159E - E) = 0.0000E$ (to 4 decimal places)

6.9.4 Competing against more than one known competitor

If there is more than one known competitor, the method of calculating the expected profit is very similar to that used for one competitor.

The result would be a number of cumulative probability distributions, the probability of being less than all competitors would be the product of the probabilities of being less than each individually.

If two such competitors are considered the result may be as shown in Fig. 6.54.

When the probability of tendering lower than all the known competitors has been calculated, the expected profit for each tender/estimate ratio can be worked out and the estimate showing the greatest expected profit can be selected as shown in Fig. 6.55.

Fig. 6.54

	Probability that tender will be less than		
T/E	*Competitor 1*	*Competitor 2*	*Competitors 1 and 2*
0.979	1.000	1.000	1.000
0.989	0.994	0.990	0.984
0.999	0.981	0.975	0.956
1.009	0.968	0.961	0.930
1.019	0.949	0.941	0.893
1.029	0.924	0.918	0.848
1.039	0.886	0.878	0.778
1.049	0.830	0.824	0.684
1.059	0.749	0.746	0.559
1.069	0.637	0.636	0.405
1.079	0.500	0.502	0.251
1.089	0.356	0.359	0.128
1.099	0.237	0.238	0.056
1.109	0.150	0.155	0.023
1.119	0.088	0.100	0.009
1.129	0.044	0.046	0.002
1.139	0.019	0.025	0.000 (to 4 dec. places)
1.149	0.006	0.010	0.000 (to 4 dec. places)
1.159	0.000	0.005	0.000 (to 4 dec. places)

6.9.5 When names of competing firms cannot be ascertained

When the competitors are not known but the number is known a slight adjustment to the method is necessary. When using the 'Code of Procedure for Selective Tendering' the maximum number of competitors will be known as this is laid down in the code. Usually this figure will be the nearest the contractor will get to the actual number of competitors.

In this situation the only difference in method is that the probability of submitting a lower tender is calculated against the average of all previous competitors. When this is ascertained the probability of a tender being lower than any number of competitors can be found as before, e.g. if a tender which includes a profit margin of 6.9%, i.e. a tender figure of 1.069*E*, has a probability of success of 0.600 against one competitor the probability against three such competitors would be $0.600 \times 0.600 \times 0.600$ or $0.63 = 0.216$.

Fig. 6.55

T/E	*Expected profit*
0.979	$1.000(0.979E - E) = -0.021E$
0.989	$0.984(0.989E - E) = -0.0108E$
0.999	$0.956(0.999E - E) = -0.0010E$
1.009	$0.930(1.009E - E) = 0.0084E$
1.019	$0.893(1.019E - E) = 0.0170E$
1.029	$0.848(1.029E - E) = 0.0246E$
1.039	$0.778(1.039E - E) = 0.0303E$
1.049	$0.684(1.049E - E) = 0.0335E$
1.059	$0.559(1.059E - E) = 0.0330E$
1.069	$0.305(1.069E - E) = 0.021E$
1.079	$0.251(1.079E - E) = 0.0198E$
1.089	$0.128(1.089E - E) = 0.0114E$
1.099	$0.056(1.099E - E) = 0.0055E$
1.109	$0.023(1.109E - E) = 0.0025E$
1.119	$0.009(1.119E - E) = 0.0011E$
1.129	$0.002(1.129E - E) = 0.0002E$
1.139	$0.000(1.139E - E) = 0.000E$
1.149	$0.000(1.149E - E) = 0.000E$
1.159	$0.000(1.159E - E) = 0.000E$

6.9.6 Tendering against an unknown number of competitors

If statistics are to be used in this case an estimate of the number of competitors must be made. It is considered bad practice to ask unlimited numbers of contractors to tender for the work but if this is done and the number cannot be estimated, e.g. in open-tendering, then statistics can be of very limited use.

6.9.7 Conclusions

It can be seen that by using statistics a contractor can gain an advantage over competitors who do not use analytical techniques in helping to determine profit margins on contracts. It may be appreciated that the main concern in this section has been to indicate the value of statistical methods for use in tendering. In practice other factors have to be taken into account besides the ones considered here. Indeed there are many other influencing factors in determining profit margins when tendering, examples being the state of the market and the economic climate at the time of tendering but this does not render statistical analysis irrelevant. Any method which assists in giving a more realistic appraisal of the situation should be taken into account.

6.10 USE OF COMPUTERS

Standard software packages are available for solving many operational research problems, typical examples being packages to solve linear programming problems, transportation problems, problems using simulation and of course Network Methods and their extensions in the form of resource allocation and cost optimisation.

Probably the greatest advantage of using the computer is its speed which enables problems, which would take months by hand to be done in a small fraction of the time. Indeed many problems could not be solved at all within a practical time without the use of a computer. Many simulation problems fall into this category.

Wherever possible it is essential that an integrated data processing system is used, as in this case much of the information required to solve a problem will probably be readily available in store, the information being collected as part of a total system.

Reference was made at the beginning of this book to integrated management information systems and in large firms a computer would be essential if an integrated system is to be run efficiently.

7 Computer Applications

7.1 INTRODUCTION

Most medium-sized and large companies use computers to some extent, although in many cases they are used for accounting purposes only. The arrival of the microcomputer is causing companies to reconsider their attitude towards computers and some who previously thought they were not viable due to the cost of hardware of suitable capacity, or because of the delay factor when using bureau, are using or considering using computers for a variety of purposes. The next few years will probably show a dramatic increase in their use in both large and small companies, even those with a turnover as low as £500 000. There are many reasons for this, including the fact that microcomputers have become cheaper and much more powerful, and the availability of suitable software has increased considerably. This software includes general purpose software such as that for word processing, spreadsheets and databases, and software written for specific purposes such as estimating, accounting and costing, and the various aspects of planning.

The saving in time and the availability of information generated by computers is having a considerable influence on some construction companies and this influence is likely to increase in the near future. Those who do not take advantage of the advances in computer technology may well cease to be competitive.

This chapter introduces and illustrates a range of applications relevant to the construction manager.

7.2 INTEGRATED SYSTEMS AND DATABASES

7.2.1. Introduction

The arrival of computers made it possible to re-examine the way in which firms are organised. The optimum organisation would be one in which the data for the company is totally integrated. Database management systems are designed to help in the development of such systems.

7.2.2 Database packages

Many packages which are called 'database' packages are really information retrieval packages, which simply organise and retrieve data. There are many examples of information retrieval systems and these include telephone directories and dictionaries. They are collections of information organised for a specific purpose. However, addresses and telephone numbers have little value on their own. They are only useful when related to a name (i.e. a Key) to make them easy to find. Information retrieval systems can organise and find data very quickly compared to manual methods. (e.g. to find a person who has a particular telephone number could take hours or even days by hand, whereas a computer would find that person in seconds).

In addition to the above a true database package in computing terms should:

Merge information in such a way that it appears to the user that only one file is in use.
Eliminate redundant or unnecessary data.
Make the data independent from the application programmes.

Fig. 7.1 shows a comparison of a database approach to a conventional file approach for accessing data.

7.2.2.1 dBase II and dBase III

dBase II, which has now been improved and made more user friendly in dBase III, is a database management system for microcomputers which will run on many machines. It allows the user to manipulate and change database contents and to produce reports. It keeps records up to date and under control. It includes a procedural language which makes it possible to create screen menus, determine the format of documents (see later), write programs to enable non-technical personnel to work with the system which has been developed. The integrated system described in section 7.3 was based on dBase II.

7.3 INTEGRATED SYSTEMS

Fully integrated systems take considerable time to develop, and as they involve a complete new look at the way things are done, they often necessitate considerable changes in organisation and procedures.

A good starting point is to concentrate on a sub-system within an overall framework and to build on to this at a later stage.

Much has been written on the need to integrate production information with that required for estimating. This is very difficult when using traditional bills of quantities based on the standard method of measurement, where bill items are not compatible with operational data, particularly when this data is required for short-term planning and for setting incentive targets.

A system using dBase II, has been developed at Sheffield City Polytechnic by Ray Oxley and Steve Westgate, for use by small firms and other firms who take off quantities based on operations. The system is called 'Micro Estimating and Targetting System' (MEAT).

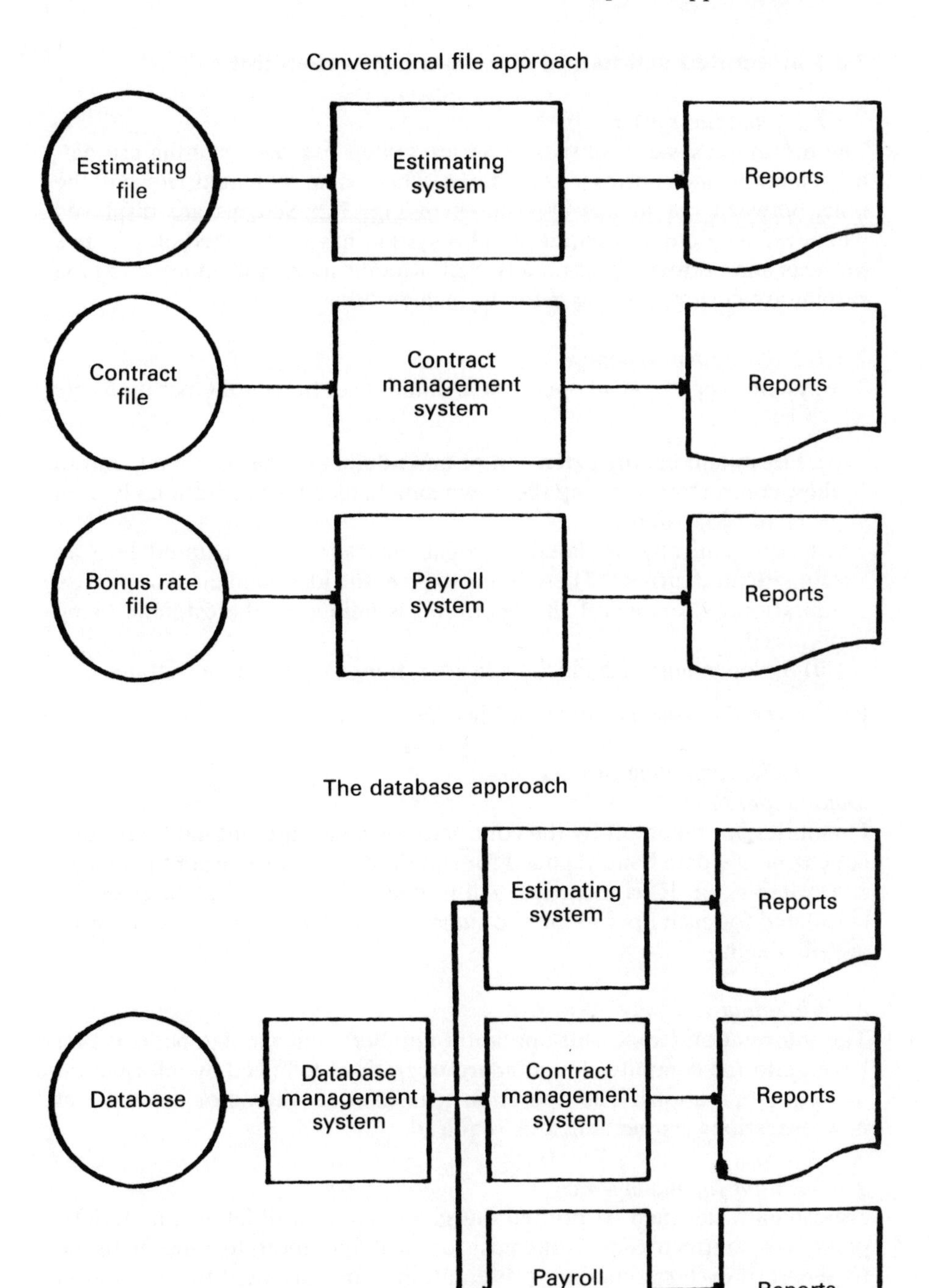

Fig. 7.1. System comparison

7.3.1 Integrated estimating and planning system (MEAT)

7.3.1.1 The total system

The prototype system consists of an integrated package covering estimating, planning and control in small firms. The system is menu driven and the links between the menus are shown in Fig. 7.2. Screens are displayed whenever data entry is required. This system has been tested on two live projects and performed extremely well. Modifications and additions to the system are currently being made to include other facilities.

7.3.1.2 The system database

The system is based on an operational database or Standard Operations file which is:

(i) Flexible to enable extension or amendment of the data bank and to allow contractors to set up their own data bank either operationally or in terms of SMM items.

(ii) Automatically up-dated as rogue operations are entered into an estimate for a project. These items to be easily identified facilitating easy alteration or removal if the operation is unique to the estimate being prepared.

(iii) Independent of financial considerations i.e. based on outputs.

Part of the database is shown in Fig. 7.3.

7.3.1.3 The estimating module

Data preparation

Quantities are taken off by the contractor using his present methods but a code from the data bank is added for standard operations or, in the case of rogue items, an 'R' is recorded. All-in material rates and plant rates are calculated for each operation, and sums to be included for sub-contractors are obtained.

Entering data

The information (trade and operation number from the database) is first keyed into the computer for standard operations followed by information for rogue operations. The operations can then be altered or deleted and new operations can be added as required.

The printout for management

An estimate can then be printed out giving the cost of labour, materials, plant, sub-contractors and total cost for each operation, together with the total cost of each resource for presentation to management for conversion into a tender.

Part of a printout for the use of management is shown in Fig. 7.4.

Conversion into a tender

At this stage, preliminaries are added and the estimate is converted into a tender by management.

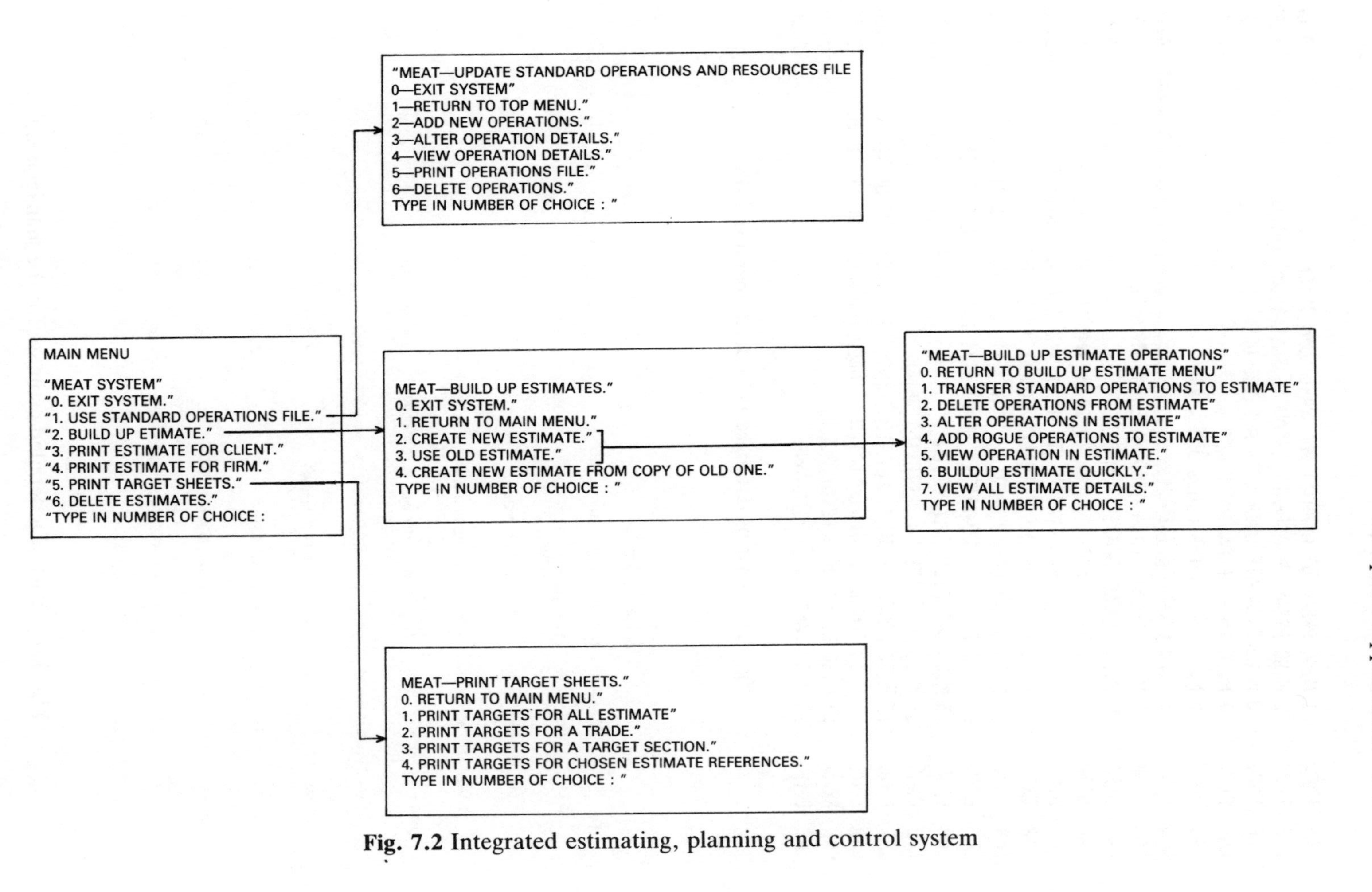

Fig. 7.2 Integrated estimating, planning and control system

LAB	EXCV	1	M.SQ	BREAK UP ASPHALT AND BASE 50MM THICK	0.66
LAB	EXCV	2	M.SQ	BREAK UP ASPHALT EXCAVATE BASE 150MM THICK	1.33
LAB	EXCV	3	M.CU	EXCAVATE OVER SITE AVE.150 DEEP	4.00
LAB	EXCV	4	M.SQ	TAKE UP PAVING FLAGS	0.60
LAB	PAVE	5	M.SQ	LAY PAVING FLAGS 900 X 600	1.50
LAB	EXCV	6	M.CU	EXCAVATE TRENCH	4.00
LAB	DISP	7	M.CU	DEPOSIT EXCAVATED MATERIAL ON SITE	1.20
LAB	DISP	8	M.SQ	REMOVE SLABS FROM SITE	0.50
LAB	DISP	9	M.CU	WHEEL MATERIAL UP TO 20 M. AND DEPOSIT IN SKIP	1.33
LAB	DISP	11	M.CU	WHEEL MATERIAL UP TO 40 M. AND DEPOSIT IN SKIP	2.66
LAB	EXCV	12	M.SQ	BREAK UP UNREINFORCED CONCRETE 150 THICK	1.20
LAB	EXCV	13	M.SQ	BREAK UP CONCRETE, LIGHTLY REINFORCED, 150 THICK	2.00
LAB	EXCV	14	M.SQ	BREAK UP CONCRETE, HEAVILY REINFORCED, 225 THICK	4.00
LAB	DISP	15	M.CU	BACKFILL EXCAVATION	1.00
LAB	CONC	16	M.CU	OVERSITE CONCRETE 150 THICK	5.33
LAB	CONC	17	M.SQ	CONCRETE IN FOUNDATIONS INCLUDING WHEELING UP TO 10M.	4.00
LAB	CONC	18	M.CU	CONCRETE IN FOUNDATIONS INCLUDING WHEELING UP TO 30M	5.33
LAB	REIN	19	M.SQ	LIGHT MESH REINFORCEMENT	0.10
LAB	DPM	20	M.SQ	DAMP PROOF MEMBRANE	0.06

Fig. 7.3. Part of database (standard operations file)

ESTIMATE FOR MSCPRO 1

REF	TRAD	NO	TARG	OUTPUT	QUANTITY	UNIT	DESCRIPTION	LAB. COST	MAT. COST	PLT. COST	SCN. COST	TOTALS
5	LAB	13	FDNS	2.00	8.30	M.SQ	BREAK UP CONCRETE, LIGHTLY REINFORCED, 150 THICK	60.09	0.00	0.00	0.00	60.09
10	LAB	6	FDNS	4.00	6.75	M.CU	EXCAVATE TRENCH	97.74	0.00	0.00	0.00	97.74
15	LAB	17	FDNS	4.00	6.55	M.SQ	CONCRETE IN FOUNDATIONS INCLUDING WHEELING UP TO 10M.	94.84	242.35	0.00	0.00	337.19
20	BLR	20	BFDN	1.28	4.00	M.SQ	HALF BRICK WALL IN FOUNDATIONS (2+1)	19.50	34.48	0.00	0.00	53.98
25	LAB	16	SLAB	5.33	5 00	M.CU	OVERSITE CONCRETE 150 THICK	96.47	185.00	0.00	0.00	281.47
30	LAB	20	SLAB	0.06	22.00	M.SQ	DAMP PROOF MEMBRANE	4.77	0.00	0.00	0.00	4.77
35	LAB	21	FIN	0.60	22.00	M.SQ	ARDIT SCREED	47.78	24.20	0.00	0.00	71.98
40	LAB	19	SLAB	0.10	22.00	M.SQ	LIGHT MESH REINFORCEMENT	7.96	33.66	0.00	0.00	41.62
45	BLR	83	BSUP	1.28	47.00	M.SQ	HALF BRICK WALL IN SUPERSTRUCTURE	229.20	405.14	0.00	0.00	634.34

Fig. 7.4. Part of a detailed estimate for the management

The printout for the client
After management have carried out the conversion and arrived at a mark-up, a printout can be obtained in the form of a priced bill for the client.
Part of a printout for issuing to the client is shown in Fig. 7.5.

REF	DESCRIPTION	QUANT	UNIT	TOTALS
	ESTIMATE FOR MSCPRO	1		
5	BREAK UP CONCRETE, LIGHTLY REINFORCED, 150 THICK	8.30	M.SQ	79.92
10	EXCAVATE TRENCH	6.75	M.CU	129.99
15	CONCRETE IN FOUNDATIONS INCLUDING WHEELING UP TO 10M.	6.55	M.SQ	448.46
20	HALF BRICK WALL IN FOUNDATIONS (2+1)	4.00	M.SQ	71.80
25	OVERSITE CONCRETE 150 THICK	5.00	M.CU	374.35
30	DAMP PROOF MEMBRANE	22.00	M.SQ	6.35
35	ARDIT SCREED	22.00	M.SQ	95.73
40	LIGHT MESH REINFORCEMENT	22.00	M.SQ	55.35
45	HALF BRICK WALL IN SUPERSTRUCTURE	47.00	M.SQ	843.68

Fig. 7.5. Part of a detailed estimate for the client

Standard estimates
Many firms carry out projects which are very similar in terms of method of construction and breakdown of operations. Standard estimate files can be set up when this occurs and new estimates can then be produced very quickly. This was considered to be a very important feature of the system.

Variations
Variations can be produced together with targets. These can be treated separately or incorporated into the main estimate.

7.3.1.4 The planning control and incentives module
This module provides information for the supervisor of the project and for the operatives.

Grouping of operations
Operations can be grouped in various ways for convenience of planning, setting targets, or control as follows:

(i) Manhours for a whole job (useful for small jobs to be carried out by one team of operatives).
(ii) Manhours for a complete trade e.g. brickwork (useful when a trade visits the site only once).
(iii) Manhours for a section of the project e.g. carpentry first-fix (useful for projects of a reasonable size when trades visit the site more than once).
(iv) Manhours for individual operations e.g. hang doors (useful when the supervisor wishes to build up his own targets, taking account of the current position on site).

Information for supervisor
Manhours included in the estimate can be provided in a number of ways as follows:

(i) Actual manhours for each operation.
(ii) Manhours for each operation adjusted to take account of the type of incentive scheme being used e.g. different rates of payback to operatives.
(iii) Manhours for each operation adjusted to allow for certain 'preliminary' items e.g. travelling time, clearing up at each stage etc.

In each case the printout gives the estimate reference number, operation description, unadjusted output per unit, adjusted and unadjusted target hours per operation and the adjusted and unadjusted total hours for the target set. A printout for foundation work for issuing to the site supervisor is shown in Fig. 7.6.

EST REF	DESCRIPTION		TARGET - HRS	
5	BREAK UP CONCRETE, LIGHTLY REINFORCED, 150 THICK		16.60	
	8.30 M.SQ 2.00 Hours/M.SQ			19.92
10	EXCAVATE TRENCH		27.00	
	6.75 M.CU 4.00 Hours/M.CU			32.40
15	CONCRETE IN FOUNDATIONS INCLUDING WHEELING UP TO 10M.		26.20	
	6.55 M.SQ 4.00 Hours/M.SQ			31.44
		TOTALS		69.80
	FACTOR 1.20	FACTOR TOTAL		83.76

Fig. 7.6. Target sheet for foundations work – for management

Information for operatives
A printout giving the estimate reference, operation description and total target hours only is provided for handing to the operatives.
A printout for foundation work for issuing to the operatives is shown in Fig. 7.7.

```
                TARGET SHEETS FOR MSCPRO TARGET SECTION FDNS

EST REF  DESCRIPTION                                          TARGET - HRS

   5   BREAK UP CONCRETE, LIGHTLY REINFORCED, 150 THICK

  10   EXCAVATE TRENCH

  15   CONCRETE IN FOUNDATIONS INCLUDING WHEELING UP TO 10M.


                                                   TOTALS              84
```

Fig. 7.7. Target sheet for foundations work – for operative

By integrating the estimating and production processes data is provided to control labour output and costs, plant output and costs, and to update the database as appropriate. The system provides all the data for an incentive scheme based on realistic outputs which should help to increase productivity. Production of data is available immediately a contract is secured and site staff are equipped with all the production data they need.

7.4 PLANNING

7.4.1 Introduction

A number of packages are available for the planning of construction work, all of which are based on network techniques in either arrow, precedence or bar chart form. The one used here to illustrate the use of a computer is Pertmaster from Abtex Computer Systems Ltd. This package, which is used by many construction companies, is available on many machines. Both arrow and precedence formats are supported. Arrow format will be used in this illustration.

Before using the package the planner must set down the logic by preparing an arrow diagram. The village hall project described in 2.4.3 will be used as a basis for this illustration, although in practice it may be considered too small to warrant the use of a computer.

7.4.2 Calendars (see Fig. 7.8)

In order to obtain printouts a calendar must be created which relates the day numbers to calendar dates. Non-working days are entered and the calendar is given a code name for identification.

7.4.3 Abbreviations (see Fig. 7.9)

An abbreviations file is also created and given a code name. Abbreviations are used because they save space in the computer memory and save time in entering the information into the computer. They can be used for parts of activity descriptions, whole descriptions, and for resources.

7.4.4 Creating a new plan

General information about the plan is first entered. This includes:

(a) A plan identifier (file name)
(b) The title of the project
(c) Other general information about the project
(d) The calendar code name
(e) The abbreviations code name

7.4.4.1 Entering activities

Activities are entered by typing the preceding and succeeding node, the duration and the description (using abbreviations if preferred).

Scheduled day numbers

These can be used for such occurrences as the start of the project, delivery of materials and components in short supply, the end of the project etc. In Fig. 7.11 the scheduled start was entered as day 26. The day number is of course related to a date by referring to the calendar. The activity 'deliver plasterboard' could have been entered as a scheduled start of day 96 for plastering instead of including it as an activity.

Start and end activities must be identified and entered with an 'S' in the preceding event node position or an 'F' in the succeeding event node position.

7.4.4.2 Milestones

Milestones are key events during the progress of the work. They are entered as separate activities by typing 'F' instead of the succeeding event number. This makes excellent use of the management by exception principle by drawing senior management's attention to these important

```
                                                           LISTING OF CALENDAR DAILY

***************************************************************************************************************************
                        Month :JAN/1987 !                               Month :FEB/1987 !                               Month :MAR/1987 !

     Sun  Mon  Tue  Wed  Thu  Fri  Sat !      Sun  Mon  Tue  Wed  Thu  Fri  Sat !      Sun  Mon  Tue  Wed  Thu  Fri  Sat !
^^^^^^^^^^^^^^^^^^^^^^^^^^^^^^^^^^^^^^^^!^^^^^^^^^^^^^^^^^^^^^^^^^^^^^^^^^^^^^^^^^^!^^^^^^^^^^^^^^^^^^^^^^^^^^^^^^^^^^^^^^^^^!
Date                     1    2    3   !Date 1    2    3    4    5    6    7    !Date 1    2    3    4    5    6    7    !
Day                      ---  ---  --- !Day  ---  21   22   23   24   25   ---  !Day  ---  41   42   43   44   45   ---  !
                                       !                                        !                                        !
Date 4    5    6    7    8    9    10  !Date 8    9    10   11   12   13   14   !Date 8    9    10   11   12   13   14   !
Day  ---  1    2    3    4    5    --- !Day  ---  26   27   28   29   30   ---  !Day  ---  46   47   48   49   50   ---  !
                                       !                                        !                                        !
Date 11   12   13   14   15   16   17  !Date 15   16   17   18   19   20   21   !Date 15   16   17   18   19   20   21   !
Day  ---  6    7    8    9    10   --- !Day  ---  31   32   33   34   35   ---  !Day  ---  51   52   53   54   55   ---  !
                                       !                                        !                                        !
Date 18   19   20   21   22   23   24  !Date 22   23   24   25   26   27   28   !Date 22   23   24   25   26   27   28   !
Day  ---  11   12   13   14   15   --- !Day  ---  36   37   38   39   40   ---  !Day  ---  56   57   58   59   60   ---  !
                                       !                                        !                                        !
Date 25   26   27   28   29   30   31  !                                        !Date 29   30   31                      !
Day  ---  16   17   18   19   20   --- !                                        !Day  ---  61   62                      !
                                       !                                        !                                        !
                                       !                                        !                                        !
                                       !                                        !                                        !
***************************************************************************************************************************
                        Month :APR/1987 !                               Month :MAY/1987 !                               Month :JUN/1987 !
     Sun  Mon  Tue  Wed  Thu  Fri  Sat !      Sun  Mon  Tue  Wed  Thu  Fri  Sat !      Sun  Mon  Tue  Wed  Thu  Fri  Sat !
^^^^^^^^^^^^^^^^^^^^^^^^^^^^^^^^^^^^^^^^!^^^^^^^^^^^^^^^^^^^^^^^^^^^^^^^^^^^^^^^^^^!^^^^^^^^^^^^^^^^^^^^^^^^^^^^^^^^^^^^^^^^^!
Date                1    2    3    4   !Date                          1    2    !Date      1    2    3    4    5    6    !
Day                 63   64   65   --- !Day                           80   ---  !Day       98   99   100  101  102  ---  !
                                       !                                        !                                        !
Date 5    6    7    8    9    10   11  !Date 3    4    5    6    7    8    9    !Date 7    8    9    10   11   12   13   !
Day  ---  66   67   68   69   70   --- !Day  ---  ---  81   82   83   84   ---  !Day  ---  103  104  105  106  107  ---  !
                                       !                                        !                                        !
Date 12   13   14   15   16   17   18  !Date 10   11   12   13   14   15   16   !Date 14   15   16   17   18   19   20   !
Day  ---  71   72   73   74   75   --- !Day  ---  85   86   87   88   89   ---  !Day  ---  108  109  110  111  112  ---  !
                                       !                                        !                                        !
Date 19   20   21   22   23   24   25  !Date 17   18   19   20   21   22   23   !Date 21   22   23   24   25   26   27   !
Day  ---  ---  ---  ---  ---  ---  --- !Day  ---  90   91   92   93   94   ---  !Day  ---  113  114  115  116  117  ---  !
                                       !                                        !                                        !
Date 26   27   28   29   30            !Date 24   25   26   27   28   29   30   !Date 28   29   30                      !
Day  ---  76   77   78   79            !Day  ---  ---  ---  95   96   97   ---  !Day  ---  118  119                      !
                                       !                                        !                                        !
                                       !Date 31                                 !                                        !
                                       !Day  ---                                !                                        !
***************************************************************************************************************************
```

Fig. 7.8. Part of a calendar

No.	Code	Description	No.	Code	Description
1	DF	- DELIVER FRAME	2	FD	- FOUNDATION
3	DP	- DELIVER PLASTERBOARD	4	EF	- ERECT FRAME
5	BD	- BRICKWORK TO DPC	6	FC	- FLOOR CONSTRUCTION
7	BE	- BRICKWORK TO EAVES	8	EP	- END PANEL
9	GL	- GLAZING	10	RC	- ROOF CONSTRUCTION
11	IP	- INTERNAL PARTITIONS	12	DR	- DRAINAGE
13	RF	- ROOF FINISH	14	RG	- RAINWATER GOODS
15	EC	- ELECTRICAL FIRST FIX	16	JF	- JOINERY FIRST FIX
17	PL	- PLASTERING	18	ES	- ELECTRICAL 2ND FIX
19	JS	- JOINERY SECOND FIX	20	MG	- MAKE GOOD PLASTER
21	FS	- FLOOR SCREED	22	SD	- FLOOR SCREED DRY
23	FF	- FLOOR FINISH	24	PD	- PAINTING AND DECORATING
25	PV	- PAVING	26	EL	- ELECTRICIAN FINISH
27	JC	- JOINERY FINISH	28	CH	- CLEAN AND HANDOVER
29	LA	- LABOURERS	30	BL	- BRICKLAYERS
31	PR	- PLASTERER	32	JO	- CARPENTERS AND JOINERS
33	GZ	- GLAZIERS	34	DL	- DRAINLAYERS
35	PA	- PAINTERS AND DECORATORS	36	SL	- SLAB LAYERS
37	PS	- PAVIORS	38	SF	- STEELFIXER
39	EX	- EXCAVATOR	40	GR	- GRADER
41	PM	- PLUMBER	42	CA	- CARPENTER
43	JB	- JCB 3CX AND OPERATOR	44	MI	- 150/100 MIXER
45	SC	- SUBCONTRACTOR	46	FL	- FORKLIFT
47	HM	- HEATING/MECHANICAL S/C	48	EO	- ELECTRICIAN
49	PC	- PRESTRESSED FLOOR S/C	50	RT	- ROOF TILING S/C
51	GS	- GLAZING S/C	52	PY	- PLASTERING S/C
53	ZX	- PAINTER S/C	54	LF	- LANDSCAPE/FINISH S/C
55	LM	- 400\300 MIXER	56	FR	- FELT ROOFING
57	LB	- BRICKLAYERS LABOURERS	58	LP	- PLUMBERS MATE
59	LE	- ELECTRICIANS MATE	60	$P	- POUNDS STERLING
61	DS	- DEMOLITION	62	RL	- PLASTERERS LABOURERS
63	DE	- DEMOLISH BUILDING	64	CR	- CRANE

Fig. 7.9. Abbreviations file

events, which give an indication of the progress of the whole project. The use of milestones is not really appropriate on the village hall project but on large projects such as major hospitals they are extremely useful. They can be used by management committees overseeing a number of projects whereby say twelve milestones can be selected on each project thus allowing overall control.

7.4.4.3 Entering resources

Resources are entered using the same screen as that used for entering activity data. Simply by pressing RES and return, the screen clears except for the node numbers of activities which can have resources. The resources are then entered using the two-character code from the abbreviations file followed by a number to represent the quantity of the particular resource, e.g. BL4 means four bricklayers, $P350 means a cash value of £350 pounds sterling. These entries represent the resource requirements per unit of time spread evenly over the activity. Entering $ before a resource causes it to be represented as a cumulative resource. Point resources can be used by prefixing the resource with an 'H', in which case the resource is assumed to be used on the last day of the activity. In the village hall example all resources are assumed to be spread evenly over the activities.

7.4.5 General reports

Before producing reports Pertmaster analyses the data and loads the calendar and abbreviations files. It then sorts the activities into the order required by the first report to be prepared.

The power of the package is in the wealth of information which can be obtained after initial entry, and the following reports can be obtained.

7.4.5.1 Standard report (see Fig. 7.10)

A full description, earliest and latest times, duration and total float are displayed with a marker on critical activities or activities with negative float. Activities can be sorted into different sequences e.g. earliest activities first, alphabetically (descriptions in alphabetical order), critical activities first etc. Selected activities can be printed by using special codes in the description to facilitate selection by Pertmaster. Day numbers are translated into calendar dates. A milestone report prints out milestones only.

7.4.5.2 Bar charts

Bar charts can be produced for a whole project or for a period of time. Both of these contain the heading with dates and day numbers and are divided into weeks. Critical and non-critical activities and total float can easily be seen. All non-working days are ignored. A project bar chart for the village hall project is shown in Fig. 7.11.

7.4.5.3 Complete resource reports

These can be provided on a daily or weekly basis. In the weekly report the resources for the week are averaged out. The reports can be at earliest or latest times, A daily resources report of the period up to March 31st, 1987 is shown in Fig. 7.12.

Pert does not undertake resource allocation but this can be done manually working from the earliest or latest start condition.

Histograms

Histograms can be produced which show the demand for a specific resource. They can be drawn for earliest or latest start condition. A histogram for bricklayers is shown in Fig. 7.13.

7.4.5.4 Cash flow curves

As a resource code which begins with a '$' always produces a cumulative graph, this is ideal for cash flow prediction. Fig. 7.14 shows a cumulative graph for the village hall project up to the end of July. If a graph is produced for earliest and latest conditions a cash flow envelope can be obtained.

PACME CONSTRUCTION LTD — VILLAGE HALL — TENBURY

STANDARD REPORT With Earliest Activities First

ANALYSIS OF HALL SCHEDULE - VERSION 7 — TIME NOW DATE 9/FEB/87

I-J	ACTIVITY DESCRIPTION	EARLIEST START	EARLIEST FINISH	LATEST START	LATEST FINISH	DURATION	FLOAT
1- 3	FOUNDATIONS	9/FEB/87	13/FEB/87	2/MAR/87	6/MAR/87	5	15
1- 2	DELIVER FRAME	9/FEB/87	20/FEB/87	23/FEB/87	6/MAR/87	10	10
1- 13	DELIVER PLASTERBOARD	9/FEB/87	27/MAY/87	9/FEB/87	27/MAY/87	70	0 *
3- 4	ERECT FRAME	23/FEB/87	23/FEB/87	9/MAR/87	9/MAR/87	1	10
4- 5	BRICKWORK TO DPC	24/FEB/87	25/FEB/87	10/MAR/87	11/MAR/87	2	10
5- 6	FLOOR CONSTRUCTION	26/FEB/87	2/MAR/87	12/MAR/87	16/MAR/87	3	10
5- 8	BRICKWORK TO EAVES	26/FEB/87	12/MAR/87	19/MAR/87	2/APR/87	11	15
6- 7	END PANELS	3/MAR/87	19/MAR/87	17/MAR/87	2/APR/87	13	10
8- 11	INTERNAL PARTITIONS	20/MAR/87	23/MAR/87	13/MAY/87	14/MAY/87	2	32
8- 21	DRAINAGE	20/MAR/87	25/MAR/87	22/JUL/87	27/JUL/87	4	80
7- 11	GLAZING	20/MAR/87	30/MAR/87	6/MAY/87	14/MAY/87	7	27
8- 9	ROOF CONSTRUCTION	20/MAR/87	9/APR/87	3/APR/87	30/APR/87	15	10
21- 24	PAVING	26/MAR/87	7/APR/87	28/JUL/87	7/AUG/87	9	80
9- 10	ROOF FINISH	10/APR/87	29/APR/87	1/MAY/87	14/MAY/87	9	10
10- 20	RAINWATER GOODS	30/APR/87	1/MAY/87	22/JUL/87	23/JUL/87	2	56
11- 13	JOINERY FIRST FIX	30/APR/87	8/MAY/87	18/MAY/87	27/MAY/87	6	11
11- 12	ELECTRICAL 1ST FIX	30/APR/87	11/MAY/87	15/MAY/87	27/MAY/87	7	10
13- 14	PLASTERING	28/MAY/87	10/JUN/87	28/MAY/87	10/JUN/87	10	0 *
14- 15	ELECTRICIAN 2ND FIX	11/JUN/87	15/JUN/87	15/JUN/87	17/[illegible]/87	3	2
14- 16	JOINERY SECOND FIX	11/JUN/87	17/JUN/87	11/JUN/87	17/JUN/87	5	0 *
16- 20	MAKE GOOD PLASTER	18/JUN/87	18/JUN/87	23/JUL/87	23/JUL/87	1	25
16- 18	FLOOR SCREED	18/JUN/87	23/JUN/87	18/JUN/87	23/JUN/87	4	0 *
18- 19	FLOOR SCREED DRY	24/JUN/87	7/JUL/87	24/JUN/87	7/JUL/87	10	0 *
19- 20	FLOOR FINISH	8/JUL/87	23/JUL/87	8/JUL/87	23/JUL/87	12	0 *
20- 22	PAINTING AND DECORATING	24/JUL/87	4/AUG/87	24/JUL/87	4/AUG/87	8	0 *
22- 23	ELECTRICAL FINISH	5/AUG/87	5/AUG/87	7/AUG/87	7/AUG/87	1	2
22- 24	JOINERY FINISH	5/AUG/87	7/AUG/87	5/AUG/87	7/AUG/87	3	0 *
24- 25	CLEAN AND HANDOVER	10/AUG/87	14/AUG/87	10/AUG/87	14/AUG/87	5	0 *

Fig. 7.10. Standard report

PACME CONSTRUCTION LTD

VILLAGE HALL
PROJECT BARCHART

TENBURY

ANALYSIS OF HALL   SCHEDULE - VERSION 7

TIME NOW DATE 9/FEB/87

I-J	ACTIVITY DESCRIPTION	EARLIEST START	EARLIEST FINISH	LATEST START
1- 3	FOUNDATIONS	9/FEB/87	13/FEB/87	2/MAR/87
1- 2	DELIVER FRAME	9/FEB/87	20/FEB/87	23/FEB/87
1- 13	DELIVER PLASTERBOARD	9/FEB/87	27/MAY/87	9/FEB/87
3- 4	ERECT FRAME	23/FEB/87	23/FEB/87	9/MAR/87
4- 5	BRICKWORK TO DPC	24/FEB/87	25/FEB/87	10/MAR/87
5- 6	FLOOR CONSTRUCTION	26/FEB/87	2/MAR/87	12/MAR/87
5- 8	BRICKWORK TO EAVES	26/FEB/87	12/MAR/87	19/MAR/87
6- 7	END PANELS	3/MAR/87	19/MAR/87	17/MAR/87
8- 11	INTERNAL PARTITIONS	20/MAR/87	23/MAR/87	13/MAY/87
8- 21	DRAINAGE	20/MAR/87	25/MAR/87	22/JUL/87
7- 11	GLAZING	20/MAR/87	30/MAR/87	6/MAY/87
8- 9	ROOF CONSTRUCTION	20/MAR/87	9/APR/87	3/APR/87
21- 24	PAVING	26/MAR/87	7/APR/87	28/JUL/87
9- 10	ROOF FINISH	10/APR/87	29/APR/87	1/MAY/87
10- 20	RAINWATER GOODS	30/APR/87	1/MAY/87	22/JUL/87
11- 13	JOINERY FIRST FIX	30/APR/87	8/MAY/87	18/MAY/87
11- 12	ELECTRICAL 1ST FIX	30/APR/87	11/MAY/87	15/MAY/87
13- 14	PLASTERING	28/MAY/87	10/JUN/87	28/MAY/87
14- 15	ELECTRICIAN 2ND FIX	11/JUN/87	15/JUN/87	15/JUN/87
14- 16	JOINERY SECOND FIX	11/JUN/87	17/JUN/87	11/JUN/87
16- 20	MAKE GOOD PLASTER	18/JUN/87	18/JUN/87	23/JUL/87
16- 18	FLOOR SCREED	18/JUN/87	23/JUN/87	18/JUN/87
18- 19	FLOOR SCREED DRY	24/JUN/87	7/JUL/87	24/JUN/87
19- 20	FLOOR FINISH	8/JUL/87	23/JUL/87	8/JUL/87
20- 22	PAINTING AND DECORATING	24/JUL/87	4/AUG/87	24/JUL/87
22- 23	ELECTRICAL FINISH	5/AUG/87	5/AUG/87	7/AUG/87
22- 24	JOINERY FINISH	5/AUG/87	7/AUG/87	5/AUG/87
24- 25	CLEAN AND HANDOVER	10/AUG/87	14/AUG/87	10/AUG/87

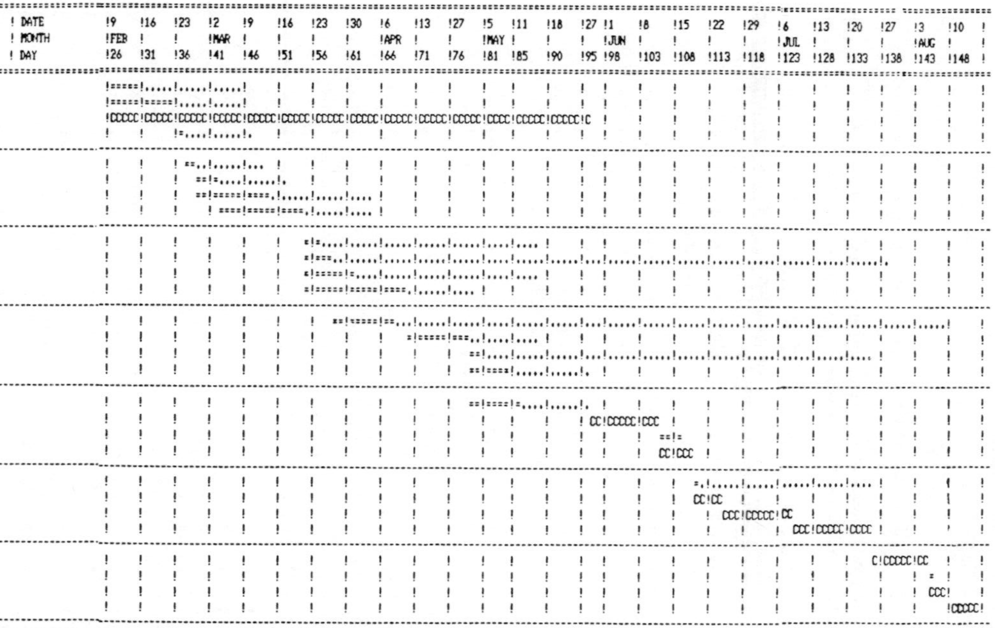

Barchart Key   CCC : Critical Activities   === : Non Critical Activities   ..... : Float (if shown)

Fig. 7.11. Project bar chart

FACME CONSTRUCTION LTD

TENBURY

GENERAL RESOURCES REPORT (Daily)

ANALYSIS OF HALL SCHEDULE - VERSION 7 ASSUMING ALL ACTIVITIES ARE AT THEIR EARLY START TIME NOW DATE 9/FEB/87

1 =POUNDS STERLING
2 =JCB 3CX AND OPERATOR
3 =400\300 MIXER
4 =LABOURERS
5 =150/100 MIXER
6 =BRICKLAYERS
7 =BRICKLAYERS LABOURERS
8 =CARPENTERS AND JOINERS
9 =PLUMBER
10 =PLUMBERS MATE
11 =FELT ROOFING
12 =ELECTRICIAN
13 =ELECTRICIANS MATE
14 =PLASTERER
15 =PLASTERERS LABOURERS
16 =FLOOR FINISH
17 =PAINTERS AND DECORATORS

	1	2	3	4	5	6	7	8	9	10	11	12	13	14	15	16	17	18	19	20	21	22	23
9/FEB/87	1150	1	1	8																			
10/FEB/87	1150	1	1	8																			
11/FEB/87	1150	1	1	8																			
12/FEB/87	1150	1	1	8																			
13/FEB/87	1150	1	1	8																			
16/FEB/87	500																						
17/FEB/87	500																						
18/FEB/87	500																						
19/FEB/87	500																						
20/FEB/87	500																						
23/FEB/87	300																						
24/FEB/87	450				1	4	2																
25/FEB/87	450				1	4	2																
26/FEB/87	1200		1	8	1	4	2																
27/FEB/87	1200		1	8	1	4	2																
2/MAR/87	1200		1	8	1	4	2																
3/MAR/87	760				1	4	2	2															
4/MAR/87	760				1	4	2	2															
5/MAR/87	760				1	4	2	2															
6/MAR/87	760				1	4	2	2															
9/MAR/87	760				1	4	2	2															
10/MAR/87	760				1	4	2	2															
11/MAR/87	760				1	4	2	2															
12/MAR/87	760				1	4	2	2															
13/MAR/87	310							2															
16/MAR/87	310							2															
17/MAR/87	310							2															
18/MAR/87	310							2															
19/MAR/87	310							2															
20/MAR/87	1030		1	8	2	5	3	2	1	1													
23/MAR/87	1030		1	8	2	5	3	2	1	1													
24/MAR/87	760		1	8	1	1	1	2	1	1													
25/MAR/87	760		1	8	1	1	1	2	1	1													
26/MAR/87	520				1	1	1	2	1	1													
27/MAR/87	520				1	1	1	2	1	1													
30/MAR/87	520				1	1	1	2	1	1													
31/MAR/87	320				1	1	1	2															

Fig. 7.12. Daily resources report

```
PACME CONSTRUCTION LTD                    VILLAGE HALL                                TENBURY
                                        HISTOGRAM REPORT

ANALYSIS OF HALL    SCHEDULE - VERSION  7                          REPORT START DATE 9/FEB/87

          ANALYSIS OF THE REQUIREMENT FOR BRICKLAYERS ASSUMING ALL ACTIVITIES ARE AT THEIR EARLY START


5    !                    **
4    !      *************  **
3    !      *************  **
2    !      *************  **
1    !      *************  *************
     ---------------------------------------------------------------------------------------
     ^      ^      ^      ^      ^      ^      ^      ^      ^      ^      ^
     26  30  34  38  42  46  50  54  58  62  66  70  74  78  82  86  90  94  98  102 106 110 114 118 122 126 130 134 138 142
     9/FEB/87    25/FEB/87   13/MAR/87   31/MAR/87   16/APR/87   12/MAY/87   1/JUN/87    17/JUN/87   3/JUL/87    21/JUL/87
```

Fig. 7.13. Histogram showing bricklayer requirements

Fig. 7.14. Cash flow forecast

7.4.5.5 Disk report

This files the information on disk for interfacing with other software packages such as databases.

7.4.5.6 Screen reports

Bar charts, histograms and activity listings can be displayed on the screen. It is then simple to move around the project examining the information in detail as required.

7.4.6 Up-dating plans

This facility allows the planner to up-date plans in order to control the project. It also allows him to try out different strategies and to see the result immediately.

To up-date a plan the durations are altered to reflect the current situation. A new plan can then be generated showing the up-to-date position. Activities can be deleted or added to take account of site instructions, and resources can be altered if required. Changes can also be made to the plan headings. Up-dated reports for the village hall project are outlined below:

Fig. 7.15 shows a bar chart for the period commencing day 126 i.e. 100 days after the start of the project. The durations of the activities have been amended as follows.

(1) Completed activity durations have been reduced to zero.
  (a) 'Rainwater goods' has been reduced to a time period of 1.
  (b) 'Floor finish' has been reduced to a time period of 6.
  (c) All other activities are unchanged.
(2) Fig. 7.16 shows a standard report.
(3) Fig. 7.17 shows a histogram for painters and decorators.

7.4.7 Other facilities

Many other facilities are available and these include provision for:

(1) Setting up a library of skeleton plans for similar projects.
(2) Testing the plan against different calendars (e.g. working a 6-day week) to see what will happen.
(3) Providing a histogram for the entire program.
(4) Printing the actual figures on the cash flow curves.
(5) Listing the activities and relationships.
(6) Zooming in on selected activities on the screen, adding new activities and building up plans on the screen.
(7) Producing the arrow diagram on the printer.

7.4.8 Conclusion

The greatest benefits of computers for planning are gained on large and/or complex projects when a number of reports are required, because they can all be produced from the same basic data. Alternative strategies can be explored and the results seen immediately, thus saving considerable sums of money on site. The immediate response gained from office and site-based microcomputers, and the direct computer links which can now be achieved, makes computer-based planning much more attractive than it was in the pre-microcomputer days.

7.5 LINE OF BALANCE

7.5.1 Introduction

Software produced by Vega Software Ltd includes an educational package called Prolin which has been developed to exploit the line-of-balance method of programming. The model offers a rapid means of computing, displaying and modifying line-of-balance schedules. It runs on RML 380Z and 480Z machines.

The screen graphics are excellent for problem solving, allowing the user to experiment with different solutions.

```
PACME CONSTRUCTION LTD                        VILLAGE HALL                                          TENBURY
                                             PROJECT BARCHART
ANALYSIS OF HALL    SCHEDULE - VERSION 9                                           TIME NOW DATE 9/JUL/87
==========================================================================================================
                                                                     ! DATE      !9     !20  !27  !3    !
I-J    ACTIVITY DESCRIPTION        EARLIEST    EARLIEST    LATEST    ! MONTH     !JUL   !    !    !AUG  !
                                   START       FINISH      START     ! DAY       !126   !133 !138 !143  !
==========================================================================================================
 10- 20 RAINWATER GOODS            9/JUL/87    9/JUL/87    16/JUL/87             !=..... !    !    !    !
  8- 21 DRAINAGE                   9/JUL/87    14/JUL/87   15/JUL/87             !====...!.   !    !    !
 19- 20 FLOOR FINISH               9/JUL/87    16/JUL/87   9/JUL/87              !CCCCCC !    !    !    !
 21- 24 PAVING                     15/JUL/87   27/JUL/87   21/JUL/87             !   ===!=====!=....!   !
----------------------------------------------------------------------------------------------------------
 20- 22 PAINTING AND DECORATING    17/JUL/87   28/JUL/87   17/JUL/87             !     C!CCCCC!CC   !   !
 22- 23 ELECTRICAL FINISH          29/JUL/87   29/JUL/87   31/JUL/87             !      !    ! =..!    !
 22- 24 JOINERY FINISH             29/JUL/87   31/JUL/87   29/JUL/87             !      !    ! CCC!    !
 24- 25 CLEAN AND HANDOVER         3/AUG/87    7/AUG/87    3/AUG/87              !      !    !    !CCCCC!
----------------------------------------------------------------------------------------------------------

==========================================================================================================
 Barchart Key     CCC : Critical Activities    === : Non Critical Activities    ..... : Float (if shown)
```

Fig. 7.15. Project bar chart showing progress after 100 days

PACME CONSTRUCTION LTD VILLAGE HALL TENBURY

STANDARD REPORT With Earliest Activities First

ANALYSIS OF HALL SCHEDULE - VERSION 9 TIME NOW DATE 9/JUL/87

I-J	ACTIVITY DESCRIPTION	EARLIEST START	EARLIEST FINISH	LATEST START	LATEST FINISH	DURATION	FLOAT
10- 20	RAINWATER GOODS	9/JUL/87	9/JUL/87	16/JUL/87	16/JUL/87	1	5
8- 21	DRAINAGE	9/JUL/87	14/JUL/87	15/JUL/87	20/JUL/87	4	4
19- 20	FLOOR FINISH	9/JUL/87	16/JUL/87	9/JUL/87	16/JUL/87	6	0 *
21- 24	PAVING	15/JUL/87	27/JUL/87	21/JUL/87	31/JUL/87	9	4
20- 22	PAINTING AND DECORATING	17/JUL/87	28/JUL/87	17/JUL/87	28/JUL/87	8	0 *
22- 23	ELECTRICAL FINISH	29/JUL/87	29/JUL/87	31/JUL/87	31/JUL/87	1	2
22- 24	JOINERY FINISH	29/JUL/87	31/JUL/87	29/JUL/87	31/JUL/87	3	0 *
24- 25	CLEAN AND HANDOVER	3/AUG/87	7/AUG/87	3/AUG/87	7/AUG/87	5	0 *

Fig. 7.16. Standard report after 100 days

```
PACME CONSTRUCTION LTD                         VILLAGE HALL                                      TENBURY
                                             HISTOGRAM REPORT

ANALYSIS OF HALL    SCHEDULE - VERSION  9                                      REPORT START DATE 9/JUL/87

            ANALYSIS OF THE REQUIREMENT FOR PAINTERS AND DECORATORS ASSUMING ALL ACTIVITIES ARE AT THEIR EARLY START

4     !    ********
3     !    ********
2     !    ********
1     !    ********
      ----------------------------------------------------------------------------------------------------
      ^           ^           ^           ^           ^           ^           ^           ^           ^           ^           ^
      126 130 134 138 142 146 150 154 158 162 166 170 174 178 182 186 190 194 198 202 206 210 214 218 222 226 230 234 238 242
      9/JUL/87    27/JUL/87   12/AUG/87   4/SEP/87    22/SEP/87   8/OCT/87    26/OCT/87   11/NOV/87   27/NOV/87   15/DEC/87
```

Fig. 7.17. Histogram for painters and decorators

7.5.2 Application to a housing project

The following example is based upon the superstructure of the housing project described in 1.10.4.

Preliminary decisions

Prior to entering data decisions must be taken on the following:

(a) *General data*
Job title.
Number of units.
Number of activities to complete one unit.
Required handover rate in units per day.
Number of hours in a working day.
Number of working days in a week.

(b) *Activity data*
Activity descriptions.
Number of manhours per unit.
Number of men per gang.
Number of gangs for the first half of the units.
Number of gangs for the second half of the units.
The buffer to be used.

7.5.2.1 Data entry

Prolin creates a new data file by loading the demonstration file and editing it via the screen editing process to represent the new project. The new plan is then saved, after which the data can be printed. A printout of the initial data for the housing project is shown in Fig. 7.18.

It can be seen that the theoretical number of gangs required is calculated by Prolin and this gives a guide on the action which might be taken to modify the plan.

7.5.2.2 Adapting the plan

To inspect and adapt the plan, it is first called from disk and displayed on the screen. A photograph of the screen based on the data in Fig. 7.18 is shown in Fig. 7.19. The relationship between the lines-of-balance can clearly be seen and by reference to the production plan, decisions can be taken on possible changes in activity data, e.g. the number of gangs used on each activity. To modify the plan the 'modify' option is chosen and any activity can then be changed. The modified line-of-balance diagram can then be displayed on the screen.

By the iterative process the diagram can be progressively improved until an acceptable solution is found. At this stage the new plan can be saved and printed in text and numeric form. The user can then draw out the solution developed if required.

Fig. 7.20 shows a photograph of the screen display and Fig. 7.21 shows the printout for the housing project in which four gangs were used for the first half and five gangs for the second half of activity 'Flashgs & glazing'.

PROLIN--GENERAL PROJECT DATA

Job title	HOUSING AT TENBURY
Job duration(working days)	135
Actual h/o rate(units/day)	Ø.536
Required h/o rate(units/day)	Ø.5ØØ
First unit handover on day	8Ø
Number of units	3Ø
No. of activities per unit	6
Hours per working day	7
Days per working week	5

BASIC ACTIVITY DATA

Activity name	No.	Work content (manhrs)	Men per gang	Time buffer (days)
BRICKWORK 1ST & 2ND LIFT	1	332	2	3
1ST FL JSTS & GAR ROOF	2	4Ø	3	3
BRICKWORK TO COMPLETION	3	222	2	3
ROOF CARCASS	4	78	3	3
ROOF COVERINGS	5	4Ø	3	3
FLASHGS & GLAZING	6	54	1	3

PRODUCTION PLAN

Act no.	Days per unit	Start days for first unit	middle unit	last unit	No. of gangs 1st half	2nd half	Theoretical gangs needed
1	24	1	28	58	12	12	11.86
2	2	3Ø	56	85	1	1	Ø.95
3	16	35	62	92	8	8	7.93
4	4	57	83	111	2	2	1.86
5	2	64	91	119	1	1	Ø.95
6	8	69	96	125	4	4	3.86

Fig. 7.18. Initial data for housing project

This is a very good model and clearly demonstrates the superiority of the computer over hand analysis for this type of problem.

7.5.3 Other modules

The total system developed by Vega Software is known as CAMM and includes modules on Project Planning Techniques, Discounting and Investment Appraisal, Optimum Policy Models and System Simulation.

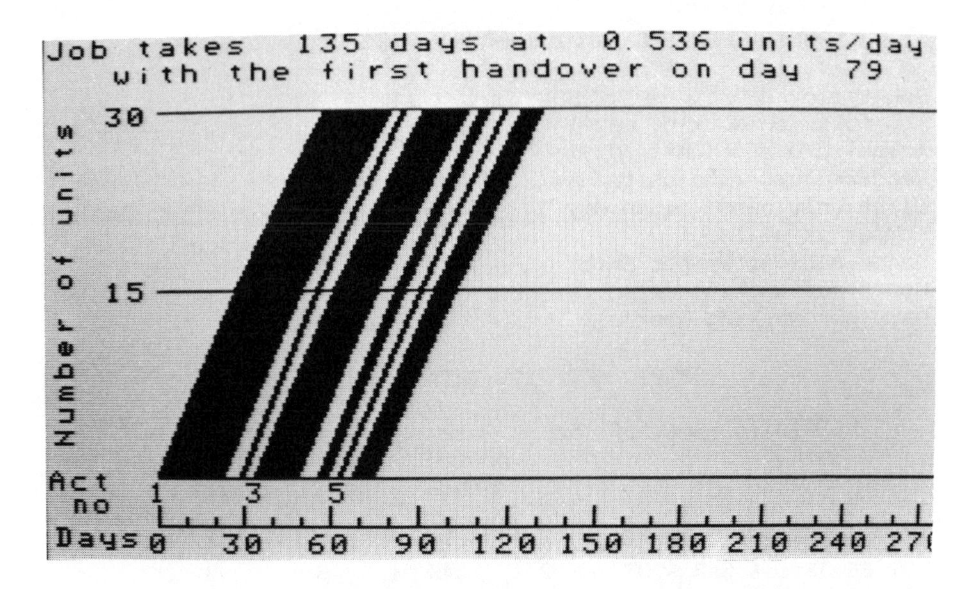

Fig. 7.19. Screen display based on initial data

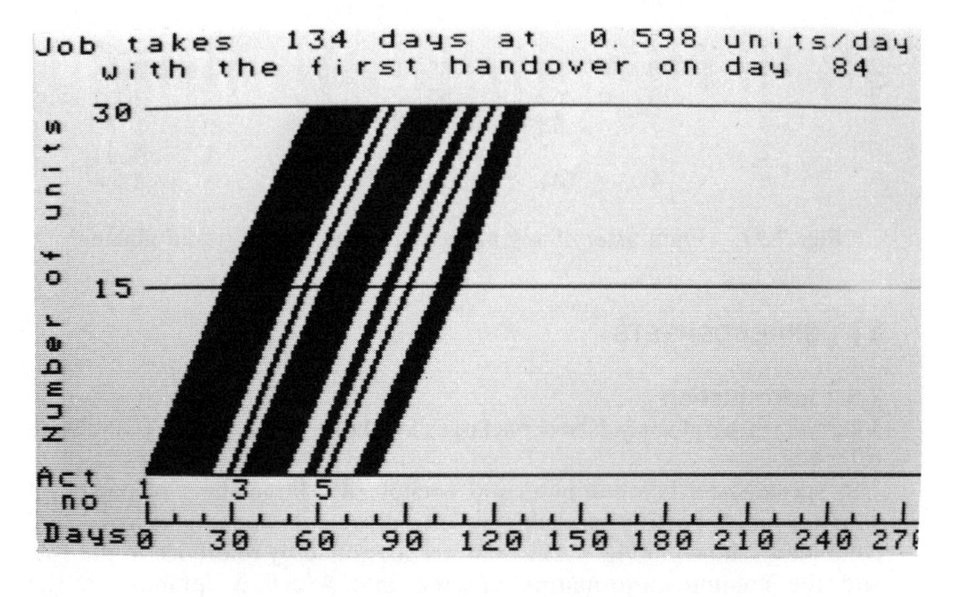

Fig. 7.20. Screen display after changing last activity

PROLIN--GENERAL PROJECT DATA

Job title HOUSING AT TENBURY	
Job duration(working days)	134
Actual h/o rate(units/day)	0.598
Required h/o rate(units/day)	0.500
First unit handover on day	85
Number of units	30
No. of activities per unit	6
Hours per working day	7
Days per working week	5

BASIC ACTIVITY DATA

Activity name	No.	Work content (manhrs)	Men per gang	Time buffer (days)
BRICKWORK 1ST & 2ND LIFT	1	332	2	3
1ST FL JSTS & GAR ROOF	2	40	3	3
BRICKWORK TO COMPLETION	3	222	2	3
ROOF CARCASS	4	78	3	3
ROOF COVERINGS	5	40	3	3
FLASHGS & GLAZING	6	54	1	3

PRODUCTION PLAN

Act no.	Days per unit	Start days for first unit	middle unit	last unit	No. of gangs 1st half	2nd half	Theoretical gangs needed
1	24	1	28	58	12	12	11.86
2	2	30	56	85	1	1	0.95
3	16	35	62	92	8	8	7.93
4	4	57	83	111	2	2	1.86
5	2	64	91	119	1	1	0.95
6	8	74	101	124	4	5	3.86

Fig. 7.21. Data after changing last activity 'flashings and glazing'

7.6 SPREADSHEETS

7.6.1 Introduction

There are several spreadsheet packages available which are very similar in principle.

A spreadsheet is a computerised version of a large sheet of paper, of which a small portion is visible on the screen. It consists of rows and columns as shown in Fig. 7.22. Cells are identified by reference to the row and the column. Information entered into a cell is retained in the computer's memory, and the cells can be related to each other. If the

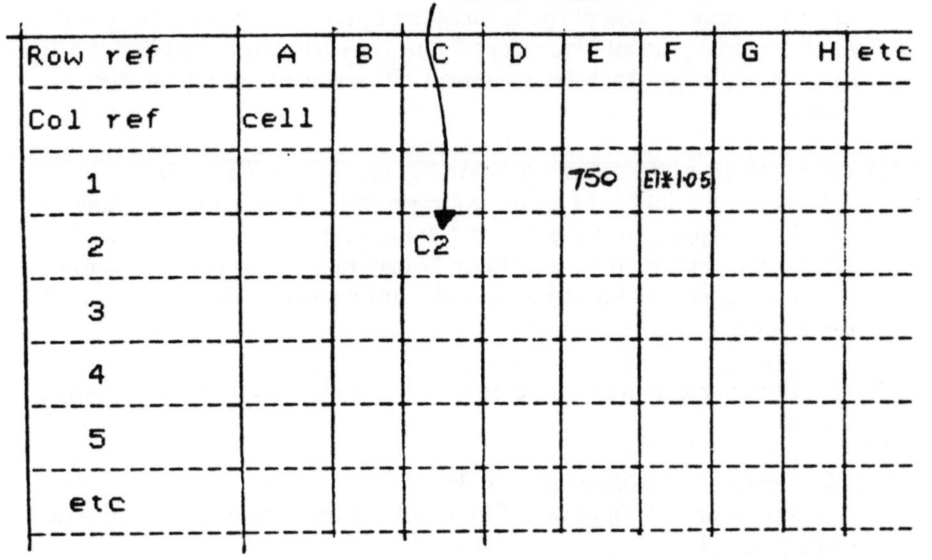

Fig. 7.22. Small part of a spreadsheet

information in one cell changes, the information in all relative cells also changes. In Fig. 7.22 cell E1 contains a value of 750, F1 will contain a value of 750 × 1.05 = 787.5. If E1 is altered to 800, F1 will change to 800 × 1.05 = 840. The screen acts as a window viewing a small part of the spreadsheet and can be moved from side to side and up and down to view all cells. Data and text can be changed very easily e.g. deleted, inserted, altered etc.

7.6.2 Links with other software

Some spreadsheet programs can be linked with:

(1) Other spreadsheet models and can extract information from them.
(2) Graphics to give histograms, bar charts, pie charts etc.
(3) Word processors enabling them to be incorporated into documents.
(4) Databases allowing them to use the database information or allow the data base to use information from the spreadsheet.

7.6.3 Uses in the construction industry

There are many uses for spreadsheets, which are ideal for solving problems and carrying out calculations on data which can be presented in tabular form. Spreadsheets are particularly useful for financial modelling and presenting contract and company cash flows. Spreadsheets could have been used for the following calculations in this book.

(1) *Line-of-balance calculations* in Fig. 1.36
This could be set up as a standard model requiring general information

such as handover rates, number of hours worked per week, number of units and activity data in the form of the operations, productive hours, minimum gang size and number of gangs in each unit to be filled in, all calculations could then be produced automatically using the spreadsheet.

(2) *Calculations on the effects of incentives on costs and earnings*
A standard model could be set up to produce all the tables starting at Fig. 3.20, requiring only hours at 100 rating, percentage added (to set target), percentage payback, basic hourly rate, all-in rate and output rating assumed for the estimate. The graphs could also be produced using some packages.

(3) All tabular calculations and graphs in the Budgetary and Cost Control section.

(4) *The queuing calculations* (Fig. 6.36)
A standard model could be set up requiring only the inter-arrival time to be entered.

(5) *The decision rule table* (Fig. 6.46)
A standard model could be set up requiring only the factor *f*, the operation and estimated duration to be entered.

(6) *Any other calculations*
Where they can be based on formulae in single column or tabular form, e.g. a model for an all-in rate calculation could be set up requiring only changed data to be entered each year.

7.6.4 Project budget and cash flow

The example discussed in 5.1.5.3 will be used as an example on the detailed application of spreadsheets.

Lotus 1 2 3 will be used. This is a very powerful spreadsheet/graphics/database package which is currently used throughout industry. The package consists of a maximum of 2048 rows and 256 columns giving a massive spreadsheet potential. Graphs can be produced in colour when a colour printer is used and a number of different fonts are available.

A model was first set up, and headings, retention, operations, duration, planned expenditure and income, and the program entered using the monetary values in each week, the floats being represented by a zero. The planned expenditure and income and monthly planned expenditure and income was then calculated using simple '@ sum' functions.

The remainder of the calculations were then carried out using formulae to relate the cells. A printout showing all activities at their earliest time and 5% retention is shown in Fig. 7.23. A printout of some of the formulae which determined this chart is shown in Fig. 7.24. The graph showing cash flow at earliest times is shown in Fig. 7.25. The power of the package is the manipulation which can be carried out at tremendous speed. By simply omitting the '1' from 'Expenditure and income at earliest time' and inserting a '1' to 'Expenditure and income at latest time' the whole spreadsheet can be changed. Any slight adjustments resulting from operations starting part way through a week can then be made. The result is shown in Fig. 7.26 and the graph produced is shown in Fig. 7.29. These changed calculations and the production of the graph would have taken at least one and a half hours manually and there is a high probability of mathematical errors being made.

Again simply by altering the retention figure to 3% a complete new spreadsheet and graph can be produced as shown in Figs. 7.27 and 7.30 showing the cash flow with all activities at their latest times. By altering the '1' back to its original position a further spreadsheet and graph can be produced showing the cash flow with activities at their earliest time, as shown in Figs. 7.30 and 7.31.

It will be observed that many other alternatives could be tried requiring very little effort but with very fast results.

It can be seen that spreadsheets have much to offer the construction manager.

7.7 WORD PROCESSING

7.7.1 Introduction

Word processing can make a valuable contribution to the efficiency of a construction company. Using word processing it is possible to do everything which can be done on a typewriter and much more. Some advantages over a standard electric typewriter are the simple quick manipulation of text as follows:

1. Margins, spacing between lines, and length of lines can be altered in seconds.
2. Whole documents, paragraphs or blocks of text can be reformed, saved, moved, copied, deleted, altered or called from disk.
3. Spelling can be corrected automatically.
4. Simple mailing facilities can be implemented.
5. Line height and character width can be changed and many other facilities exploited.

7.7.2 The use of word processing to the construction manager

Whilst the construction manager may not physically use word processing (although some do) it has much to offer because of the above capabilities. Typical uses could be as follows:

1. Setting up a library of standard descriptions for building up work packages based on method statements for targetting purposes.
2. In the case of small firms, setting up a standard specification form to

PROGRAMME FOR SMALL OFFICE BUILDING
EXPENDITURE AND INCOME AT EARLIEST TIME 1 RETENT. 5 %
EXPENDITURE AND INCOME AT LATEST TIME

OPERATION	DURN DAYS	PLAN'D EXPEND.	PLAN'D INCOME	WEEKS 1	2	3	4	5	6	7	8	9	10	11	12	13	14	15	16
SITE SET UP	3	210	240	210															
				240															
SETTING OUT	4	60	80	60															
				80															
EXCAVATE TO REDUCE LEVEL	10	1000	1200	100	500	400													
				120	600	480													
PILING	12	3600	4800			300	1500	1500	300										
						400	2000	2000	400										
DRAINS	6	300	360			50	250	0	0	0	0	0	0	0					
						60	300	0	0	0	0	0	0	0					
PAD FOUNDATION	9	630	720						280	350									
									320	400									
STRIP FOUNDATION	5	500	600						400	100	0	0	0	0					
									480	120	0	0	0	0					
INSITU CONCRETE FRAME	22	22000	33000								5000	5000	5000	5000	2000				
											7500	7500	7500	7500	3000				
GROUND FLOOR SLAB	5	800	900												480	320	0		
															540	360	0		
BRICK INFILL PANELS	15	8700	9750												1740	2900	2900	1160	
															1950	3250	3250	1300	
PRECAST STAIRS AND FLOORS	3	2400	2700													2400	0	0	
																2700	0	0	
HARDWOOD FRAMES	10	1000	1200															300	500
																		360	600
PRECAST ROOF SLAB	2	2000	2500														2000	0	0
																	2500	0	0
BLOCK PARTITIONS GROUND FLOOR	7	1750	2100														1250	500	0
																	1500	600	0
GLAZING	6	240	300																
EXTERNAL DOORS	4	1600	1800																
ROOF COVERING	5	1000	1200														600	400	0
																	720	480	0
EXTERNAL PLUMBING	6	180	210														90	90	0
																	105	105	0
BLOCK PARTITIONS FIRST FLOOR	7	2100	2450															900	1200
																		1050	1400
PLUMBING FIRST FIX	5	900	1050																300
																			350
ELECTICAL FIRST FIX	8	400	480																
EXTERNAL PAINTING	12	240	300																40
																			50
EXTERNAL WORKS	20	7564	9000																756
																			900
JOINERY FIRST FIX	8	800	880																
PLASTERING	18	1800	2160																
FLOOR SCREEDING	10	600	700																
QUARRY TILING	1	100	140																
PLUMBING SECOND FIX	10	1000	1200																
JOINERY SECOND FIX	10	1000	1200																
INTERNAL PAINTING	14	420	504																
VINYL TILING	6	240	300																
ELECTRICAL SECOND FIX	6	360	540																
CLEANING AND HANDOVER	1	50	60																
PRELIMINARIES		13175	13950	425	425	425	425	425	425	425	425	425	425	425	425	425	425	425	425
				450	450	450	450	450	450	450	450	450	450	450	450	450	450	450	450
PLANNED EXPENDITURE		78719		795	925	1175	2175	1925	1405	875	5425	5425	5425	5425	4645	6045	7265	3775	3221
PLANNED INCOME			98574	890	1050	1390	2750	2450	1650	970	7950	7950	7950	7950	5940	6760	8525	4345	3750
PLANNED EXPENDITURE MONTHLY							5070				9630				20920				20306
PLANNED INCOME MONTHLY							6080				13020				29790				23380
RETENTION ON PLANNED INCOME							304				651				1490				1169
INCOME LESS RETENTION TIME ADJUSTED							0				5776				12369				28301
ACCUMULATIVE EXPENDITURE							5070				14700				35620				55926
ACCUMULATIVE INCOME TIME ADJUSTED							0				5776				18145				46446
VALUATION NUMBER							1	1			2	2			3	3			4
DIFFERENCE BETWEEN EXP. AND INC.							-5070	-5070			-14700	-8924			-29844	-17475			-37781
GRAPH PLOTTINGS																			
0	1	1	2	2	3	3	4	4	5	5	6	6	7	7	7.75	8	8	9	9
0	-5070	-5070	-14700	-8924	-29844	-17475	-37781	-9481	-22029	183	-3678	10183	5963	10086	7921	7921	12515	12515	17391

Fig. 7.23. Activities at earliest times – retention 5%

17	18	19	20	21	22	23	24	25	26	27	28	29	30	31	32	33	34	35	36	37	38	39	59	60
200																								
240																								
0																								
0																								
0																								
0																								
120	120																							
150	150																							
1200	400																							
1350	450																							
0	0																							
0	0																							
0	0																							
0	0																							
0	0																							
0	0																							
600																								
700																								
	100	250	50																					
	120	300	60																					
100	100	0	0	0	0	0	0	0	0	0	0	0	0	0										
125	125	0	0	0	0	0	0	0	0	0	0	0	0	0										
1891	1891	1891	1135	0	0	0	0	0	0	0	0	0	0	0										
2250	2250	2250	1350	0	0	0	0	0	0	0	0	0	0	0										
			400	400	0	0																		
			440	440	0	0																		
			400	500	500	400																		
			480	600	600	480																		
						60	300	240																
						70	350	280																
								100	0	0														
								140	0	0														
								100	500	400														
								120	600	480														
								100	500	400														
								120	600	480														
										30	150	150	90											
										36	180	180	108											
													80	160										
													100	200										
													120	240										
													180	360										
														50										
														60										
425	425	425	425	425	425	425	425	425	425	425	425	425	425	425										
450	450	450	450	450	450	450	450	450	450	450	450	450	450	450										
4536	3036	2566	2410	1325	925	885	725	965	1425	1255	575	575	715	875										
5265	3545	3000	2780	1490	1050	1000	800	1110	1650	1446	630	630	838	1070										
			12548				3860				4220			2165										
			14590				4340				4836			2538										
			730				217				242			127										
			22211				13861				4123				4594				2411					
			68474				72334				76554			78719										
			68657				82517				86640				91234				96110					
4			5	5			6	6			7	7		7.75	8	8			9	9			10	10
-9481			-22029	183			-3678	10183			5963	10086		7921	7921	12515			12515	17391			17391	19855
10	10																							
17391	19855																							

1	A	B	C	D	E	F	G	H	I	J	K	L	M	N	O	P	Q	R	S	T
2		PROGRAMME FOR SMALL OFFICE BUILDING																		
3		EXPENDITURE AND INCOME AT EARLIEST	TIME	1		RETENT.	5 %													
4		EXPENDITURE AND INCOME AT LATEST	TIME																	
5																				
6		OPERATION	DURN	PLAN'D	PLAN'D	WEEKS														
7			DAYS	EXPEND.	INCOME	1	2	3	4	5	6	7	8	9	10	11	12	13	14	15
8																				
9		SITE SET UP	3	210	240	210														
10						240														
11		SETTING OUT	4	60	80	60														
12						80														
13		EXCAVATE TO REDUCE LEVEL	10	1000	1200	100	500	400												
14						120	600	480												
15		PILING	12	3600	4800			300	1500	1500	300									
16								400	2000	2000	400									
17		DRAINS	6	300	360			50*D3	250*D3	0	0	0	0	0	300*D4	250*D4				
18								60*D3	300*D3	0	0	0	0	0	360*D4	300*D4				
19		PAD FOUNDATION	9	630	720						280	350								
20											320	400								
21		STRIP FOUNDATION	5	500	600						400*D3	100*D3	0	0	400*D4	100*D4				
22											480*D3	120*D3	0	0	480*D4	120*D4				
23		INSITU CONCRETE FRAME	22	22000	33000								5000	5000	5000	5000	2000			
24													7500	7500	7500	7500	3000			
25		GROUND FLOOR SLAB	5	800	900												480*D3	320*D3	800*D4	
26																	540*D3	360*D3	900*D4	
27		BRICK INFILL PANELS	15	8700	9750												1740	2900	2900	1160
28																	1950	3250	3250	1300
29		PRECAST STAIRS AND FLOORS	3	2400	2700													2400*D3	0	2400*D4
30																		2700*D3	0	2700*D4

Fig. 7.24. Formulae which determine whether activity is at earlier or latest time

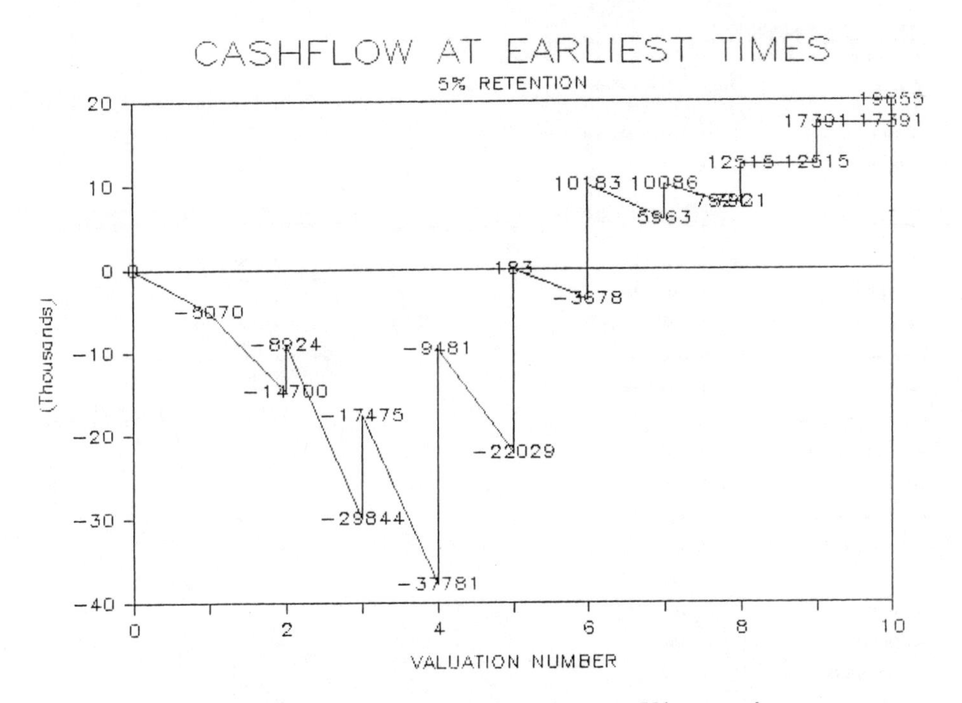

Fig. 7.25. Activity at earliest times – 5% retention

be sent out with quotations. This could then be edited to suit the particular project in question.

3. Setting up standard documents, e.g. agenda and minutes for meetings and the many other forms which may simply need editing when being re-used.
4. Again in the case of small firms, setting up a library of descriptions for standard quotations or standard paragraphs for repetitive work.
5. Keeping employee data up to date. Simple amendments can be made as circumstances change.
6. Standard headings and endings for letters and other documents.
7. Sending out general information to clients and sub-contractors using 'mail-merge'.
8. Producing draft reports.

There are many more uses for word processing and its use can save much time for secretaries and typists. The text produced by word processing can result in extremely professional looking documents.

PROGRAMME FOR SMALL OFFICE BUILDING
EXPENDITURE AND INCOME AT EARLIEST TIME RETENT. 5 %
EXPENDITURE AND INCOME AT LATEST TIME 1

OPERATION	DURN DAYS	PLAN'D EXPEND.	PLAN'D INCOME	WEEKS 1	2	3	4	5	6	7	8	9	10	11	12	13	14	15	16	17
SITE SET UP	3	210	240	210																
				240																
SETTING OUT	4	60	80	60																
				80																
EXCAVATE TO REDUCE LEVEL	10	1000	1200	100	500	400														
				120	600	480														
PILING	12	3600	4800			300	1500	1500	300											
						400	2000	2000	400											
DRAINS	6	300	360			0	0	0	0	0	0	0	250	50						
						0	0	0	0	0	0	0	300.	60						
PAD FOUNDATION	9	630	720						280	350										
									320	400										
STRIP FOUNDATION	5	500	600						0	0	0	0	400	100						
									0	0	0	0	480	120						
INSITU CONCRETE FRAME	22	22000	33000								5000	5000	5000	5000	2000					
											7500	7500	7500	7500	3000					
GROUND FLOOR SLAB	5	800	900												0	0	800			
															0	0	900			
BRICK INFILL PANELS	15	8700	9750												1740	2900	2900	1160		
															1950	3250	3250	1300		
PRECAST STAIRS AND FLOORS	3	2400	2700													0	0	2400		
																0	0	2700		
HARDWOOD FRAMES	10	1000	1200															300	500	200
																		360	600	240
PRECAST ROOF SLAB	2	2000	2500														0	0	1000	1000
																	0	0	1250	1250
BLOCK PARTITIONS GROUND FLOOR	7	1750	2100														0	0	1250	500
																	0	0	1500	600
GLAZING	6	240	300																	120
																				150
EXTERNAL DOORS	4	1600	1800																	1200
																				1350
ROOF COVERING	5	1000	1200														0	0	0	600
																	0	0	0	720
EXTERNAL PLUMBING	6	180	210														0	0	0	90
																	0	0	0	105
BLOCK PARTITIONS FIRST FLOOR	7	2100	2450															0	0	900
																		0	0	1050
PLUMBING FIRST FIX	5	900	1050																300	600
																			350	700
ELECTICAL FIRST FIX	8	400	480																	
EXTERNAL PAINTING	12	240	300																0	0
																			0	0
EXTERNAL WORKS	20	7564	9000																0	0
																			0	0
JOINERY FIRST FIX	8	800	880																	
PLASTERING	18	1800	2160																	
FLOOR SCREEDING	10	600	700																	
QUARRY TILING	1	100	140																	
PLUMBING SECOND FIX	10	1000	1200																	
JOINERY SECOND FIX	10	1000	1200																	
INTERNAL PAINTING	14	420	504																	
VINYL TILING	6	240	300																	
ELECTRICAL SECOND FIX	6	360	540																	
CLEANING AND HANDOVER	1	50	60																	
PRELIMINARIES		13175	13950	425	425	425	425	425	425	425	425	425	425	425	425	425	425	425	425	425
				450	450	450	450	450	450	450	450	450	450	450	450	450	450	450	450	450
PLANNED EXPENDITURE		78719		795	925	1125	1925	1925	1005	775	5425	5425	6075	5575	4165	3325	4125	4285	3475	5635
PLANNED INCOME			98574	890	1050	1330	2450	2450	1170	850	7950	7950	8730	8130	5400	3700	4600	4810	4150	6615
PLANNED EXPENDITURE MONTHLY							4770				9130				21240				15210	
PLANNED INCOME MONTHLY							5720				12420				30210				17260	
RETENTION ON PLANNED INCOME							286				621				1511				863	
INCOME LESS RETENTION TIME ADJUSTED							0				5434				11799				28700	
ACCUMULATIVE EXPENDITURE							4770				13900				35140				50350	
ACCUMULATIVE INCOME TIME ADJUSTED							0				5434				17233				45933	
VALUATION NUMBER							1	1			2	2			3	3			4	4
DIFFERENCE BETWEEN EXP. AND INC.							-4770	-4770			-13900	-8466			-29706	-17907			-33117	-4418

GRAPH PLOTTINGS

0 1 1 2 2 3 3 4 4 5 5 6 6 7 7 7.75 8 8 9 9 10
0 -4770 -4770 -13900 -8466 -29706 -17907 -33117 -4418 -14338 2060 -2201 8734 2623 7164 -914 -914 5818 5818 17391 17391

Fig. 7.26. Activities at latest times – retention 5%

18	19	20	21	22	23	24	25	26	27	28	29	30	31	32	33	34	35	36	37	38	39	59	60
120																							
150																							
400																							
450																							
400																							
480																							
90																							
105																							
1200																							
1400																							
100	250	50																					
120	300	60																					
0	0	0	0	0	0	0	0	0	0	0	40	100	100										
0	0	0	0	0	0	0	0	0	0	0	50	125	125										
0	0	0	0	0	0	0	0	0	0	1891	1891	1891	1891										
0	0	0	0	0	0	0	0	0	0	2250	2250	2250	2250										
		0	0	300	500																		
		0	0	330	550																		
		400	500	500	400																		
		480	600	600	480																		
					60	300	240																
					70	350	280																
							0	0	100														
							0	0	140														
							100	500	400														
							120	600	480														
							100	500	400														
							120	600	480														
									30	150	150	90											
									36	180	180	108											
												80	160										
												100	200										
												120	240										
												180	360										
													50										
													60										
425	425	425	425	425	425	425	425	425	425	425	425	425	425										
450	450	450	450	450	450	450	450	450	450	450	450	450	450										
735	675	875	925	1225	1385	725	865	1425	1355	2466	2506	2706	2866										
155	750	990	1050	1380	1550	800	970	1650	1586	2880	2930	3213	3445										
		9920				4260				6111			8078										
		11510				4780				7086			9588										
		576				239				354			479										
		16397				10935				4541				6732				9109					
		60270				64530				70641			78719										
		62330				73264				77805				84537				96110					
		5	5			6	6			7	7		7.75	8	8			9	9			10	10
		-14338	2060			-2201	8734			2623	7164		-914	-914	5818			5818	17391			17391	19855
10																							
855																							

PROGRAMME FOR SMALL OFFICE BUILDING
EXPENDITURE AND INCOME AT EARLIEST TIME RETENT. 3 %
EXPENDITURE AND INCOME AT LATEST TIME 1

OPERATION	DURN DAYS	PLAN'D EXPEND.	PLAN'D INCOME	WEEKS 1	2	3	4	5	6	7	8	9	10	11	12	13	14	15	16	
SITE SET UP	3	210	240	210																
				240																
SETTING OUT	4	60	80	60																
				80																
EXCAVATE TO REDUCE LEVEL	10	1000	1200	100	500	400														
				120	600	480														
PILING	12	3600	4800			300	1500	1500	300											
						400	2000	2000	400											
DRAINS	6	300	360			0	0	0	0	0	0	0	250	50						
						0	0	0	0	0	0	0	300	60						
PAD FOUNDATION	9	630	720						280	350										
									320	400										
STRIP FOUNDATION	5	500	600						0	0	0	0	400	100						
									0	0	0	0	480	120						
INSITU CONCRETE FRAME	22	22000	33000								5000	5000	5000	5000	2000					
											7500	7500	7500	7500	3000					
GROUND FLOOR SLAB	5	800	900												0	0	800			
															0	0	900			
BRICK INFILL PANELS	15	8700	9750												1740	2900	2900	1160		
															1950	3250	3250	1300		
PRECAST STAIRS AND FLOORS	3	2400	2700													0	0	2400		
																0	0	2700		
HARDWOOD FRAMES	10	1000	1200															300	500	20
																		360	600	24
PRECAST ROOF SLAB	2	2000	2500														0	0	1000	100
																	0	0	1250	125
BLOCK PARTITIONS GROUND FLOOR	7	1750	2100														0	0	1250	50
																	0	0	1500	60
GLAZING	6	240	300																	12
																				15
EXTERNAL DOORS	4	1600	1800																	120
																				135
ROOF COVERING	5	1000	1200														0	0	0	60
																	0	0	0	72
EXTERNAL PLUMBING	6	180	210														0	0	0	9
																	0	0	0	10
BLOCK PARTITIONS FIRST FLOOR	7	2100	2450															0	0	90
																		0	0	105
PLUMBING FIRST FIX	5	900	1050																300	60
																			350	70
ELECTICAL FIRST FIX	8	400	480																	
EXTERNAL PAINTING	12	240	300																0	
																			0	
EXTERNAL WORKS	20	7564	9000																0	
																			0	
JOINERY FIRST FIX	8	800	880																	
PLASTERING	18	1800	2160																	
FLOOR SCREEDING	10	600	700																	
QUARRY TILING	1	100	140																	
PLUMBING SECOND FIX	10	1000	1200																	
JOINERY SECOND FIX	10	1000	1200																	
INTERNAL PAINTING	14	420	504																	
VINYL TILING	6	240	300																	
ELECTRICAL SECOND FIX	6	360	540																	
CLEANING AND HANDOVER	1	50	60																	
PRELIMINARIES		13175	13950	425	425	425	425	425	425	425	425	425	425	425	425	425	425	425	425	42
				450	450	450	450	450	450	450	450	450	450	450	450	450	450	450	450	45
PLANNED EXPENDITURE		78719		795	925	1125	1925	1925	1005	775	5425	5425	6075	5575	4165	3325	4125	4285	3475	563
PLANNED INCOME			98574	890	1050	1330	2450	2450	1170	850	7950	7950	8730	8130	5400	3700	4600	4810	4150	661
PLANNED EXPENDITURE MONTHLY							4770				9130				21240				15210	
PLANNED INCOME MONTHLY							5720				12420				30210				17260	
RETENTION ON PLANNED INCOME							172				373				906				518	
INCOME LESS RETENTION TIME ADJUSTED							0				5548				12047				29304	
ACCUMULATIVE EXPENDITURE							4770				13900				35140				50350	
ACCUMULATIVE INCOME TIME ADJUSTED							0				5548				17596				46900	
VALUATION NUMBER							1	1			2	2			3	3			4	
DIFFERENCE BETWEEN EXP. AND INC.							-4770	-4770			-13900	-8352			-29592	-17544			-32754	-345

GRAPH PLOTTINGS

0	1	1	2	2	3	3	4	4	5	5	6	6	7	7	7.75	8	8	9	9	1
0	-4770	-4770	-13900	-8352	-29592	-17544	-32754	-3451	-13371	3372	-888	10276	4165	8802	724	724	7597	7597	18376	1837

Fig. 7.27. Activities at latest times – 3% retention

18	19	20	21	22	23	24	25	26	27	28	29	30	31	32	33	34	35	36	37	38	39	59	60		
120																									
150																									
400																									
450																									
400																									
480																									
90																									
105																									
1200																									
1400																									
100	250	50																							
120	300	60																							
0	0	0	0	0	0	0	0	0	0	0	40	100	100												
0	0	0	0	0	0	0	0	0	0	0	50	125	125												
0	0	0	0	0	0	0	0	0	0	1891	1891	1891	1891												
0	0	0	0	0	0	0	0	0	0	2250	2250	2250	2250												
		0	0	300	500																				
		0	0	330	550																				
		400	500	500	400																				
		480	600	600	480																				
					60	300	240																		
					70	350	280																		
							0	0	100																
							0	0	140																
							100	500	400																
							120	600	480																
							100	500	400																
							120	600	480																
									30	150	150	90													
									36	180	180	108													
												80	160												
												100	200												
												120	240												
												180	360												
													50												
													60												
425	425	425	425	425	425	425	425	425	425	425	425	425	425												
450	450	450	450	450	450	450	450	450	450	450	450	450	450												
2735	675	875	925	1225	1385	725	865	1425	1355	2466	2506	2706	2866												
3155	750	990	1050	1380	1550	800	970	1650	1586	2880	2930	3213	3445												
		9920				4260				6111			8078												
		11510				4780				7086			9588												
		345				143				213			288												
		16742				11165				4637				6873				9300							
		60270				64530				70641			78719												
		63642				74806				79443				86316				97095							
		5	5			6	6			7	7		7.75	8	8			9	9					10	10
		-13371	3372			-888	10276			4165	8802		724	724	7597			7597	18376					18376	19855

10
19855

PROGRAMME FOR SMALL OFFICE BUILDING
EXPENDITURE AND INCOME AT EARLIEST TIME 1 RETENT. 3 %
EXPENDITURE AND INCOME AT LATEST TIME

OPERATION	DURN DAYS	PLAN'D EXPEND.	PLAN'D INCOME	WEEKS 1	2	3	4	5	6	7	8	9	10	11	12	13	14	15	16	17
SITE SET UP	3	210	240	210																
				240																
SETTING OUT	4	60	80	60																
				80																
EXCAVATE TO REDUCE LEVEL	10	1000	1200	100	500	400														
				120	600	480														
PILING	12	3600	4800			300	1500	1500	300											
						400	2000	2000	400											
DRAINS	6	300	360			50	250	0	0	0	0	0	0	0						
						60	300	0	0	0	0	0	0	0						
PAD FOUNDATION	9	630	720						280	350										
									320	400										
STRIP FOUNDATION	5	500	600						400	100	0	0	0	0						
									480	120	0	0	0	0						
INSITU CONCRETE FRAME	22	22000	33000								5000	5000	5000	5000	2000					
											7500	7500	7500	7500	3000					
GROUND FLOOR SLAB	5	800	900												480	320	0			
															540	360	0			
BRICK INFILL PANELS	15	8700	9750												1740	2900	2900	1160		
															1950	3250	3250	1300		
PRECAST STAIRS AND FLOORS	3	2400	2700													2400	0	0		
																2700	0	0		
HARDWOOD FRAMES	10	1000	1200															300	500	200
																		360	600	240
PRECAST ROOF SLAB	2	2000	2500														2000	0	0	0
																	2500	0	0	0
BLOCK PARTITIONS GROUND FLOOR	7	1750	2100														1250	500	0	0
																	1500	600	0	0
GLAZING	6	240	300																	12[illegible]
																				15[illegible]
EXTERNAL DOORS	4	1600	1800																	120[illegible]
																				135[illegible]
ROOF COVERING	5	1000	1200														600	400	0	0
																	720	480	0	0
EXTERNAL PLUMBING	6	180	210														90	90	0	0
																	105	105	0	0
BLOCK PARTITIONS FIRST FLOOR	7	2100	2450															900	1200	0
																		1050	1400	0
PLUMBING FIRST FIX	5	900	1050																300	60[illegible]
																			350	70[illegible]
ELECTICAL FIRST FIX	8	400	480																	
EXTERNAL PAINTING	12	240	300																40	10[illegible]
																			50	125
EXTERNAL WORKS	20	7564	9000																756	189[illegible]
																			900	2250
JOINERY FIRST FIX	8	800	880																	
PLASTERING	18	1800	2160																	
FLOOR SCREEDING	10	600	700																	
QUARRY TILING	1	100	140																	
PLUMBING SECOND FIX	10	1000	1200																	
JOINERY SECOND FIX	10	1000	1200																	
INTERNAL PAINTING	14	420	504																	
VINYL TILING	6	240	300																	
ELECTRICAL SECOND FIX	6	360	540																	
CLEANING AND HANDOVER	1	50	60																	
PRELIMINARIES		13175	13950	425	425	425	425	425	425	425	425	425	425	425	425	425	425	425	425	425
				450	450	450	450	450	450	450	450	450	450	450	450	450	450	450	450	450
PLANNED EXPENDITURE		78719		795	925	1175	2175	1925	1405	875	5425	5425	5425	5425	4645	6045	7265	3775	3221	4536
PLANNED INCOME			98574	890	1050	1390	2750	2450	1650	970	7950	7950	7950	7950	5940	6760	8525	4345	3750	5265
PLANNED EXPENDITURE MONTHLY							5070				9630				20920				20306	
PLANNED INCOME MONTHLY							6080				13020				29790				23380	
RETENTION ON PLANNED INCOME							182				391				894				701	
INCOME LESS RETENTION TIME ADJUSTED							0				5898				12629				28896	
ACCUMULATIVE EXPENDITURE							5070				14700				35620				55926	
ACCUMULATIVE INCOME TIME ADJUSTED							0				5898				18527				47423	
VALUATION NUMBER							1	1			2	2			3	3			4	4
DIFFERENCE BETWEEN EXP. AND INC.							-5070	-5070			-14700	-8802			-29722	-17093			-37399	-8503

GRAPH PLOTTINGS

0	1	1	2	2	3	3	4	4	5	5	6	6	7	7	7.75	8	8	9	9	10
0	-5070	-5070	-14700	-8802	-29722	-17093	-37399	-8503	-21051	1628	-2232	11920	7700	11910	9745	9745	14436	14436	18376	18376

Fig. 7.28. Activities at earliest times – retention 3%

18	19	20	21	22	23	24	25	26	27	28	29	30	31	32	33	34	35	36	37	38	39	59	60
120																							
150																							
400																							
450																							
0																							
0																							
0																							
0																							
0																							
0																							
100	250	50																					
120	300	60																					
100	0	0	0	0	0	0	0	0	0	0	0	0	0										
125	0	0	0	0	0	0	0	0	0	0	0	0	0										
1891	1891	1135	0	0	0	0	0	0	0	0	0	0	0										
2250	2250	1350	0	0	0	0	0	0	0	0	0	0	0										
		400	400	0	0																		
		440	440	0	0																		
		400	500	500	400																		
		480	600	600	480																		
					60	300	240																
					70	350	280																
							100	0	0														
							140	0	0														
							100	500	400														
							120	600	480														
							100	500	400														
							120	600	480														
									30	150	150	90											
									36	180	180	108											
												80	160										
												100	200										
												120	240										
												180	360										
													50										
													60										
425	425	425	425	425	425	425	425	425	425	425	425	425	425										
450	450	450	450	450	450	450	450	450	450	450	450	450	450										
3036	2566	2410	1325	925	885	725	965	1425	1255	575	575	715	875										
3545	3000	2780	1490	1050	1000	800	1110	1650	1446	630	630	838	1070										
		12548				3860				4220			2165										
		14590				4340				4836			2538										
		438				130				145			76										
		22679				14152				4210				4691				2462					
		68474				72334				76554			78719										
		70102				84254				88464				93155				97095					
		5	5			6	6			7	7		7.75	8	8			9	9			10	10
		-21051	1628			-2232	11920			7700	11910		9745	9745	14436			14436	18376			18376	19855
10																							
19855																							

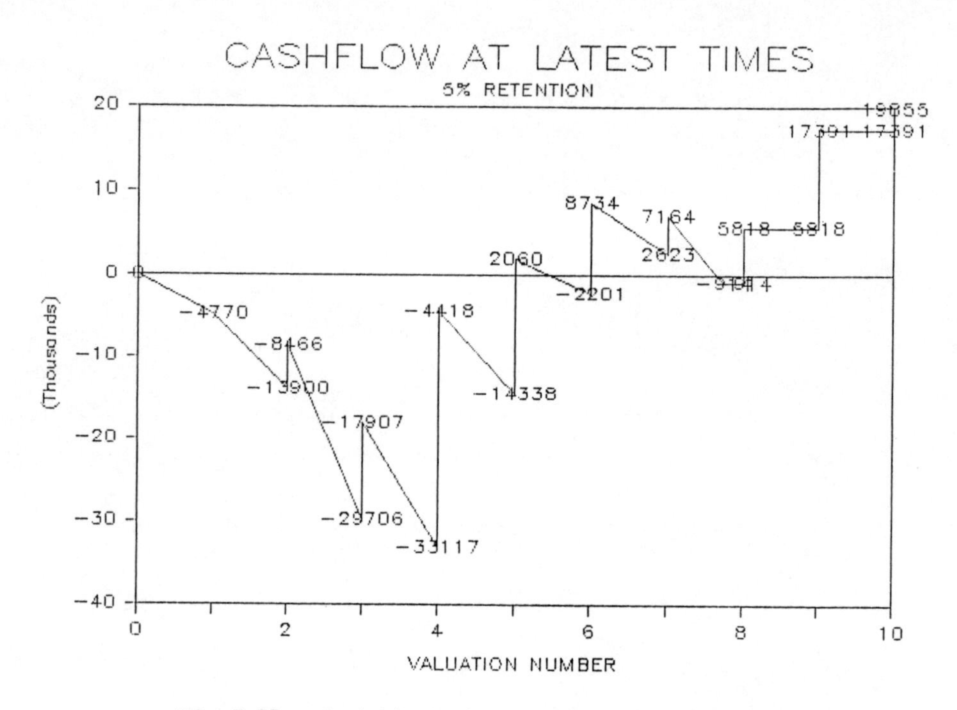

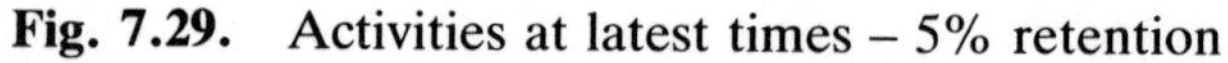
Fig. 7.29. Activities at latest times – 5% retention

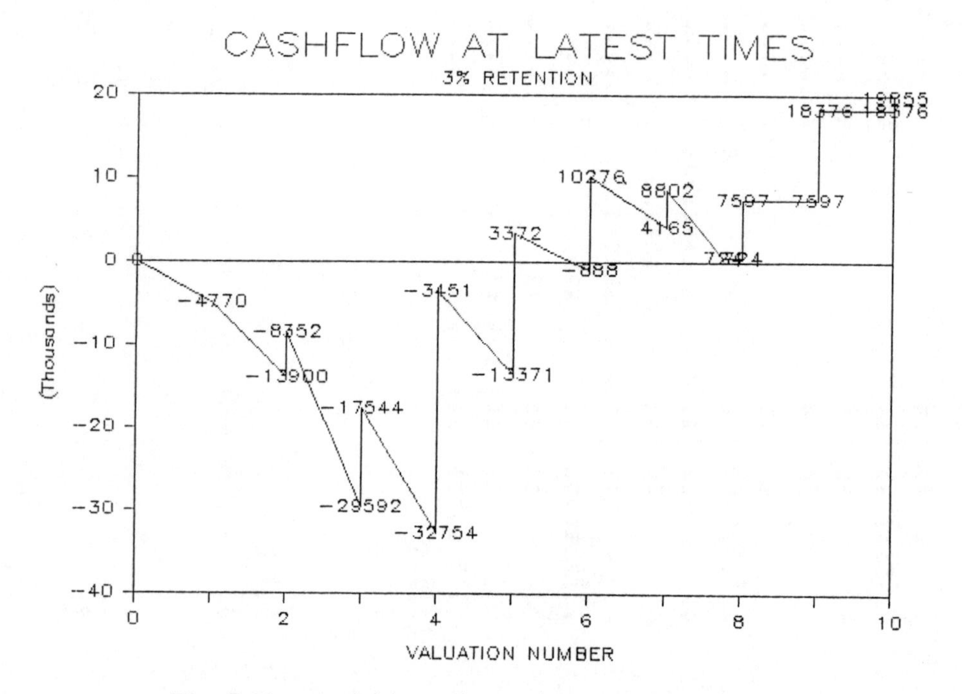

Fig. 7.30. Activities at latest times – 3% retention

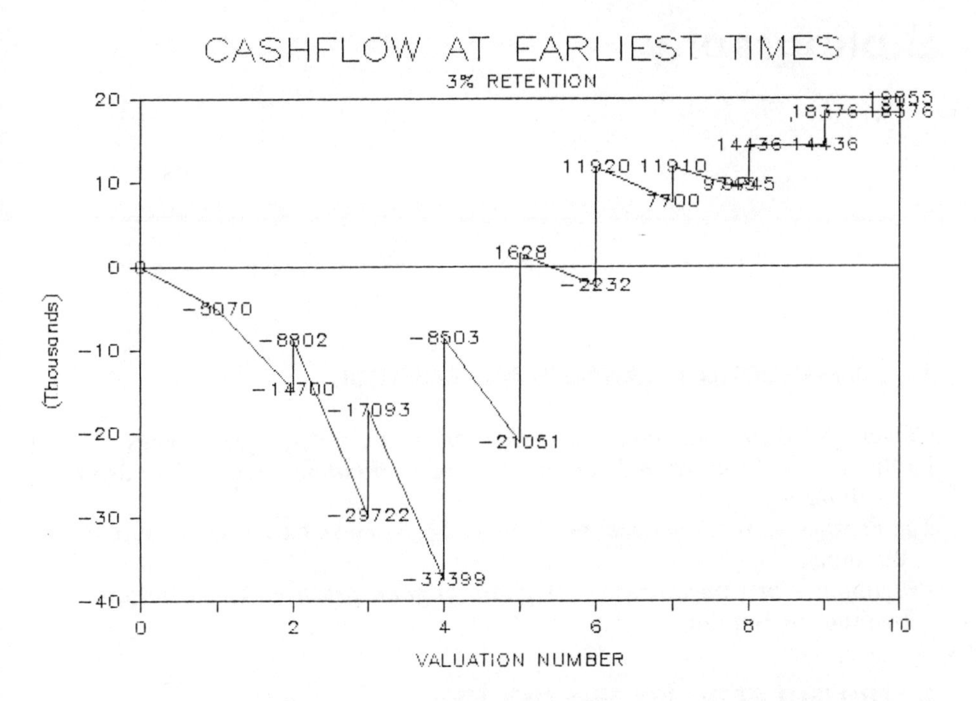

Fig. 7.31. Activities at earliest times – 3% retention

Bibliography

1 CONSTRUCTION PLANNING AND CONTROL

Calvert, R. E. *Introduction to Building Management.* Georges Newnes.

Hollins, R. J. *Production Planning Applied to Building.* George Goodwin for *Building.*

The Practice of Site Management, Vols 1, 2 & 3. The Chartered Institute of Building.

Programmes in Construction – A guide to good practice. The Chartered Institute of Building.

2 PROJECT NETWORK TECHNIQUES

Antill, J. M. and Woodhead, R. W. *Critical Path Methods in Construction Practice.* Wiley.

Burman, P. J. *Precedence Networks for Project Planning and Control.* McGraw Hill Inc.

BS 4335 (1972) *Glossary of Terms used in Project Network Techniques.* London: British Standards Institution.

O'Brian, J. J. *CPM in Construction Management.* McGraw Hill Inc.

3 WORK STUDY

BS 3138 *Glossary of Terms in Work Study.* London: British Standards Institution.

Outline of Work Study, Parts 1, 2, 3. The British Institute of Management.

Geary, R. *Work Study Applied to Building. Building.*

Turner, G. J. and Elliot, K. R. J. *Project Planning and Control in the Construction Industry.* Cassell.

Currie, R. M. *Work Study.* British Institute of Management.

Oxley, R. *Incentives in the Construction Industry – Effects on Earning and Costs. The Practice of Site Management*, Vol. 3. The Chartered Institute of Building, 1985.

4 STATISTICS

Hays, Samuel. *An Outline of Statistics.* 7th Edn. Longmans, Green.
Rhodes, E. C. *Elementary Statistical Methods.* Routledge & Kegan Paul.
Oliver, F. R. *What do Statistics Show?.* Hodder & Stoughton.
Moroney, M. J. *Facts from Figures.* Penguin Books.
Reichman, W. J. *Use and Abuse of Statistics.* Pelican Books.
Rose, T. G. *Business Charts.* Pitman.
Moore, P. G. *Principles of Statistical Techniques.* Cambridge University Press.
McIntosh, J. D. *Concrete and Statistics.* C.R. Books Ltd.
Moore, P. G. *Statistics and the Manager.* Macdonald.

5 BUDGETARY AND COST CONTROL

Brandwood, F. *Builders Cost Control, Bonusing and Accounts.* Gee & Co.
Institute of Cost and Works Accountants, *Introduction to Budgetary Control, Standard Costing, Material Control and Production Control.* Gee & Co.
Scott, J. A. *Budgetary Control and Standard Costs.* Pitman.
Gobourne, J. *Cost Control in the Construction Industry.* Newnes Butterworth.
Nazem, S. M. *Planning Contractors' Capital.* Building Technology & Management.
Cooke, B. & Jepson, W. B. *Cost and Financial Control for Construction Firms.* Macmillan Press Ltd.
Pilcher, R. *Project Cost Control in Construction.* Collins.

6 OPERATIONAL RESEARCH

Theil, Henry, Boot, John C. G. and Kloek, Tuen. *Operational Research and Quantitative Economics – An Elementary Introduction.* McGraw Hill.
Rubey, Harry and Milner, Walker W. *Construction and Professional Management – An Introduction.* Macmillan.
Duckworth, Eric. *A Guide to Operational Research.* University Paperbacks.
Sargeaunt, M. J. *Operational Research for Management.* Heinemann.
Rowe, A. J. and Jackson, J. R. (1956–7) 'Research problems in production routing and scheduling.' *Journal of Industrial Economics.* 116–21.

7 COMPUTER APPLICATIONS

Construction Computing – Quarterly – The Chartered Institute of Building.
Oxley, R. and Poskitt, J. 'Systems analysis and design applied to the construction industry'. *Building Technology and Management*, December

1969, and *The Illustrated Carpenter and Builder*, 19th March 1971.
Paul Barton (Editor). 'Information Systems in Construction Management, Principles & Applications'. Batsford.
Cooke & Balakrishnan. *Computer Spreadsheet Applications in Building & Surveying*. Macmillan.
Brandon, P. and Moore R. G. *Microcomputers in Building Appraisal*, Granada.
Brandon, P., Moore R. G. and Main P. *Computer Programs for Building Cost Appraisal*, Collins.

Index